Advanced Topics in Science and Technology in China

Volume 62

Zhejiang University is one of the leading universities in China. In Advanced Topics in Science and Technology in China, Zhejiang University Press and Springer jointly publish monographs by Chinese scholars and professors, as well as invited authors and editors from abroad who are outstanding experts and scholars in their fields. This series will be of interest to researchers, lecturers, and graduate students alike.

Advanced Topics in Science and Technology in China aims to present the latest and most cutting-edge theories, techniques, and methodologies in various research areas in China. It covers all disciplines in the fields of natural science and technology, including but not limited to, computer science, materials science, the life sciences, engineering, environmental sciences, mathematics, and physics.

This book series is indexed by the SCOPUS database.

If you are interested in publishing your book in the series, please contact Dr. Mengchu Huang (Email: mengchu.huang@springer.com).

More information about this series at http://www.springer.com/series/7887

Jialin Zhou · Erwin Oh

Full-Scale Field Tests of Different Types of Piles

Project-Based Study

ZHEJIANG UNIVERSITY PRESS
浙江大学出版社

Springer

Jialin Zhou
School of Engineering and Built Environment
Griffith University
Gold Coast, QLD, Australia

Erwin Oh
School of Engineering and Built Environment
Griffith University
Gold Coast, QLD, Australia

ISSN 1995-6819 ISSN 1995-6827 (electronic)
Advanced Topics in Science and Technology in China
ISBN 978-981-33-6182-9 ISBN 978-981-33-6183-6 (eBook)
https://doi.org/10.1007/978-981-33-6183-6

Jointly published with Zhejiang University Press
The print edition is not for sale in China. Customers from China please order the print book from: Zhejiang University Press.

This Springer imprint is published by the registered company Springer Nature Singapore Pte Ltd.
The registered company address is: 152 Beach Road, #21-01/04 Gateway East, Singapore 189721, Singapore

Preface

This book provides full-scale field tests of different types of pile foundations. For the testing, the traditional static load tests which consider various loading orientations are provided. In detail, the compressive, uplift and horizontal static load tests with project case studies are demonstrated. For the test in which loading capacity is limited, the improved static load test is introduced. This improved testing method can be used for huge capacity determination of super-long and large diameter pile foundations. Moreover, the dynamic load tests, inclinometer monitoring, and the tests that aim to determine the load transfer mechanism of pile foundation are detailed in this book.

This book also covers the up-to-date popular topic with detailed project studies. This includes the academic investigation of post-grouting technology effect on drilled shaft piles, the research of displacement and non-displacement precast pile foundation, the study of fiber-reinforced polymer (FRP) material used in the geotechnical environment such as deep excavation pit in tunneling project, and the research of super-long and large diameter pile foundations. The study of the composite pile foundation provided in this book contains two types: the FRP bar reinforced concrete piles and FRP laminar confined concrete piles; these new types of composite piles provide practical information to the geotechnical engineer, and they could be an alternative for the traditional pile in some specific environmental conditions. Most importantly, this book provides pile failure tests considering different failure conditions, the failure caused by inappropriate pile construction, eccentric loading, plunging failure of piling system are illustrated. All these investigations provide essential and academic information for researchers as well as engineers in the role of Civil and Geotech.

Not only the different types of piles are studied, but also the relevant theory and literatures are reviewed in this book. In addition, the diagrams are plotted in an easy way and the explanation of the diagrams and tables are described in detail. The research methods corresponding to the practical projects are illustrated in detail as well, hence it can be a perfect book for the bachelor and master's degree students whose objectives are related to the civil and geotechnical engineering.

Gold Coast, Australia Jialin Zhou

Contents

Abbreviations

AASHTO	American Association of State Highway and Transportation Officials
AFRP	Aramid Fibre-reinforced Polymer
BFRP	Basalt Fibre-reinforced Polymer
CAPWAP	Case Pile Wave Analysis Program
CD	Consolidated Drained Tri-axial Tests
CFRP	Carbon Fibre-reinforced Polymer
CPT	Cone Penetration Test
CU	Consolidated Undrained Tri-axial Tests
DCM	Deep Cement Mixing
ER	Equipment Energy Ratio
FHWA	Federal Highway Administration
FRP	Fibre-reinforced Polymer
GFRP	Glass Fibre-reinforced Polymer
LVDT	Linear Variable Differential Transformer
O-cell	Osterberg Cell
OCR	Over-consolidation Ratio
PDA	Pile Driving Analyser
PT(s)	Proof Test(s)
PHC	Pretensioned Spun High-strength Concrete
SPE	Soil Plug Effect
ROV	Remote-operated Vehicles
RSP	Static Resistance on Pile during Driving
RTL	Total Static and Dynamic Resistance during Driving
SLT	Static Load Test
SMW	Soil Mixing Wall
SPT	Standard Penetration Test

SRP	Structurally Reinforced Plastic
TBM	Tunnel Boring Machine
TDM	T-shaped Deep Mixed
UU	Unconsolidated Undrained Tri-axial Tests

Notations

N-value	Standard Penetration Test (SPT) Blow Count
N_{60}	Corrected SPT N-value
$(N_1)_{60}$	Normalised Blow Count
C_N	Overburden Correction Factor
I_D	Relative Density
σ'_{v0}	Overburden Pressure
C_u	Undrained Shear Strength
P_a	Atmospheric Pressure
I_{P}	Plasticity Index
q_{u}	Unconfined Compressive Strength
γ	Unit Weight
e	Void Ratio
σ'	Vertical Effective Stress
C_{c}	Compression Index
C_{e}	Expansion Index
m_{v}	Coefficient of Volume Compressibility
E'_{oed}	One-dimensional Elastic Modulus
w	Moisture Content
ρ	Material Density
C_{z}	Coefficient of Curvature
I_{L}	Liquidity Index
U_{v}	Degree of Consolidation
C_{v}	Coefficient of Consolidation
c	Cohesion of Soil
c'	Effective Cohesion of Soil
φ'	Effective Friction Angle of Soil
T_{v}	Time Factor
σ	Normal Stress of Soil
φ	Friction Angle of Soil
μ	Coefficient of Friction
$\overline{N}_O$	Average Corrected SPT N-value for Stratum Overlying Bearing Stratum

$\overline{N}_B$	Average Corrected SPT *N*-value of Bearing Stratum
D_B	Pile Embedment Depth into Bearing Stratum
K_δ	Lateral Earth Pressure Coefficient at Depth *d*
C_F	Correction Factor for K_δ
α_t	Depth–Width Relationship Factor
N'_q	Capacity Factor
q_c	Cone Tip Resistance of Cone Penetration Test (CPT)
f_s	Shaft Resistance
q_t	Toe or Tip Resistance
B	Bjerrum-Burland Beta Coefficient
K_s	Earth Pressure Coefficient
Δ	Interface Friction Angle between Pile and Soil
N_t	Toe Bearing Capacity Coefficient
p_t	Effective Overburden Pressure at Pile Toe
s_f	Offset Settlement
Q_u	Ultimate Bearing Capacity of Pile
Q_b	Base Resistance of Pile Foundation
Q_s	Shaft Resistance of Pile Foundation
R_s	Compressive Settlement Ratio
R_d	Uplift Displacement Ratio
R_h	Horizontal Displacement Ratio
S_t	Vertical Settlement at Certain Loading at Certain Time
H_t	Horizontal Displacement at Certain Loading at Certain Time
$S_{(Fi)}$	Vertical Settlement at End of Certain Loading Stage
$H_{(Fi)}$	Horizontal Displacement at End of Certain Loading Stage
$S_{(In)}$	Settlement at Beginning of Certain Loading Stage
$H_{(In)}$	Horizontal Displacement at Beginning of Certain Loading Stage
Q_{uk}	Ultimate Bearing Capacity of Precast Concrete Pile
Q_{sk}	Ultimate Shaft Capacity of Precast Concrete Pile
Q_{pk}	Ultimate End Capacity of Precast Concrete Pile
q_{sik}	Shaft Resistance of Precast Concrete Pile
q_{qk}	End Resistance of Precast Concrete Pile
A_p	Total Area from Pipe Pile Toe
A_j	Effective Area of Pipe Pile Toe
λ_p	Plug Effect Coefficient
A_{p1}	Hollow Area of Pipe Pile Toe
h_b	Embedment Depth of Pile in Bearing Stratum
H_{cr}	Critical Load of Horizontal Static Load Tests (SLTs)
H_u	Ultimate Load of Horizontal SLTs
$R_{P\text{-}S}$	Permanent Settlement Ratio
S_{max}	Settlement under Maximum Applied Load
S_p	Permanent Settlement
Q_{Max}	Maximum Applied Loads during SLTs
Q_a	Allowable Load of Pile

F_t	Converted Force by Pile Driving Analyser (PDA)
ε_t	Recorded Strain by Strain Transducers
V_t	Converted Velocity by PDA
a_t	Recorded Acceleration by Accelerometer
F_{stn}	Measured Statnamic Force
F_v	Damping Force from Soil
F_p	Pore Water Pressure Resistance

List of Figures

List of Tables

Chapter 1
Introduction

1.1 Background

With the soaring requirement for building space in metropolises, high-rise buildings are becoming increasingly popular. Pile foundations can resist more loading through end bearing and friction resistance than can shallow foundations; hence, the use of pile foundations is more common. As a structural element that transfers the loads from upper structures into the soil layers, piles can be generally categorized into precast piles and cast-in-situ piles. As a result of the numerous advantages—such as convenience of construction without considering the transfer of piles, cost and schedule of construction—cast-in-situ piles are the most accepted type of piles in construction projects.

For bored piles, the process of construction leads to soil deposits remaining at the base area after drilling the hole, which consequently leads to a decrease of the pile capacity and increase of settlement. Engineers currently use admixture, such as bentonite or polymeric slurry, to avoid the collapse of the soil during the drilling process. However, these admixtures create a layer between the concrete pile surface and soil layers. This new layer, created by the reaction among bentonite, soil and water, has a low friction coefficient, which results in decreased shaft resistance. To solve the above problems, the post grouting technique is the most acceptable method. This method involves pressurizing admixtures into soils through a device, such as a pipe installed inside the pile. The highly pressurized materials compact the loose soils, and the injected materials mix with the soils and reinforce the pile end or shaft.

For precast piles, piles are driven into the soil layers and the soils are forced to compact. Based on the amount of soil compacted, these precast piles can be categorized into small and large displacement piles. Normally, open-ended pipe piles are

J. Zhou and E. Oh, *Full-Scale Field Tests of Different Types of Piles*,
Advanced Topics in Science and Technology in China 62,
https://doi.org/10.1007/978-981-33-6183-6_1

considered small displacement piles, and solid piles are identified as large displacement piles. A non-displacement pile generally refers to a bored pile (or cast-in-place pile) because the soil is removed during construction. There are many investigates in terms of the small and large displacement piles, but the research of the non-displacement precast piles is very limited.

When based on the selected type of materials, the pile foundation can be categorized into timber, steel, concrete or composite piles. Past studies investigated Fiber-reinforced polymer (FRP) composite piles with consideration of their damage or corrosive effect on steel, timber, fiber and concrete materials. Different to the previous research on FRP piles that are made of fiber and polymer, this book provides the investigation on concrete piles with inside FRP bar reinforcement and outside FRP laminar reinforcement.

Within this context, there are different types of piles used under different construction technology. To investigate these piles under different effects, such as post grouting or composite confinement, it is vital to determine the ultimate bearing capacity and settlement behavior of the pile foundation. Compared with experimental tests in lab, field tests can provide the most accurate information about pile behavior. For compressively loaded piles, the test method can include Static Load Tests (SLTs), dynamic load tests, Osterberg Cell (O-cell) tests and statnamic tests. However, for tests that apply lateral and uplift loading, SLTs are the only option. In practice, static field tests are mostly the Proof Tests (PTs), which aims to assure the designed pile foundation possessing adequate capacity within required settlement. Field tests for loading the pile to failure are quite limited. The obtained result sometimes may provide limited information, and is hard to be used for capacity determination, thus interpretations based on the data obtained from field tests shall be used.

This book focuses on field tests of different types of piles associated with various up-to-date technologies. Static and dynamic load tests are performed to determine the ultimate compressive bearing capacity and settlement behaviors of pile foundations. However, the traditional SLT can only provide limited loads especially when testing the super-long and large diameter pile foundations. So, this book also introduces an improved SLT that can provide large compressive loads and the load transfer mechanism research is involved. As aforementioned above, the PTs are the mostly performed tests, thus this book not only covers the PTs, but also includes failure field tests.

1.2 Objectives

The objectives of this book are to investigate bored piles with different grouting technology; small, large and no displacement precast piles; FRP composite piles, super-long and large diameter piles and piles applied with ultimate loading. A full-scale field tests are illustrated including static compressive, uplift and horizontal load tests, dynamic load tests, inclinometer monitoring as well as tests for determination of the stress in pile foundation.

For cast-in-situ concrete piles, this book aims to:

1. Determine the compressive and uplift ultimate bearing capacity of non-grouted, base grouted only, and base-and-shaft grouted piles.
2. Research the shaft resistance mechanism of grouted piles through conducting static and dynamic load tests.
3. Propose empirical formula for vertical settlement and uplift displacement prediction.
4. Investigate the behavior of pile foundation under different types of failure conditions, and compare failure load test with proof test.
5. Study the capacity as well as the load transfer mechanism of the super-long and large diameter pile foundations.

For precast concrete piles, this book aims to:

1. Determine the ultimate bearing capacity of small and large displacement piles (concrete pipe and rectangular piles).
2. Research the geotechnical mechanism of concrete pipe and rectangular piles.
3. Investigate the behavior of non-displacement piles through performing uplift and horizontal SLTs.
4. Propose empirical formula for horizontal displacement prediction.

For composite piles, this book aims to:

1. Determine the ultimate bearing capacity of precast pipe and rectangular piles confined by Carbon FRP (CFRP), and determine the capacity improvement compared with piles without FRP confinement effect.
2. Research the deflection behavior of concrete bored pile with inside Glass FRP (GFRP) reinforcement as a replacement for traditional steel bar reinforcement.

For the super-long and large-diameter concrete piles, this book aims to:

1. Research the load transfer mechanism, shaft resistance development and shear stress distribution of the super-long and large diameter pile foundations.
2. Introduce an improved static load test which can apply large amount of compressive load.

For piles applied with ultimate loads, this book aims to:

1. Determine the concrete pile behavior under failure caused by inadequate concrete strength.
2. Investigate the failure of concrete piles because of the application of eccentric loading, via performing field tests.
3. Study the plunging failure of a pile foundation.

1.3 Book Organization

After the background and objectives have been illustrated in this chapter, the related general academic principles and practices is provided in Chap. 2. Chapter 2 first reviews the methods to determine the soil parameters and subsurface exploration, and then reviews the categorization of piles. This is followed by discussion of the analytical methods used to design the pile capacity, and review of field tests to indicate which test methods should be selected to obtain the real pile capacity onsite.

Based on the research methodology presented in Chap. 2, Chap. 3 focuses on studying concrete bored piles with consideration of the construction process, and provides one case study. The post grouting technique effect is researched through comparing base-and-shaft grouted, base grouted and non-grouted piles. The ultimate compressive and uplift bearing capacities of these piles are determined through interpretations of the result obtained from SLTs. Further, the shaft resistances of these piles are analyzed and compared with the results obtained from dynamic load tests.

Chapter 4 investigates small, large and non-displacement concrete precast piles through two case studies. For the small and large displacement piles, this chapter focuses on determining compressive behaviors by considering various pile lengths and concrete strengths. For the non-displacement precast piles, this chapter focuses on determining the piles' uplift and horizontal ultimate bearing capacity. This chapter also analyses the uplift load transfer mechanisms of these non-displacement precast piles.

Different to the investigations of bored piles (Chap. 3) and precast piles (Chap. 4), Chap. 5 focuses on the composite piles with use of FRP materials. Chapter 5 presents two case studies. The first case study researches the ultimate bearing capacity of concrete piles with CFRP laminar confinement, and the second case study examines the deflection behavior of passive bored piles in which steel reinforcement is replaced by GFRP bars during the construction process.

Chapter 6 illustrates the improved compressive static load test which is used for testing the super-long and large dimeter piles. Detailed pile design as well as installation of the steel rebar meter (also known as strain gauge or sister rebar) are demonstrated based on a case study. The load transfer mechanism, shaft resistance development and shear stress distribution are discussed via comparing the results from three super-long and large diameter piles.

As a result of SLTs mostly being used as PTs, which requires extrapolation of the results, previous research on SLTs until failure is limited, and data demonstrating plunging failure is rare. Thus, in Chap. 7, SLTs under different failure conditions are performed, with consideration of the aspects of rigidity of concrete, soils and eccentric loading.

Chapter 8 provides the capacity and settlement discussion of the different types of concrete pile foundations obtained from performed field tests. Also, the methods to obtain the empirical formula that are used for prediction of vertical and lateral displacement of soil-pile system are proposed. Finally, Chap. 9 presents the study findings, conclusions and recommendations based on all the case studies undertaken in Chaps. 3 to 7. It also discusses potential avenues for further research.

Chapter 2
General Principles and Practices

2.1 General Introduction

All types of engineers are required to have sophisticated understanding and knowledge of subsurface conditions to undertake their projects. Soil analysis is more complicated than analysis of other materials because of soil's non-continuum characteristic. Materials such as steel and concrete are relatively uniform solids, and their materials analysis can be based with a high degree of confidence on the assumption of the materials' solid mechanics and strength (Samtani and Nowatzki 2006a). In contrast, soil materials can widely differ over time and space. Further, many factors can influence the behavior of soil, such as soil particles and climatic issues.

The constitutive behavior of soil, which indicates the strength and stiffness properties, can be obtained through conducting laboratory and in-situ tests. Commonly, the laboratory tests include consolidation, direct shear and triaxial tests. Given that the change of soil in volume is a time-dependent process, determining the degree of consolidation, U_v, and coefficient of consolidation, C_v, are very important for analysis of the total settlement of soil. Further, the shear strength of soils is another significant aspect used for foundation design. This chapter first reviews the literature referring to direct shear, oedometer and triaxial tests. The parameters of cohesion, c, and friction angle φ, of the soil can be obtained from both direct shear and triaxial tests. Direct shear tests are more usual and are easy to handle, yet triaxial tests allow the soil to be consolidated or drained first; therefore, engineers must assess the projects' requirements to appropriately select and perform the tests. Note that, for the laboratory index, this book does not review factors such as moisture content (w), density (ρ), unit weight (γ), void ratio (e), coefficient of uniformity (C_u) and curvature (C_z), plasticity (I_P) and liquidity index (I_L).

Laboratory tests, sometimes however, are valueless for quantifying the mechanical behavior of an element of soil. For example, it is difficult and expensive to determine accurate data from an undisturbed deposit, such as sand and sensitive clays. Further, the response of a small element of soil that contains features within the macro-fabric

J. Zhou and E. Oh, *Full-Scale Field Tests of Different Types of Piles*,
Advanced Topics in Science and Technology in China 62,
https://doi.org/10.1007/978-981-33-6183-6_2

cannot represent the complete soil behavior (Knappett and Craig 2012). The principal in-situ tests which include the Standard Penetration Test (SPT), Cone Penetration Test (CPT), field vane test and pressuremeter test could cover the shortage of the lab tests. These tests are related to the corresponding response of soil under direct loading to determine the soil properties onsite instead of soil transfer. Boreholes are required to conduct SPTs, others such as CPT and field vane test that do not require boreholes are considered as 'direct-push' technologies.

Analytical methods indicate the procedure to obtain the capacity of a pile for design, based on the parameters acquired from laboratory and in-situ tests. Meyerhof (1976), Nordlund (1963) and Brown et al. (2001) proposed methods for estimating of the bearing capacity of a single pile in cohesionless soils. For a single pile in cohesive soils, Tomlinson (1980) and Fellenius (1991) provided the total and effective methods to determine the capacity. All of these methods are now widely accepted by engineers, and are discussed follows by the review of subsurface explorations in Sect. 2.2.

It should be noted that analytical methods are used for primary design, and this is the theoretical value of pile capacity. Static Load Test (SLT) is the most accurate method to determine the load capacity of pile foundations. SLT involves loading from the pile head with specific time intervals, and monitoring the displacement of the pile head. These tests can be categorized into vertical compressive and uplift testing and lateral testing. Detailed information is provided in the standards AS 2159 (Standards Australia 2009), ASTM-D-1143/D-07 (ASTM International 1994), ASTM-D-3966/D-07 (ASTM International 1995a) and ASTM-D-3689/D-07 (ASTM International 1995b). However, performance of these compressive, uplift and lateral SLTs until failure load is expensive and time consuming; hence, the most common method is to twice apply the designed load and determine the pile head displacement, which is called a 'proof test'. The purpose of this test is to confirm that the foundation movement is within the standard requirement to safely support the designed loading. However, this test cannot directly provide the ultimate bearing capacity of a pile.

Various methods can be used for interpretation of SLTs to determine the ultimate bearing capacity of piles. Section 2.5 reviews the methods for extrapolations of SLTs, such as those by Davisson (1972), Butler and Hoy (1976), DeBeer (1970), Brinch-Hansen (1963) and Chin (1970). Further, this same section provides the theory, test setup, and data analysis methods of the dynamic load test, Osterberg cell test and statnamic test that are also used to determine the capacity of single piles.

2.2 Soil Stratum Determination

2.2.1 Laboratory Tests

2.2.1.1 Direct Shear Tests

The shear strength parameters of soils are important for foundation design, and the soil strength contains frictional strength and cohesive strength. Frictional strength is dependent on the stress state, such as overburden pressure and the friction angle between soil particles. Based on the experiments, Coulomb (1776) proposed a formula to determine the shear resistance of soil, which is related to normal stress, σ and the internal friction angle, φ, as provided in Eq. 2.1. It can be seen that soil strength is related to friction between soil particles because tan φ is equal to the coefficient of friction, μ. However, soil strength also relates to drainage of water, as Terzaghi (1944) indicated. So, the soil strength is also represented by the effective stress and friction angle provided in Eq. 2.2:

$$\tau_f = c + \sigma \tan\varphi \tag{2.1}$$

$$\tau_f = c' + \sigma' \tan\varphi' \tag{2.2}$$

where:

τ_f = shear resistance of soil.
c and c' = total and effective cohesion.
σ and σ' = total and effective normal stress.
φ and φ' = total and effective friction angle.

The parameters of shear strength can be determined by conducting laboratory direct shear and triaxial tests. This section provides the setup, procedures and data analysis of direct shear tests based on AS 1289.6.2.2 (Standards Australia 1998a) and ASTM-D-3080–98 (ASTM International 2007a). The basic theory involves applying horizontal shear force to the sample, which is vertically pressured and confined in a metal box (Fig. 2.1). The horizontal and vertical displacement of the sample are recorded under a certain vertical stress, and then the peak and ultimate shearing strength can be determined through plotting horizontal shear stress versus displacement, as illustrated in Fig. 2.2. These tests can also indicate the soil displacement behavior until failure condition. Through changing the value of vertical normal stress

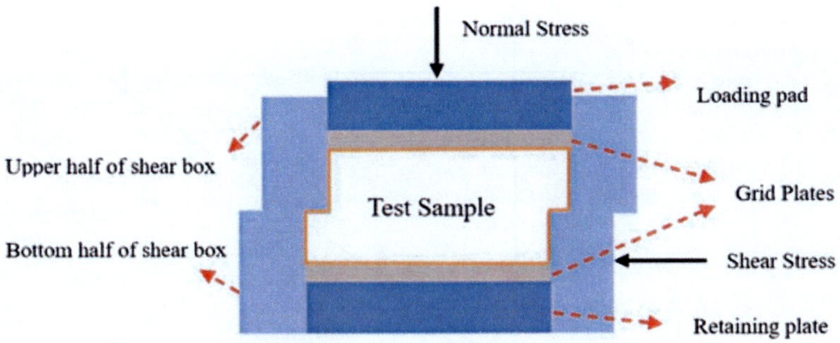

Fig. 2.1 Direct shear apparatus-shear box

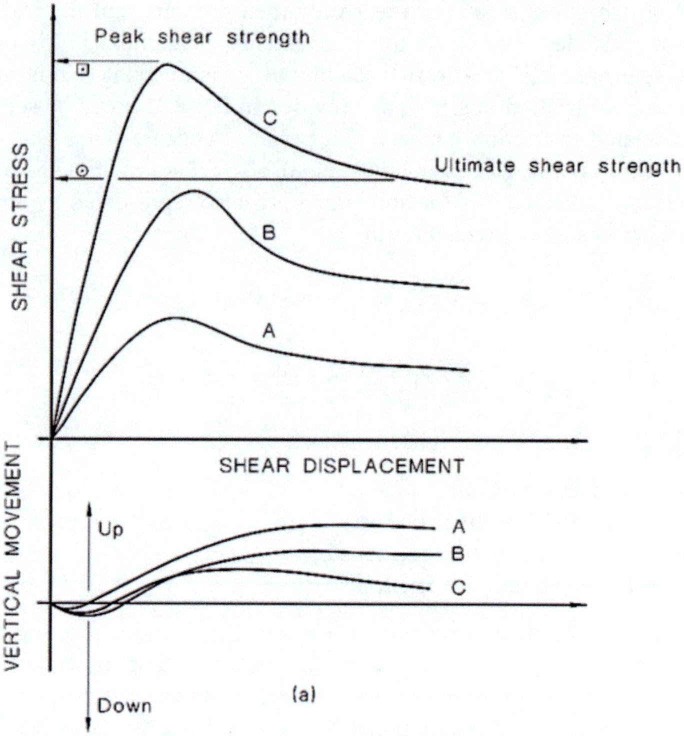

Fig. 2.2 Typical results from drained direct shear test (AS 1289.6.2.2, 1998)

and repeating the test, the peak and ultimate friction angle can be obtained after plotting normal stress versus shear stress, as shown in Fig. 2.3. It should be noted that, for the pure saturated sand sample, the cohesion value is zero.

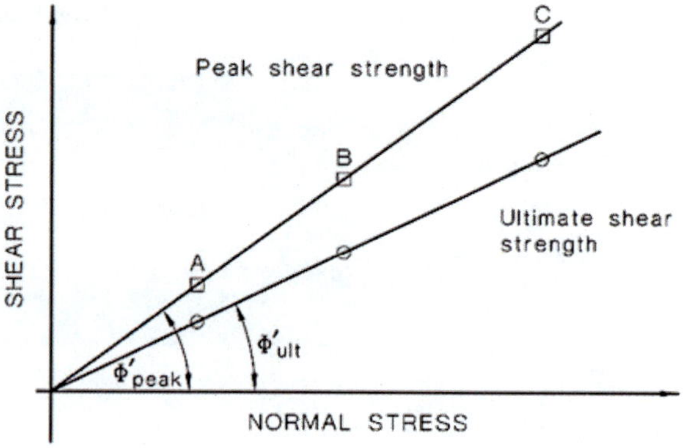

Fig. 2.3 Determination of friction angle from direct shear test (AS 1289.6.2.2, 1998)

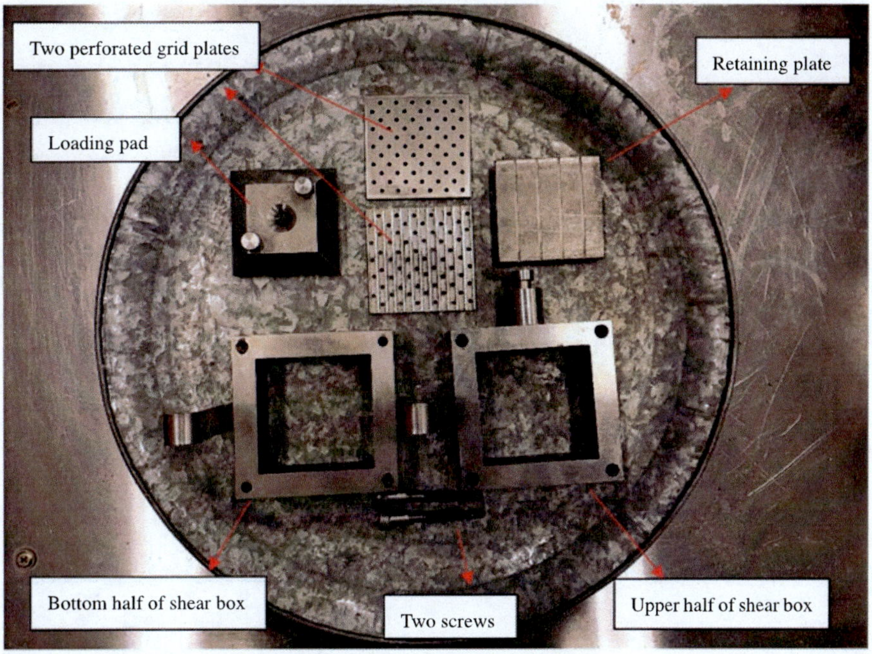

Fig. 2.4 Components of shear box

To perform the tests, it is first necessary to prepare the sample in the shear box. Figure 2.4 shows the components of this box. The upper half of the shear box is placed on the lower half, ensuring that all the inscribed holes match. Then, the two clamping screws are placed in their proper holes (clearance holes) to hold the two shear box halves in place.

The retaining plate at the bottom of the shear box is fixed, and then the perforated grid plate is fixed, noting that the serration orientation should be perpendicular to the shear force. A spoon is used to transfer the sample into the shear box several times, noting that several blows need to be applied with a tamping tool to mechanically dense the sand (20 times should be adequate, keeping the tamping light and uniform). The height clearance from the lid of the chamber to the final sand level should be measured four times near each wall of the shear box. This measured height should be more than 5 mm.

The perforated grid plate is fixed on top (minding the serration) with the loading pad, and then the assembly is placed into the shear box container, as shown in Fig. 2.5, slotting the end onto the push rod. Later, it is necessary to ensure that the lever beam is horizontal at a free position, and the required surcharge weight is added to produce the required normal stress, and then the jack screw is slowly released.

The horizontal and vertical Linear Variable Differential Transformers (LVDTs) and load cell are fixed, ensuring that the equipment is alive. The connecting bolts are removed to avoid the loading being resisted by bolts, and the lifting screws are

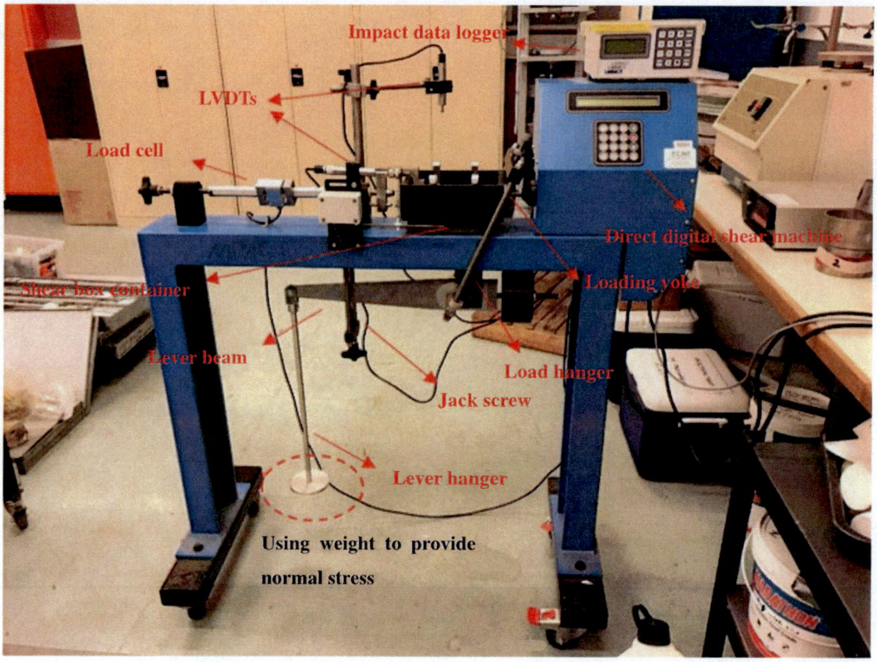

Fig. 2.5 Direct shear apparatus

wound down a 'half a turn' each to separate the upper and lower halves by about 0.5 mm. These are then backed off a few turns.

Trial direct shear tests were conducted at Griffith University. The vertical stresses applied were 50, 150 and 250 kPa. The plotted horizontal movement versus shear stress diagram, horizontal versus vertical displacement diagram, and normal stress versus shear stress diagram are provided in Figs. 2.6, 2.7 and 2.8, respectively. Table 2.1 summarizes the data obtained from these tests. The parameters of peak and ultimate friction angles were determined as 42° and 38°, and the cohesion was determined as 5.25 and 2.59 kN/m^2 respectively.

2.2.1.2 Consolidation Tests

A fully saturated soil with a property of low permeability will gradually decrease its volume under loading because of change in effective stress. This soil behavior may be due to a reduction in pore water pressure or drainage of pore water pressure, and this process is called 'consolidation'. In contrast, the process of swelling is an increase in soil volume under loading because of negative excess pore water pressure. This section reviews the one-dimensional consolidation of soil, where loading increment is applied vertically, and the lateral strain is restricted.

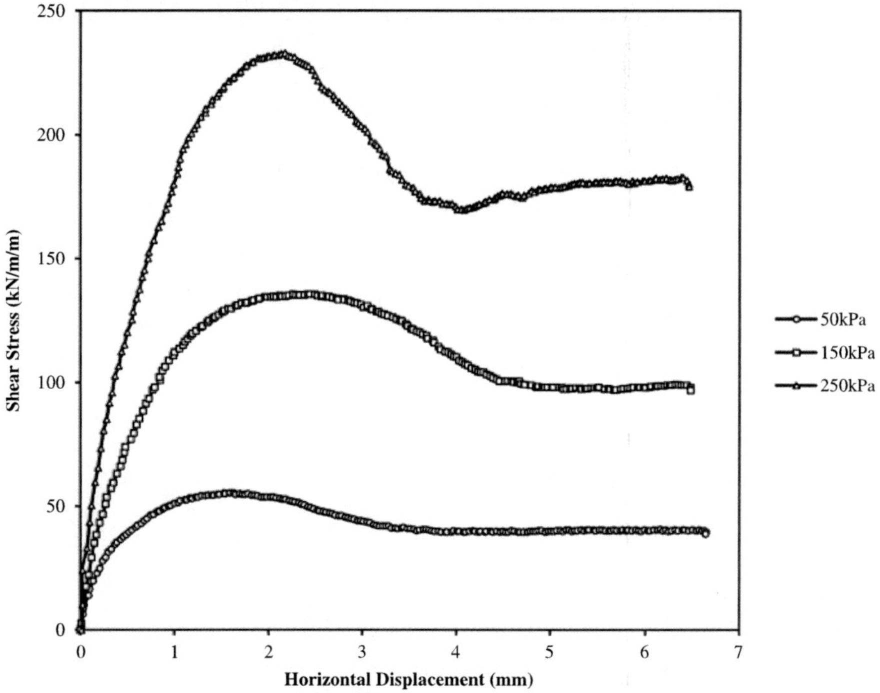

Fig. 2.6 Horizontal movement versus shear stress

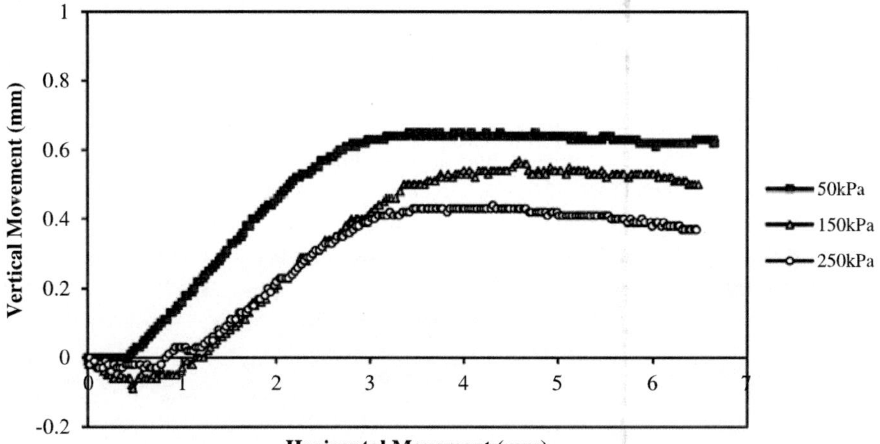

Fig. 2.7 Horizontal movement versus vertical movement

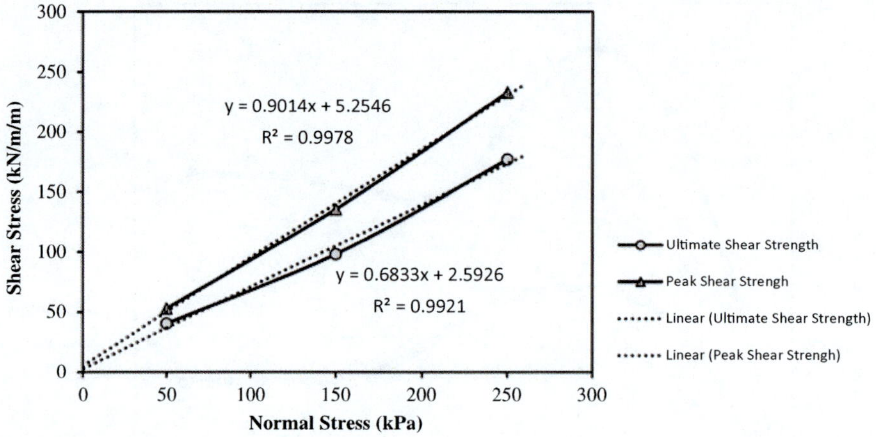

Fig. 2.8 Normal stress versus shear stress

Table 2.1 Ultimate and peak shear stresses and normal stresses

Ultimate shear stress	Maximum shear stress	Normal stress
40.28	52.78	50
98.06	135.56	150
176.94	233.06	250

As shown in Fig. 2.9, the height of an element will decrease from initial height of H_0 to H_1 when vertical stress is applied, and, at this stage, the void ratio, *e*, can be determined. By plotting *e* under each load stage versus the vertical effective stress, σ', and in logarithmic scales, the compressibility characteristics can be preliminarily determined, as shown in Fig. 2.10.

The *e*-log σ' curve illustrates a one-dimensional compression line (1DCL), as shown in Fig. 2.10. The slope of this virgin compression line is called a 'compression index', with a symbol of C_c, and the slope of unload–reload is called the 'expansion index', with a symbol of C_e or C_r (re-compression index). The compression index can be determined as follows:

$$C_c = \frac{e_0 - e_1}{\sigma_1^{'} - \sigma_0^{'}} \quad (2.3)$$

where e_0 and $\sigma_0^{'}$ are the initial void ratio and initial stress.

The coefficient of volume compressibility, m_v, is defined as the volume change per unit increase in effective stress per unit volume. It is related to e_0 and e_1, as illustrated in Eq. 2.4. The constrained modulus or one-dimensional elastic modulus, $E_{oed}^{'}$ is the reciprocal of m_v:

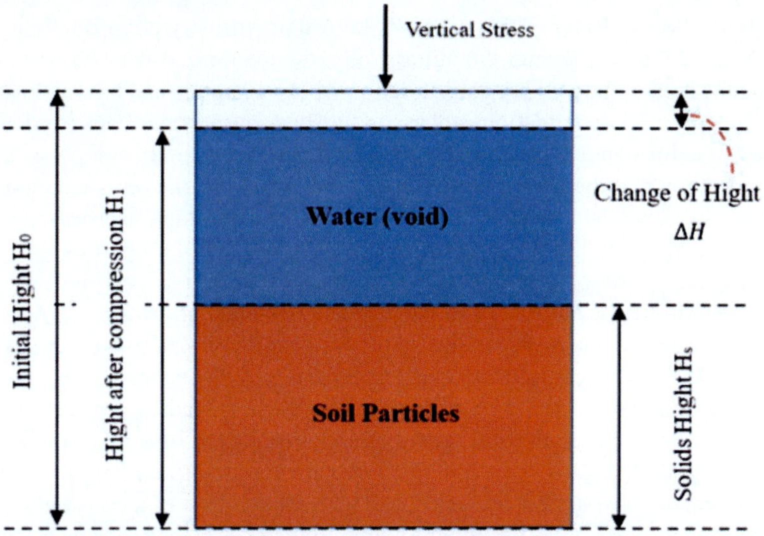

Fig. 2.9 Basic theory of element under stress

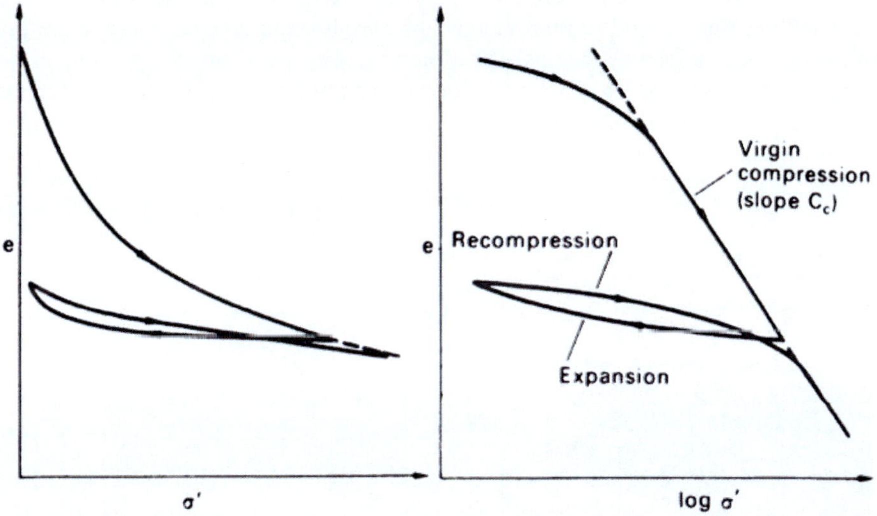

Fig. 2.10 Void ratio–effective stress relationship (Craig 1983)

$$m_v = \frac{1}{1 + e_0}\left(\frac{e_0 - e_1}{\sigma_1^{'} - \sigma_0^{'}}\right) \quad (2.4)$$

The consolidation of soil mass involves a time-dependent change in volume. The settlement of soil under stress refers to the readjustment of soil particles resulting from dissipation of water. This process is known as 'primary consolidation'. It has been reported that, the greater the initial void ratio, the more water will be squeezed out, and the greater the primary consolidation will be (Samtani and Nowatzki 2006a). After the primary consolidation process is finished, the soil particles reorient their position or deform under constant load, which leads to secondary compression. This process is known as the 'creep behavior' of soil. Given that the second compression is also a time-dependent process, as with the primary consolidation process, it is vital to plot the diagram with time or lg time to determine the soil properties.

Traditionally, consolidation tests which are also known as oedometer tests are used to determine the soil characteristics. The consolidation tests are governed by AS 1289.6.6.1 (Standards Australia 1998b) and ASTM-D-2435–06 (ASTM International 2007b). The equipment used to perform oedometer tests contains a load device (Fig. 2.11a) that can provide the required vertical loads; a consolidation cell (Fig. 2.11b) used to hold the soil specimen that is confined by a ring; a monitoring device, such as dial gauges or LVDTs to obtain the deformed height of the soil element; and a stopwatch to record time. The configuration of a consolidation cell may be different, yet normally contains two porous stones, a load cap and a confining ring.

To obtain the plotted compression versus time diagram depicted in Fig. 2.12, an oedometer test should be performed. The procedure for this test is as follows:

A cutting ring is used to prepare the soil sample and make the top and bottom of the sample confined by the consolidation ring, also, avoid any compaction on the sample. Later, filter paper is placed on the top and bottom as protection to stop the drainage plates from being contaminated. The filter paper size must be appropriate, and the paper should avoid contacting the ring.

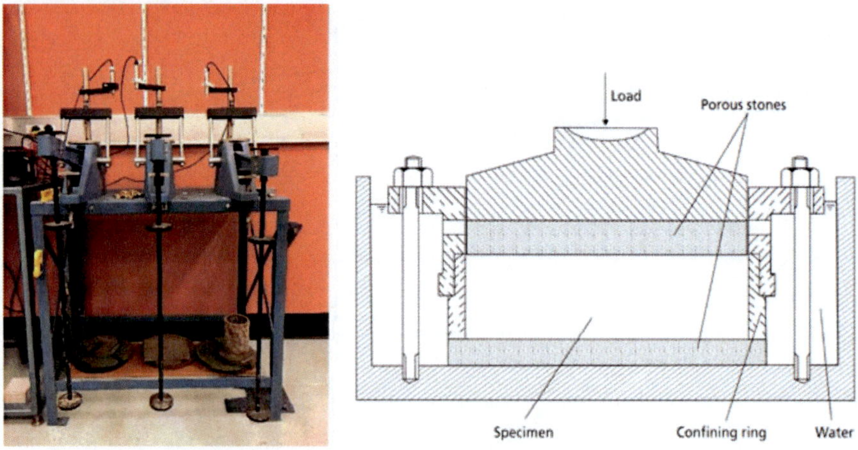

(a) Loading device (b) Consolidation cell (Craig, 1983)

Fig. 2.11 Apparatus of consolidation test

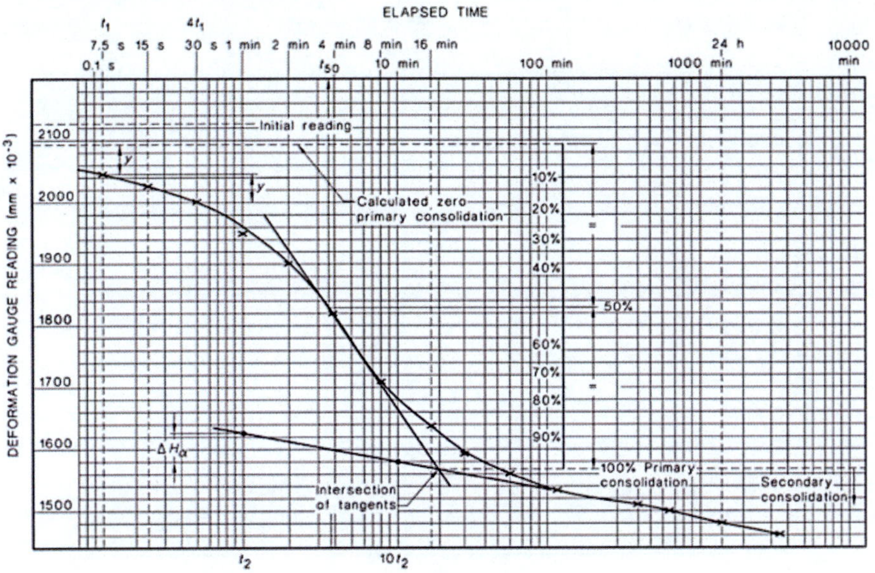

Fig. 2.12 Typical compression versus time (AS 1289.6.6.1, 1998)

The porous disc is then located, which should be soaked in water for 24 h aforehand, and then the prepared samples (with ring and filter paper) are carefully placed on this porous disc. Another porous disc is added on top of the samples, and the cutting ring is clamped. Later, the top cap is positioned. The cell is placed on the oedometer, and the loading yoke is placed onto the top cap. There must be contact without any extra loading applied. Later, the cell is filled with water to allow the sample to be fully saturated. The dial gauge or LVDT is set up to record the vertical settlement of the sample under loading.

A one-dimensional consolidation trial test of fine clay was conducted at Griffith University, and the results are provided in Fig. 2.13. Based on the Casagrande (1936) method or log time method, the theoretical curve contains three parts. The first part is approximately parabolic, the second part is closely linear, and the third part is an asymptote referring to the horizontal axis at $U_v = 100\%$.

As depicted in Fig. 2.13, the dial gauge reading of a_0 is related to the initial reading, and as corresponds to $U_v = 0$. Through finding the intersection of the two tangent lines, as shown in Fig. 2.13, the a_{100} can be determined, and the gauge reading that represents a_{50} and t_{50} can be determined later. The compression between a_s to a_{100} is known as primary consolidation. The time factor, T_v, can be determined once the t_{50} value is determined, and the C_v can then be determined via Eq. 2.5. Note that the d can be taken as half of the average thickness of the specimen because the oedometer test is a two-way drainage condition.

The theory curve that plots the diagram of root time versus dial gauge reading is linear up to 60% consolidation, and $\sqrt{t_{90}}$ is used for extrapolation of the linear part

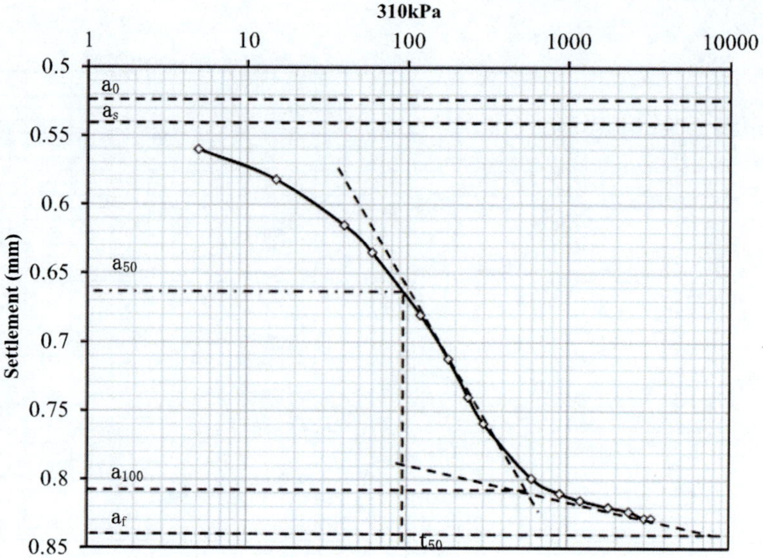

Fig. 2.13 The log time method

of the curve (Taylor 1948). As shown in Fig. 2.14, similar to the Casagrande method, the settlements between a_0 and a_s represent initial compression, the settlements between a_s and a_{100} give primary consolidation and the settlements between a_{100} and a_f illustrate secondary compression. The coefficient of consolidation is governed by Eq. 2.6:

$$C_v = \frac{0.196d^2}{t_{50}} \tag{2.5}$$

$$C_v = \frac{0.848d^2}{\sqrt{t_{90}}} \tag{2.6}$$

Referring to these two methods, the initial compression ratio, primary consolidation ratio and second consolidation ratio can be expressed by Eqs. 2.7–2.9. Using the data from this test, the relationship between the void ratios versus vertical pressure can also be found with Fig. 2.15:

$$\text{Initial compression ratio: } r_0 = (a_0 - a_s)/(a_0 - a_f) \tag{2.7}$$

$$\text{Primary compression ratio: } r_p = (a_s - a_{100})/(a_0 - a_f) \tag{2.8}$$

$$\text{Secondary compression ratio: } r_s = 1 - (r_0 + r_p) \tag{2.9}$$

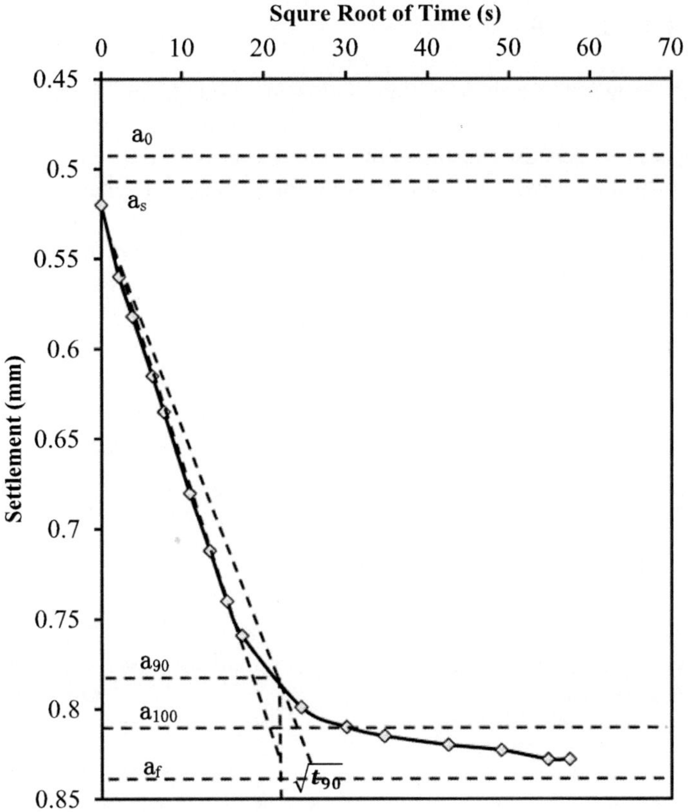

Fig. 2.14 The root time method

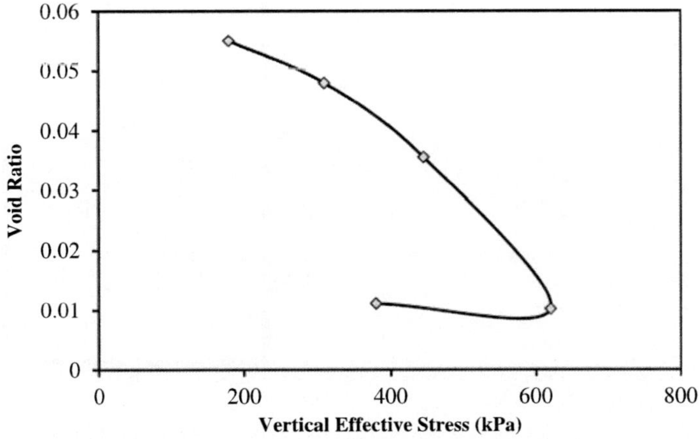

Fig. 2.15 Void ratios versus vertical pressure of consolidation tests

2.2.1.3 Triaxial Tests

The disadvantage of direct shear tests is that the shear failure surface is predetermined as a horizontal plane. Further, direct shear tests are mostly performed on a drained specimen because the apparatus cannot control drainage of water. For clay that contains water, triaxial tests are the traditional method to determine the strength parameters.

With a soil element in failure state, the principal stresses $\sigma_1^{'}$ and $\sigma_3^{'}$ are applied vertically and horizontally, as shown in Fig. 2.16. Shear stress and normal stress act on the failure plane. Through geometrical relationship in triangle *abc*, and establish force equations of equilibriums:

$$\sigma_3^{'} d_s \sin\theta - \sigma_f^{'} d_s \sin\theta + \tau_f d_s \cos\theta = 0 \tag{2.10}$$

$$\sigma_1^{'} d_s \cos\theta - \sigma_f^{'} d_s \cos\theta - d_s \tau_f \sin\theta = 0 \tag{2.11}$$

By moving Eq. 2.11 into Eq. 2.10, stresses $\sigma_f^{'}$ and τ_f are provided in Eqs. 2.12 and 2.13:

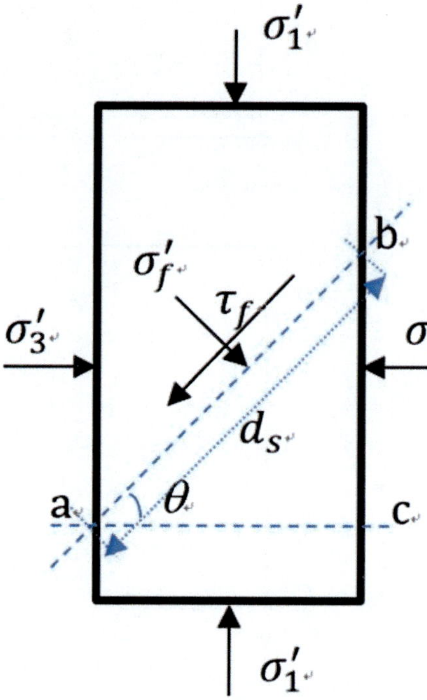

Fig. 2.16 Theory of soil element at failure state

$$\sigma'_f = \frac{1}{2}\left(\sigma'_1 + \sigma'_3\right) + \frac{1}{2}(\sigma'_1 - \sigma'_3)\cos 2\theta \tag{2.12}$$

$$\tau_f = \frac{1}{2}(\sigma'_1 - \sigma'_3)\cos 2\theta \tag{2.13}$$

The relationships between σ'_f and τ_f and σ'_1 and σ'_3 can be provided by a Mohr–Coulomb diagram, as shown in Fig. 2.17. The failure envelope represents the function, as previously provided in Eq. 2.2. Through the geometrical relationship, it can be found that:

$$\mathrm{AD} = \mathrm{RD}\sin\varphi' = \left[c'\cot\varphi' + \frac{1}{2}\left(\sigma'_1 + \sigma'_3\right)\right]\sin\varphi' = \frac{1}{2}(\sigma'_1 - \sigma'_3) \tag{2.14}$$

Equation 2.14 can be rearranged into Eqs. 2.15 and 2.16 for clay as follows:

$$\sigma'_1 = \sigma'_3\tan^2(45^0 + \frac{\varphi'}{2}) + 2c'\tan(45^0 + \frac{\varphi'}{2}) \tag{2.15}$$

$$\sigma'_3 = \sigma'_1\tan^2(45^0 - \frac{\varphi'}{2}) - 2c'\tan(45^0 - \frac{\varphi'}{2}) \tag{2.16}$$

For cohesionless soil, such as sand, $c = 0$, the principle stresses can be determined as follows:

$$\sigma'_1 = \sigma'_3\tan^2(45^0 + \frac{\varphi'}{2}) \tag{2.17}$$

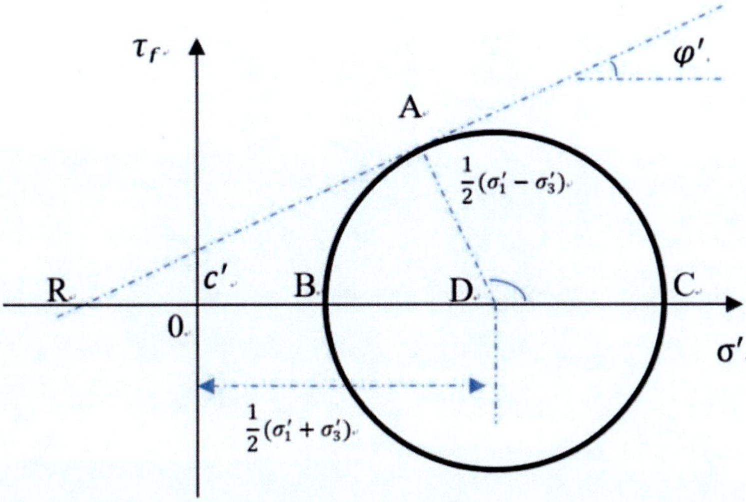

Fig. 2.17 Mohr–Coulomb failure criterion

$$\sigma_3^{'} = \sigma_1^{'}\tan^2(45^0 - \frac{\varphi^{'}}{2}) \tag{2.18}$$

As reviewed above, for one type of soil, there will be a point that represents the failure condition under particular principle stresses from the Mohr–Coulomb diagram (Point A in Fig. 2.17). By changing the state of stress, a series of points can be determined, all of which are in the failure envelope function (when soil fails). Triaxial tests can determine certain parameters (such as cohesion and friction angle) through applying different stresses vertically ($\sigma_1^{'}$) and horizontally ($\sigma_3^{'}$) until the test sample reaches failure and plotting Mohr–Coulomb diagrams. The primary advantage of triaxial tests is that they are suitable for all types of soils because they allow saturated soil with relatively low permeability to be consolidated. Compared with direct shear tests, this test has a device to control the drainage of soil, and the amount of water drained out of soil can be measured.

Triaxial tests are governed by AS 1289.6.4.1 (Standards Australia 2016a) and AS 1289.6.4.2 (Standards Australia 2016b), and the equipment is provided in Fig. 2.18. The soil sample is first prepared in a cylinder and protected by a rubber membrane. Later, hydraulic pressure is applied on this sample, and this pressure is simultaneously applied horizontally and vertically. Then, the vertical deviation stress is applied through the loading ram onto the loading cap until shear failure occurs.

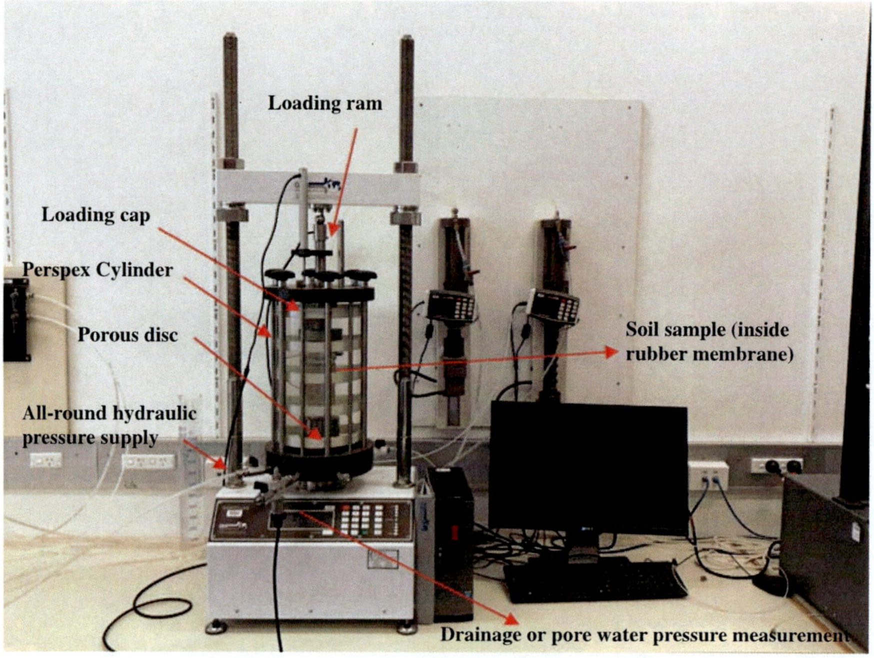

Fig. 2.18 The triaxial apparatus

For the Unconsolidated Undrained (UU) test, the drainage and consolidation process are not allowed during the application of confining or axial pressure. The performance of a UU test can determine the undrained shear strength, and can model the response of soil applied with a rapid axial load. Different to this test, the Consolidated Drained (CD) test can allow the consolidation of soil while confining pressure before the axial pressure is applied. During this test, the loading rate is slow enough to ensure there is no excess pore water pressure built. The CD test simulates the long-term drained condition thus the effective strength of soil can be determined. This equipment can also perform the Consolidated Undrained (CU) test. Similar to the CD test, the sample is consolidated first, yet the vertical pressure is applied relatively quickly. The amount of pore pressure can be measured during CU tests, so these tests can obtain the effective and total strength parameters of soil.

A UU trail test was conducted in a laboratory at Griffith University, with confinement pressures of 150, 300 and 600 kPa. While the vertical deviation pressure was applied, the strain data were determined, and the results are provided in Fig. 2.19. The normal stresses that represented the failure of soil were determined as 789, 1,160 and 1,608 kPa. When these values were plotted into the Mohr–Coulomb diagram, the function of the failure envelope could be determined (Fig. 2.20) and the parameters of soil strength could be obtained later. As shown in Fig. 2.20, the intersection between this function (failure envelope) and y-axis gave the cohesion, and the slope of this function was the friction angle.

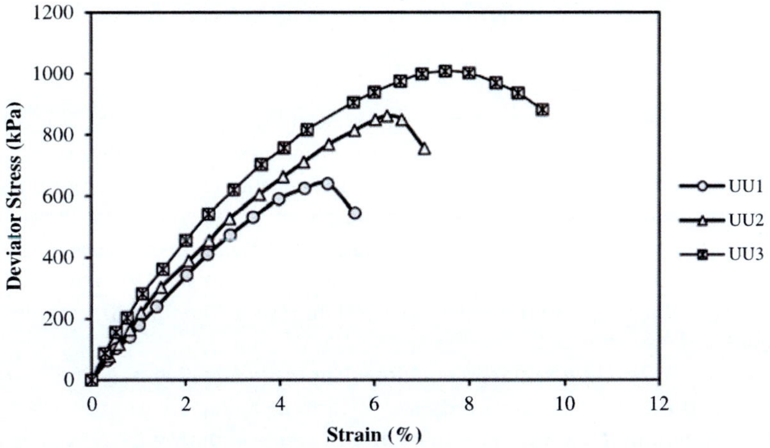

Fig. 2.19 Set of three triaxial tests

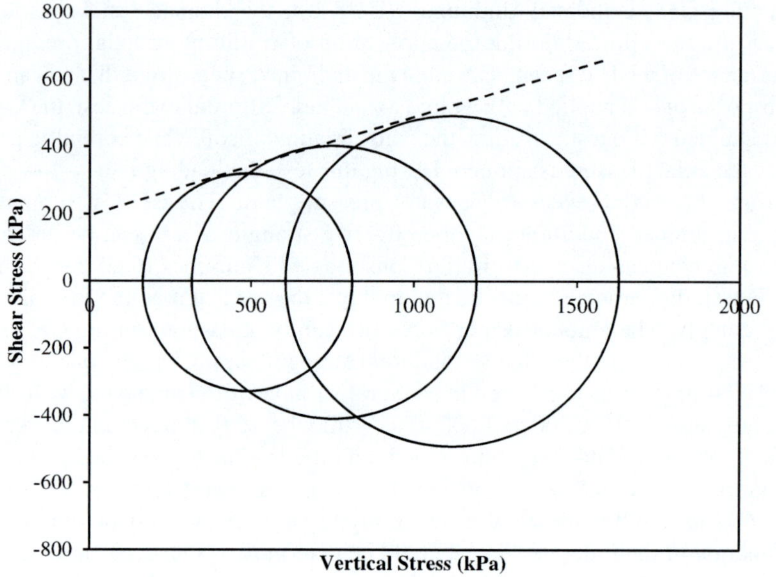

Fig. 2.20 Mohr–Coulomb circle of triaxial tests

2.2.2 *In-Situ Tests*

2.2.2.1 Standard Penetration Test (SPT)

In 1902, the SPT was introduced by the Raymond Pile Company, and it is now one of the most widely used in-situ tests because of its low cost, simplicity and ability to be rapidly conducted. As shown in Fig. 2.21, in this test, a split barrel sampler is driven into the soil by a drop hammer (weight of 63.5 kg) with falling height of 30 in (760 mm) to achieve to a 6 in (150 mm) penetration, this process is known as 'seating' process. After the second and third penetration finished (6 in, respectively), the number of drops (penetration depth of 300 mm) of the hammer is recorded. This is known as the uncorrected N-value or SPT blow count.

There are three common types of drop hammer (safety, donut and automatic hammer) and four types of drill rods for performing the tests, which consequently lead to the amount of energy being transferred to the soil during the SPT. The actual energy transferred is less than the theoretical energy, which is influenced by numerous factors, such as eccentric loading and frictional losses. The constitutive properties of the soils cannot vary with the equipment used; thus, the N-value should be corrected to N_{60}, representing a standardized Energy Ratio (ER) of 60%. This corrected N-value or N_{60} can be determined through Eq. 2.19 with consideration of borehole diameter, rod length and energy ratios.

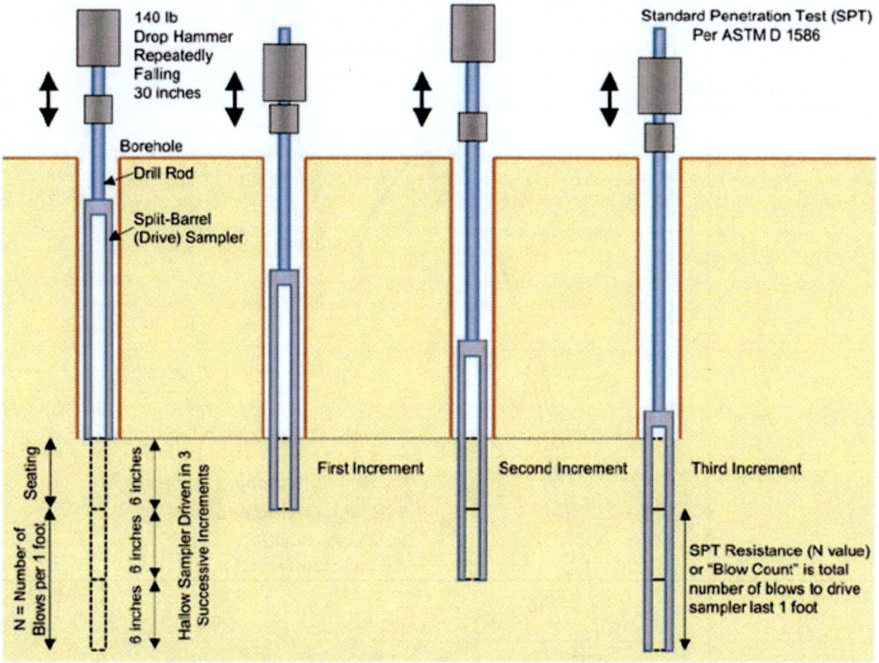

Fig. 2.21 Standard Penetration Test schematic (Mayne et al. 2002)

$$N_{60} = N\left(\frac{ER}{60}\right) \tag{2.19}$$

The N-values of similar materials increase with increasing overburden pressure (σ'_{vo}) at the test depth. For coarse-grained soils, such as sand and gravel, the N_{60} is conventionally normalized to $(N_1)_{60}$, which refers to 1 atmosphere effective overburden stress. This is governed by Eq. 2.20, while the overburden correction factor is provided in Eqs. 2.21 (Liao and Whitman 1986) and 2.22 (Peck et al. 1974), and Figs. 2.22 (Skempton 1986) and 2.23 (Samtani and Nowatzki 2006a).

$$(N_1)_{60} = C_N N_{60} \tag{2.20}$$

$$C_N = \left(\frac{p_a}{\sigma'_{V0}}\right)^n \tag{2.21}$$

$$C_N = 0.77log\left(\frac{40}{\sigma'_{V0}}\right) \leq 2 \tag{2.22}$$

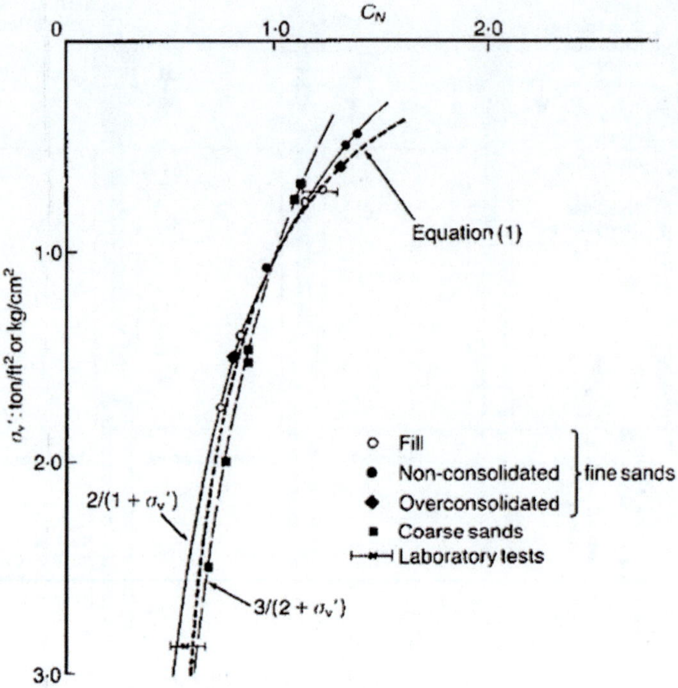

Fig. 2.22 Overburden correction factor (Skempton 1986)

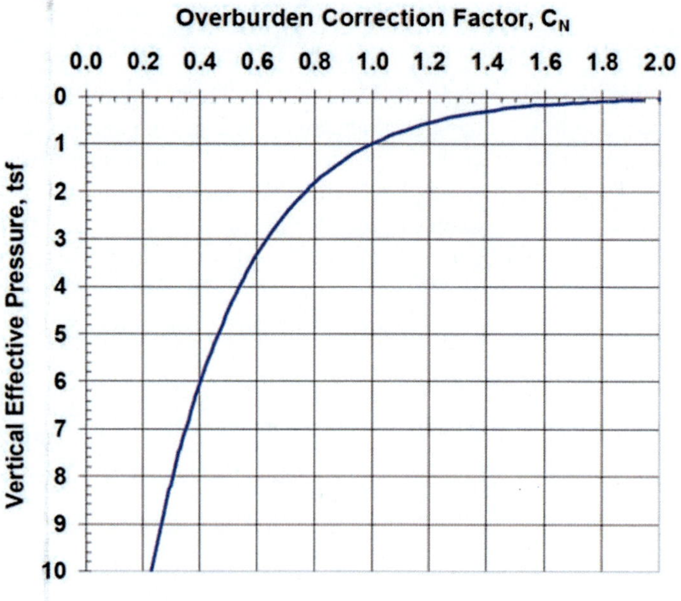

Fig. 2.23 Variation of overburden correction factor C_N (Samtani and Nowatzki 2006a)

Table 2.2 Empirical values for φ, γ & I_D of granular soils based on corrected N-value (Bowles 1977)

Description	Very loose	Loose	Medium	Dense	Very dense
Relative density I_D	0–0.15	0.15–0.35	0.35–0.65	0.65–0.85	0.85–1.00
Corrected SPT N	0–4	4–10	10–30	30–50	50 +
Friction angle φ	25–30°	27–32°	30–35°	35–40°	38–43°
Unit weight γ (kN/m^3)	11.0–15.7	14.1–18.1	17.3–20.4	17.3–22.0	20.4–23.6

Table 2.3 Empirical values for unconfined compressive strength (q_u) and consistency of cohesive soils based on uncorrected N-value (Bowles 1977)

Consistency	Very soft	Soft	Medium	Stiff	Very stiff	Hard
q_u (kPa)	0–24	24–48	48–96	96–192	192–384	384 +
SPT N-value	0–2	2–4	4–8	8–16	16–32	32 +
Unit weight γ (saturated) kN/m^3	15.8–18.8	15.8–18.8	17.3–20.4	18.8–22.0	18.8–22.0	18.8–22.0

where:

$(N_1)_{60}$ = SPT N-value corrected for energy and overburden pressure.
N_{60} = SPT N-value corrected for 60% energy transfer.
C_N = overburden correction factor.
P_a = atmospheric pressure.
σ'_{V0} = effective vertical stress at the sample depth.
n = exponent typically equal to 1 in clays and 0.5 in sandy materials.

The corrected and uncorrected N-values can be used to determine not only the relative density, I_D, and undrained shear strength, C_u, but also the angle of internal friction, φ; unit weight, γ; and unconfined compression strength, q_u, as provided in Tables 2.2 and 2.3. It should be noted that, for soils that contain gravel-size particles, Table 2.2 may be unreliable, and the correlations can only be used as a rough estimation. Further, Table 2.3 should only be used for preliminary design purposes because the unconfined compressive strength of cohesive soils is crude and unreliable (Hannigan et al. 2006).

2.2.2.2 Cone Penetrometer Test (CPT and CPTu)

The cone penetrometer is one of the most versatile tools available for soil exploration (Lunne et al. 1997). Tests are mostly performed using a CPT test rigs truck, as shown in Fig. 2.24. This rig has a hydraulic jack inside the truck, and the weight of the truck itself is used as a reaction mass to push the CPT rod into the ground. During this test, a penetrometer consisting of a friction sleeve and cone is used to determine the tip (q_c) and shaft (f_s) resistance during penetration. Performing CPT aims to

Fig. 2.24 Schematic of CPT truck (Samtani and Nowatzki 2006a)

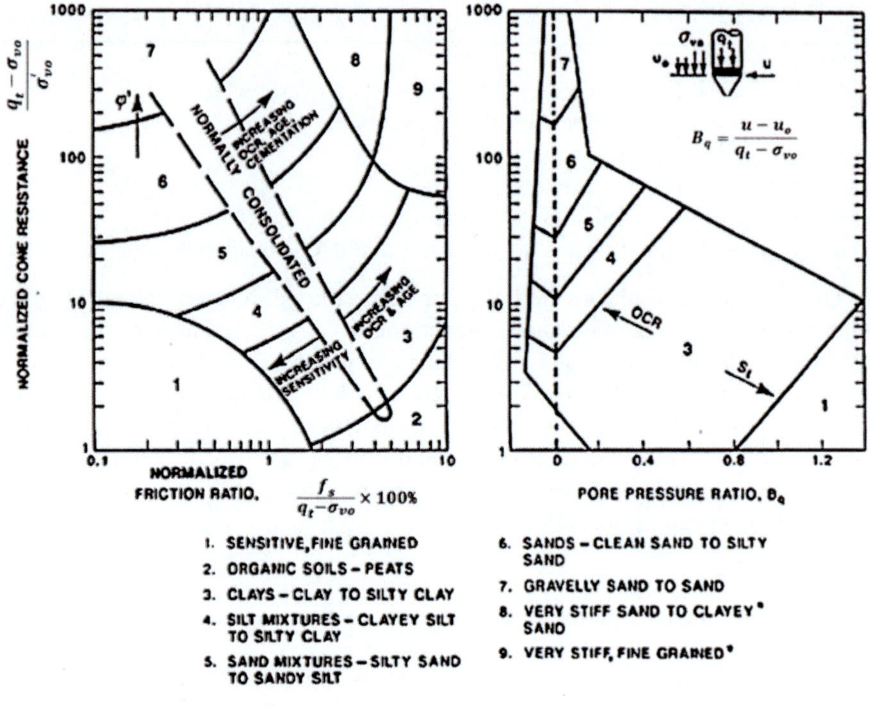

Fig. 2.25 Soil behaviour type classification chart (Robertson 1990)

identify the stratigraphy and provide soil layer information. Based on the calculation as illustrated in Fig. 2.25, the property of soil can be obtained. It can also be used for deep foundation design because the CPT process is similar to driven pile installation.

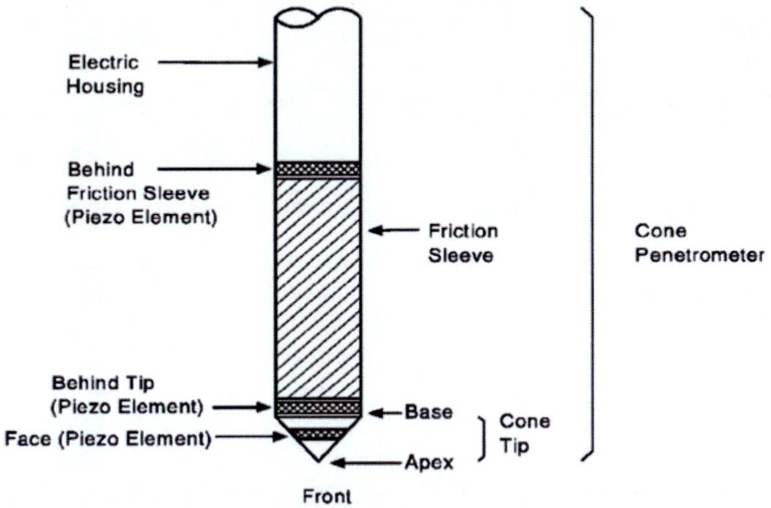

Fig. 2.26 Piezocone measurement (CPT_u) (Samtani and Nowatzki 2006a)

The cone tip resist, q_c, is determined by F_c/A_c (force and area of the piezocone), and the friction resistance, f_s, obtained from the friction sleeve is determined by F_s/A_s (force and area of the sleeve). For a standard CPT, the soil tip resistance q_t is equal to cone tip resistance q_c. As shown in Fig. 2.25 (left side of the diagram), after determined the normalized cone resistance $((q_t - \sigma_{vo})/\sigma_{vo}^{'})$ and friction ratio $(f_s/(q_t - \sigma_{vo}))$, the soil classification can be accomplished after finding the zone in figure.

However, for fine-grained soils, the excess pore water pressure should be considered; thus, a more sophisticated device is required. As shown in Fig. 2.26, a piezocone can be used because this device can measure the excess pore water pressure (CPT_u). If this equipment is used, pore water ratio $((u - u_o)/(q_t - \sigma_{vo}))$ is used for soil classification as illustrated from the right side of Fig. 2.25. CPT_u tests are faster and cheaper than drilling and sampling. They also provide more consistent and reliable results than SPT N-values due to the load condition is easier to control. The disadvantage of CPT_u is that it requires skilled operator to run the test, and no soil samples are obtained.

Typical result of CPT_u includes the shaft and tip resistance as well as water pressure shown in Fig. 2.27. Also, CPT_u data can be used to interpret the peak angle of shear resistance ($\varphi_{\max}^{'}$), relative density (I_D), undrained shear strength (C_u) and OCR based on proposed equations and diagrams (Jamiolkowski et al. 2003; Mayne 2007). There has been intensive research and tests performed to predict the capacity of piles based on CPT data. Abu-Farsakh et al. (2017) indicated that the Schmertmann (1978) method; De Kuiter and Beringen (1979) method; LCPC method (Bustamante and Gianeselli 1982); and recently developed method by Lehane et al. (2013) can all provide better results for pile capacity than can the other CPT methods.

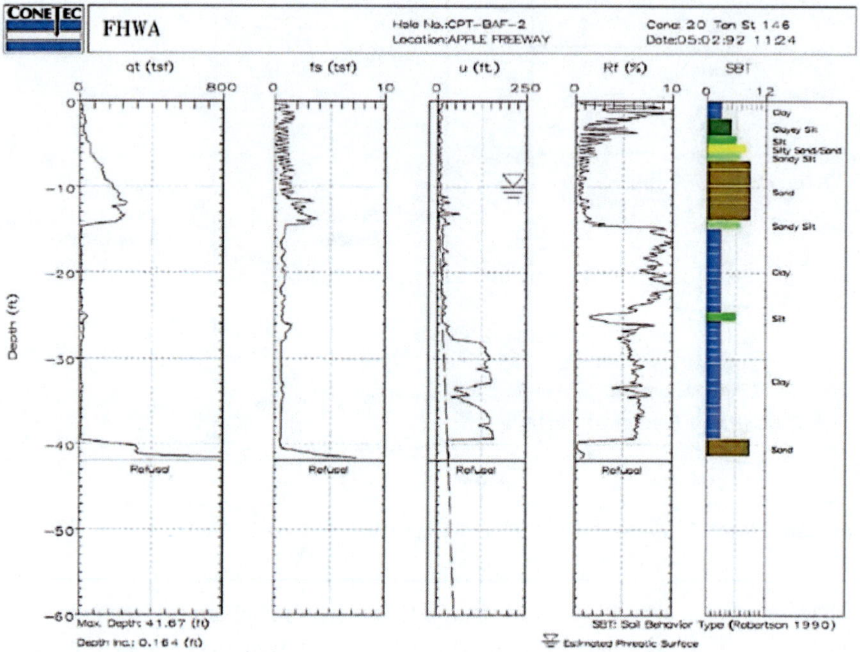

Fig. 2.27 Piezocone results for Apple Freeway Bridge (Samtani and Nowatzki 2006a)

2.3 Types of Piles

2.3.1 General Pile Categorization

Pile foundation or deep foundation is the slender structural element that transfers loading from the upper structure to the soil (Tomlinson 2001). Compared with shallow foundations, there are situations in which the use of pile foundation is preferred, such as when the pile is designed to resist large concentrated loads, when the near surface of soil layers is not stiff, and when displacements should be kept small for settlement-sensitive buildings. There are numerous types of pile foundation, and they can be broadly categorized into concrete, steel, timber and composite piles, based on the materials used. A detailed classification system of pile foundations is provided in Fig. 2.28, referring to the installation methods and equipment used for installation.

Timber piles have been used for over a millennium. Timber is an ideal selection for piling when used for cohesion pilling or for resisting light loads. The advantages of timber include its high strength to weight ratio, ease of handling and ability to be cut during construction (Tomlinson and Boorman 2001). Timber piles are suited for a modest load because this material is vulnerable and can be easily damaged at the pile head and toe when driven hard. In addition, when pile foundations are required to be driven into dense sand or gravel or to be used as end bearing piles on rocks,

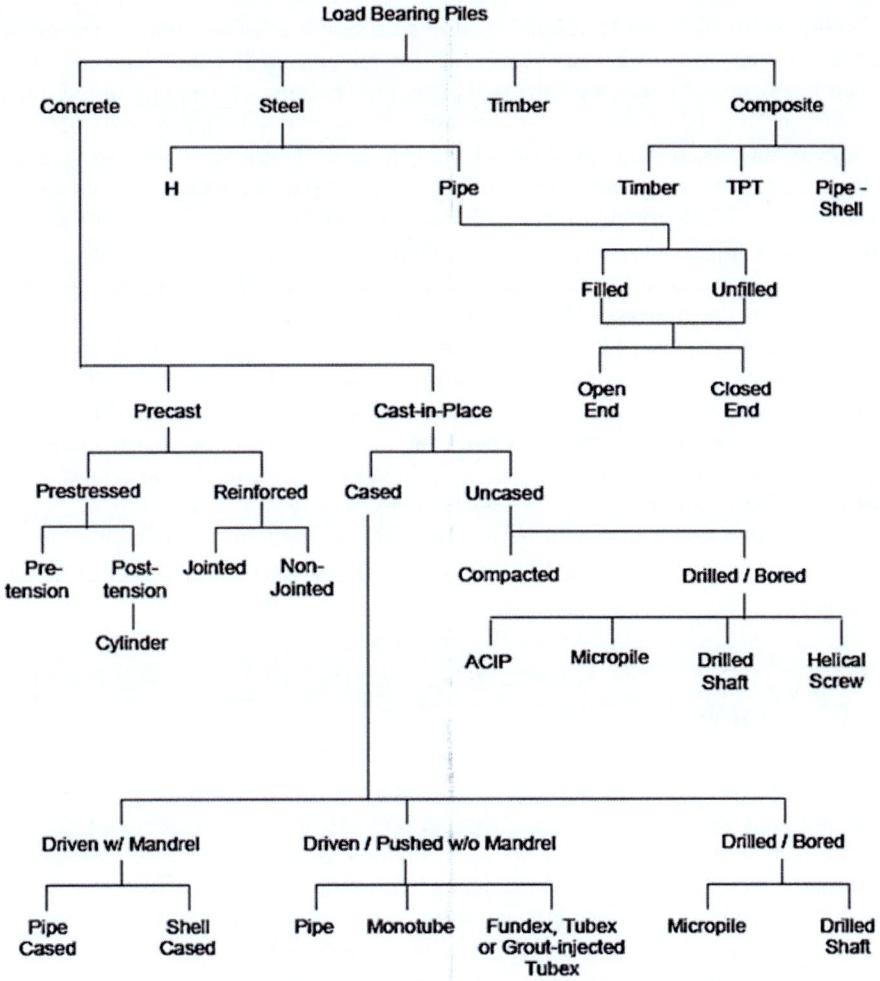

Fig. 2.28 Pile classification chart (Hannigan et al. 2006)

the use of a timber pile can cause excessive bending, and cracking may occur. Under such situations, a steel pile is preferred because of its properties of a relatively small cross-section area and high strength. The configurations of the steel pile include H-piles, pipe piles and screw piles. H-piles with high strength and improved corrosion resistance in atmospheric and marine environments are detailed in ASTM A588 and ASTM A690, respectively. However, these H-piles are difficult to obtain.

The most common pile foundations used currently are concrete piles. These piles can be categorized into bored piles (cast-in-situ piles) and precast piles. Precast piles can be classified into prestressed concrete piles and reinforced concrete piles. The configurations of precast piles and reinforced concrete piles are similar. The difference between these two types of piles is that pretensioned or post-tensioned steel reinforcements are used in prestressed piles, and, because of using these reinforcements, prestressed concrete piles are more resistant to weathering and corrosion under continued compression loads (Hannigan et al. 2006). From another perspective, prestressing or tensile stress results in damage to prestressed piles during driving and handing. Compared with precast piles, the advantage of bored concrete piles is that there is no need to consider the piles' transfer to the construction site, and no need to predetermine pile lengths. However, casting onsite may lead to dust pollution when mixing concrete and welding the reinforcement cage.

The pile foundation can also be categorized into compressive, uplift and lateral loaded piles based on the loading conditions. Figure 2.29 depicts the loading situations of pile foundations. Normally, a pile foundation can reach the bearing stratum and loads can be vertical and horizontal (Fig. 2.29a, d). In addition to the individual types of loading or combination loading, pile foundations can also be needed when there is a scour around effect, liquefaction effect or seismic event (Fig. 2.29e, h). Further, a pile foundation may be used as a fender system or retaining structure to resist earth pressure caused by excavation and undesirable movement caused by downdrag (Fig. 2.29i, k).

2.3.2 *Cast-In-Situ Piles*

Pile foundations perform well in high-rise buildings because of their high value of bearing capacity. However, during the process of cast-in-situ construction, some soil deposits may remain at the base area after drilling the hole. These soil deposits are caused by a collapse of soil layers and are very difficult to remove from the drilled hole. This leads to decreased end resistance capacity of the pile and increased pile settlement. As a result of these remaining soil deposits from the base of the hole, SLTs conducted in Taiyuan city, China, found that the capacity and settlement of some piles from a mansion were much lower than the designed requirements (Shi 2005). Further, the drilling operation loosening the soil underneath the base of the bored hole also leads to excessive working load settlement.

Another problem associated with the bored pile is the use of admixtures, such as bentonite or polymeric slurry (slurry support), which are used to avoid the collapse of soil layers into bored holes. The use of these materials may lead to capacity decrease of the pile shaft because these admixtures combine with soil and water to create a layer between the soil and pile, which consequently leads to a decrease in the friction resistance of the pile. It has been reported that this admixture layer or composite layer decreases 30 to 40% of the pile-bearing capacity (Li et al. 2000).

Such limitations of cast-in-situ concrete piles can be handled by grouting techniques. Grouting equipment can be categorized into a flat jack system, which consists of grout delivery pipes connected to a steel plate with a rubber membrane and sleeve-port system that comprises two to four U-tubes installed at the bottom of the pile. This

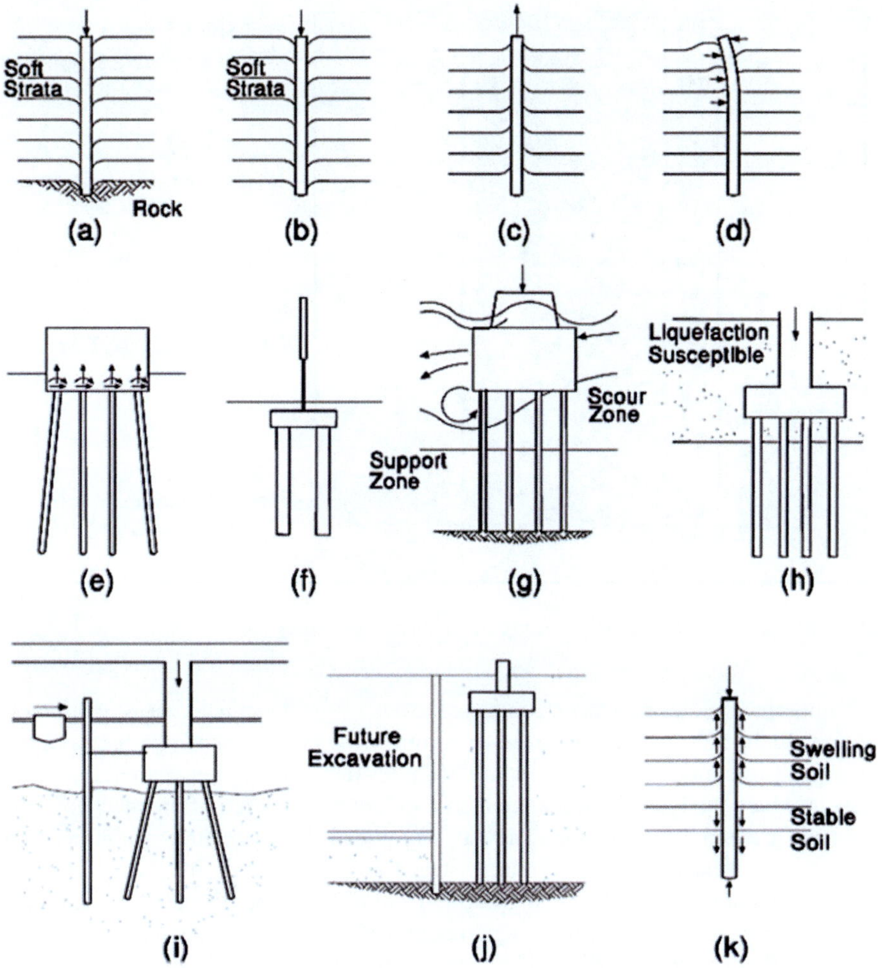

Fig. 2.29 Situations in which deep pile foundations may be needed. (modified from Vesic 1977)

U-tube is covered by rubber and can be arranged in various configurations (Dapp et al. 2006).

Cement is a widely used grouting material. It is blended into various ratios and then injected into the pile toe under high pressure, transferred by a grouting tube. Through this technology, the grouting application restores the original density of the base soil and reduces the settlement of the pile when loading transfers from upper structures (Tomlinson and Boorman 2001). Moreover, as shown in Fig. 2.30, the grouted cement can force out the soil-admixture layer that will lead to a decrease of the shaft resistance, and create a new solid cement layer, this layer bonds the soil as well as the concrete surface, and consequently improves the friction between soil layers and pile surface (Zhou et al. 2017a).

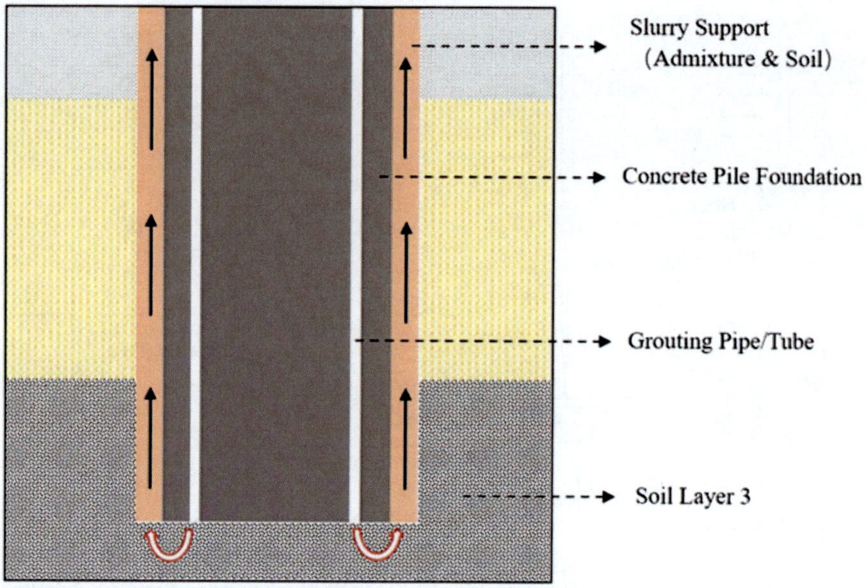

Fig. 2.30 Cast-in-situ concrete pile using slurry support technology

Numerous studies have been conducted to determine the effectiveness of this technology. Field SLTs were conducted to compare the behaviors of one base grouted pile with two conventional slurry stabilized piles (no grouting being applied). These tests provided the load transfer characteristics and clarification of the mechanism of the base grouted concrete pile. They also provided the correlation among cement consumption, the number of grouting stages and the volume of a pile shaft (Ho 2003). The diameter of piles with grouting techniques ranges from 0.4 m to approximately 2.5 m. Tests of two piles were conducted at Paksey Bridge over the Padma (Ganges) River in western Bangladesh, with tested piles of 1.5 m in diameter (Wilkins and Castelli 2004). The construction of large diameter grouted piles (diameters of 2.4 m) commenced with The Pinnacle—a 290 m high skyscraper in London, England (Patelet al. 2015).

Besides analytical methods, the finite element method has also been used to determine the bearing capacity for grouting treated piles. Recent tests and numerical simulation have shown that post grouted concrete piles can double the ultimate capacity of a defected pile, and increase about 20% of a traditional pile's capacity (Nguyen et al. 2012). Several case history projects with application of base grouting techniques were provided by Sinnreich and Simpson (2013). In their paper, they used the O-cell test method which can provide a bidirectional axial compressive load for the determination of the shaft-and-end grouted pile capacity. However, the results were inconclusive because some projects illustrated an increase of grouted pile capacity and some projects did not.

Pressure grouting piles or post grouting piles have been successfully employed around the world for about 40 years (Lai et al. 2000). However, most previous grouting pile investigations concerned base grouting; therefore, research into shaft grouting is very limited. Moreover, research has seldom considered the ultimate uplift bearing capacity of grouted bored piles under static and dynamic load tests. As Sinnreich and Simpson (2013) stated, further research into the mechanics of grouting bored piles is needed, as some results of ultimate bearing capacity between grouted and non-grouted piles have been paradoxical.

2.3.3 Precast Piles

Precast piles can be categorized into timber, steel and concrete piles based on the materials, or into H, circular, square and polygon piles based on the configurations of pile sections. Generally, they can also be categorized into small, large and non-displacement piles based on the amount of soil squeezed during a driven process. Normally, solid concrete piles and end-closed pipe piles are considered large displacement piles, and piles such as H-steel piles with a small cross-sectional area are considered small displacement piles.

For precast concrete rectangular piles (large displacement piles), the sizes can range from 200 to 700 mm in diameter and 10 to 25 m in length, and the working loads that can be resisted vary from 200 to 1,200 kN. These piles can be a normally reinforced structure or a prestressed structure. Past studies focused on the pile tests illustration (Mansur and Hunter 1970), pile capacity under different types of soil profiles (Gregersen et al. 1975; Balasubramaniam et al. 2009), drivability (Ashford and Jakrapiyanun 2001; Fellenius and Samson 1976; Hussein et al. 2006), load transfer mechanism (Fellenius 2002; Hsu, 2014), configurations of piles and strength of concrete (Liew et al. 2004). Recently, there have been intensive investigations aiming to research materials' effect on concrete piles. For example, because of a lack of data using Illinois pulverised coal combustion bottom ash in concrete structures, full-size tests were performed on concrete piles with this utilization of bottom ash, and then the results were compared with the traditional reinforced concrete pile with fly ash admixture (Kumar et al. 2004).

For pipe piles, open-ended piles possess an anticipated greater efficiency (capacity to weight ratio) than do round solid piles. Other advantages of these small displacement piles include that they are easy to drive, and the soil compaction effect is small. However, the behavior of open-ended piles is more complicated because the soil plug effect created inside the pile should be considered when investigating these piles. Recently, a new CPT-based HKU method was proposed for base capacity estimation of open-ended pipe piles with mechanical consideration of annulus resistance and plug resistance (Yu and Yang 2012). Detailed reviews of these precast piles were provided with consideration of soil profiles, methods of driven technique and configurations of precast piles (Marcos et al. 2013). Further, another aspect associated with precast piles in clay is the setup or 'freeze' of the foundation, which is a normal phenomenon whereby pile capacity increases with the increment of time. Recently, research has commenced to determine the amount of setup time of precast piles by performing dynamic load tests, statnamic load tests and SLTs (Abu-Farsakh et al. 2017).

A screw pile is also a small displacement precast pile which is mostly made of steel. This precast pile consists of a helical plate and circular hollow section. AS 2159 (Standards Australia 2009) classifies the steel screw pile as a 'preformed displacement' pile; however, this screw pile does not compact the ground as does a driven pile. The methods based on CPT value to predict the capacity of pile are inaccurate. Yttrup and Abramsson (2003) proposed a method for determining the base-bearing capacity of steel screw piles in sand. The interpretation used to determine the ultimate bearing capacity of this screw pile is based on the modified Davisson's offset method and is governed by a settlement equal to 10% of the helix diameter (Perko 2009). Recent research has found that Chin's method and Decourt's method overestimate the ultimate bearing capacity of this screw pile; therefore, the reduction factor has proposed (Malik et al. 2016).

Non-displacement piles are mostly bored and cast-in-place. As illustrated in Sect. 2.3.2, these piles are mostly manufactured by removing the soil through a drilling process by machine such as auger, and then by placing reinforcements inside the drilled hole. Finally, this hole is filled with concrete. As a result of this construction process, where no soil is compacted, these piles are considered non-displacement piles. However, there is very limited research on the process of removing the soil first and then descending the precast pile into drilled hole, which can also be considered a non-displacement pile.

2.3.4 Composite Piles

2.3.4.1 Traditional Composite Piles

Traditional piles refer to piles made of traditional construction materials. These are steel, concrete and timber piles. Piles made of more than one material are recognized as traditional composite piles. For example, piles made of steel pipes with a concrete core are considered composite piles, as shown in Fig. 2.31. In 1898, Considere found that confined cylinders exhibit an extreme increase in longitudinal compressive strength, and the first use of concrete-filled steel tubes occurred in 1901 by Sewell (Peterson 1999). The inner diameter of concrete-filled tubes or piles can range from 0.2 to 2.8 m, with wall thickness varying from 8 to 20 mm. These piles are mainly used in marine engineering and bridge engineering because of their high strength and good ductility. In particular, for bridge construction, the use of concrete-filled pipe piles is more economic because they account for 15 to 20% of the total construction cost of conventional steel bridges (Nakamura 1998).

When these piles are used in the marine condition, seawater corrosion leads to decreased tube wall thickness and delamination between the tube and concrete core.

Fig. 2.31 Concrete-filled pipe pile (Khaleghi et al. 2016)

In this context, numerous studies have been performed to investigate ways to inspect these composite piles. Compared with remote-operated vehicles (ROVs), acoustic methods—including sonic, ultrasonic and acoustic emission methods—are less time consuming. However, ROVs cannot be used to examine internal defects. Thus, many wave modes have been researched, and many studies have been conducted to investigate and detect corrosion and cracks in tubes using cylindrical Lamb waves (Alleyne and Cawley 1995; Cheng and Cheng 1999; Kundu and Ryu 2002; Rose et al. 1994; Yang et al. 2014).

Other research aspects of these concrete-filled pipe piles vary. For example, field tests on piles were performed in consideration of the fact that earthquakes lead to the soil liquefaction. Three tested piles were applied with lateral vertical and moment loadings. The tested piles included two concrete shaft piles and one concrete-filled pipe pile. The results indicated that stronger piles may tolerate larger lateral displacement than weaker piles (Naesgaard 1992). Moreover, some investigations have focused on pile behavior under special soil conditions. For example, a study of the concrete-filled pipe pile was conducted focusing on the effect of frozen soils on lateral behavior. It found that frozen soils drastically changed the ground motion characteristics and lateral behavior (Xu 2009). Other research has focused on the behavior of these piles under different loadings (Chen et al. 2011; Lao et al. 2004). Further research on the inclined steel pipe pile was conducted recently and found that p-y curves increased linearly when the lateral displacement at the pile head was smaller than 20 mm (Huo et al. 2015).

Piles with relatively high strength can be recognized as rigid piles. These piles are mostly made of cement, aggregates and reinforcement. In contrast, Deep Cement Mixing (DCM) piles, which is a product created by DCM technology that is used for soft ground treatment, can be determined as soft piles. The strength of these piles is comparatively low because there are no aggregates used. As shown in Fig. 2.32, the machine drills the hole and simultaneously sprays cement solidifying agent and

Fig. 2.32 Deep soil mixing (Madhyannapu and Puppala 2015)

water until the required depth, and then agitates the ground by reverse rotation. This process is finished when the ground is sufficiently agitated.

These soft piles can improve the stability of road embankments (Taesiri and Chantaranimi 2001) and can decrease vertical settlements (Wonglert and Jongpradist 2015). Intensive research focused on the slope stabilization of DCM piles. These previous works considered strength and deformation parameters, creep behavior and safety factors for designing DCM piles (Bergado et al. 1999; Buathong and Mairaing 2010; Laiet al. 2006; Miura et al.2001; Mokhlesur et al. 2011). A recent study referring to the spacing, depth, elastic modulus and volume of a row of DCM piles was conducted with results comparison to FEM modelling. The researchers presented recommendations of parameter determination for design to minimize the lateral displacement of soil (Boathong et al. 2014). For settlement prediction, a recent study proposed an empirical settlement formula through non-linear regression of data obtained from a field test (Yang et al. 2012); however, a limitation was that the test loading was relatively small and the settlement predictions from the unloading stages were not presented.

2.3.4.2 Fiber-Reinforced Polymer (FRP) Composite Piles

There are many uses for piles, and issues arise when piles are located in harsh environments, especially in marine or coastal conditions. Piles with traditional materials can be destroyed because of corrosion of steel, deterioration of timber and degradation of concrete, as shown in Fig. 2.33. The deterioration of timber, concrete and steel piling systems costs the United States nearly US$2 billion per year for repair

Fig. 2.33 Corrosion of steel, deterioration of timber and degradation of concrete (Pando 2003)

and replacement (Hassan and Iskander 1998). In 1998, the Federal Highway Administration (FHWA) initiated a project on the use of FRP pilings as the replacement of traditional piles in the context of a waterfront rehabilitation project, whose one goal was to replace up to 100,000 bearing piles for lightweight structures.

The development of concrete piles has a long history. Engineers are now facing a problem relating to piles because, even for piles made of concrete and steel, which possess good rigidity and high strength, damage still occurs, as shown in Fig. 2.34. These figures depict damage to concrete piles photographed from Adelaide in Australia.

FRP piles have been employed for approximately three decades. Dating back to April 1987, the first prototype recycled pile was driven at the Port of Los Angeles (Juran and Komornik 2006). This replacement of creosote-treated timber piles successfully avoided the threat of marine borers. As early as 1998, the Empire State Development Corporation undertook a waterfront rehabilitation project, known as Hudson River Park, which involved replacing up to 100,000 bearing piles for lightweight structures (Pando 2003). Moreover, concrete-filled FRP composite piles were employed by the Virginia Department of Transportation in 2000 for an entire

Fig. 2.34 Degraded concrete piles in Adelaide, South Australia, Australia

bend of the new Route 40 bridge over the Nottoway River in Sussex County, Virginia (Pando 2003; Pando et al. 2006).

Although this FRP material is expensive, the overall cost in the long term is more economical than for the traditional materials of concrete and steel because of the advantages of low management costs and a long service life. Ballinger also stated that, although the cost of FRP composite materials may be higher, the cost of labor and use of equipment for construction work may be lower because of their lighter weight (Pando et al. 2002). Moreover, aside from the cost, there is growing concern about the environmental and health effects of using treated timber and steel painted with solvent or heavy-metal coatings (Guades et al. 2012). These poisonous materials threaten marine borers, which led to engineers starting to replace timber piles with FRP piles made of fiber and plastics or resins. Robinson and Iskander (2008) stated that, by using recycled plastics to manufacture FRP piles, materials are being used that may have otherwise been landfill. Therefore, this approach can be more economical in aggressive environments when lifecycle costs are considered. According to data from the EPA (2006), less than 10% of the 13.7 million tons (equal to loading of 122 GN) of plastic containers and packaging produced annually in the United States is recovered by recycling (Robinson and Iskander 2008).

Many researchers have started to examine FRP piles. Most research has devoted attention to the drivability, driving efficiency, durability and surface friction between FRP and soils, the effect of the hammer, and the resistance of the soil. Research on the driving hammer effect was conducted using wave equation analysis that considered the effects of the weight and velocity of the hammer and pile property. This research indicated that a single-acting steam hammer was more efficient than a diesel hammer because it could drive the composite deeper with the same number of blows (Hassan and Iskander 1998). A modelling study simulated via the software program Microwave indicated that both hollow and concrete-filled FRP piles could be driven by a heavier hammer (Ashford and Jakrapiyanun 2001). Moreover, soil resistance to driving—including side friction and end bearing resistance—was also investigated using the wave equation analysis program. The results from the entire spectrum of the study demonstrated that, for drivability, there was more substantial difference in friction and end bearing conditions for concrete-filled FRP and concrete piles (Mirmiran and Shahawy 1996). The outcome of a study that researched the interface behavior between sand and FRP concluded that FRP exhibited similar relationships between peak interface friction coefficients and relative roughness for a given granular material (Fam 2000; Pando et al. 2002).

A review on the driving performance of FRP composite piles indicated that the types of composite piles are steel pipe core piles, structurally reinforced plastic (SRP) piles, concrete-filled FRP piles, fiberglass pultruded piles, fiberglass-reinforced plastic piles, hollow FRP piles and FRP sheet piles, as shown in Fig. 2.35 (Guades et al. 2012). A further study by Pando (2003) emphasized that the composite products available in the market are steel pipe core piles, SRP matrix piles, concrete-filled FRP pipe piles, fiberglass pultruded piles, plastic lumber piles, hollow FRP piles and FRP sheet piles.

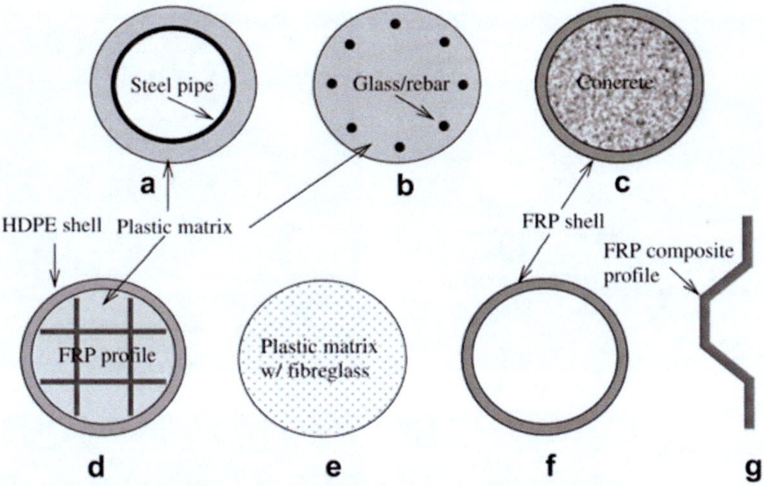

Fig. 2.35 Types of FRP composite piles (Guades et al. 2012)

Based on the results obtained from previous investigations, Robinson and Iskander (2008) conducted notable research to examine the behaviors of FRP composite piles under vertical loads to determine their stiffness, flexibility, settlement and bearing capacities. Through this research, in-situ static and dynamic tests were conducted with onsite SPTs and CPTs. Four types of FRP pile were tested in Elizabeth, New Jersey, involving concrete-filled fiberglass shell piles (Lancaster pile, Fig. 2.36a), polyethylene piles reinforced with steel bars (PPI pile, Fig. 2.36b), polyethylene piles reinforced by fiberglass bars (SEAPILES, Fig. 2.36c) and solid polyethylene piles (American Ecoboard pile, Fig. 2.36d). These tests indicated the possible applicability of plastic piles to traditional axial loading applications, and highlighted the need for further work on the long-term creep performance and durability of these piles (Robinson and Iskander 2008).

Concrete-filled FRP piles came into notice after these tests were completed and were considered the best FRP piles to resist upper loading. Though the other types of FRP piles showed inadequate capacity, using these piles to replace the traditional pile is acceptable. The reason is that the steel pipe pile is mostly used under conditions of being exposed to water, which means the FRP is actually only used for protection from corrosion. In another example, SRP, is mainly used in fendering applications and regarded as a potential load-bearing pile (Guades et al. 2012). The FHWA proposed that FRP composite piles can be used effectively as vertical load-bearing piles and represent an alternative for deep foundation construction, especially in waterfront environments and aggressive soils (Juran and Komornik 2006).

A large amount of research has focused on investigating reinforced concrete structures with externally bonded FRP applications for the purpose of strengthening and repairing. However, one of the limitations of this technology is the propensity of the FRP to prematurely de-bond at strains well below its rupture strain

(a) Lancaster pile (b) PPI pile

(c) SEAPILE pile (d) American Ecoboard pile

Fig. 2.36 Tested composite piles (Juran and Komornik 2006)

(Zhang et al. 2012a). Some researchers conducted tests to design FRP anchors, and provided the criterion for optimal assessment. Luo (2014) reported on other application projects used for slope reinforcement. For example, the technology of GFRP reinforcement anchor was adopted in the project of the Changji Expressway, built in 2008, for the purpose of red sandstone slope reinforcement. This proposed reinforcement system used GFRP bolts with 28 mm diameter, and the results demonstrated that the slope was stabilized overall. Luo (2014) also discussed another GFRP anchor application used in an underground retaining project, in which GFRP replaced the Chinese traditional HPB325, and the results were successful after the pull-out tests.

Other potential options for anchorage include applying FRP bolts or FRP anchor bolts, which are used for slope treatment in expansive soils. The reason for using FRP bolts is that the traditional main governance method for an expansive soil slope is to use a steel anchor bolt and frame beam or grid; however, these are easily corroded and have poor durability (Liu 2014). Through comparison and analysis of the FRP anchor bolt and steel bolt, the author pointed out that, although the maximum crack widths were different, the pulling resistances were almost equal.

A Tunnel Boring Machine (TBM) is one of the most efficient machines used for tunneling; however, this machine cannot easily break the walls or piles that are used to retain the earth pressure (structures with steel reinforcement). The traditional way to solve this problem is with manual work; however, this is dangerous to the workers and is time consuming. For example, the project of Yuantong subway station in Nanjing city, China, suffered human losses when breaking retaining structures in 2007, as the soil layers collapsed. In this accident, the ground cracked and extended to 150 m (Liu et al. 2014).

Application of GFRP bars is the most commonly used form of reinforcement during shield tunneling construction. The replacement of traditional steel reinforcement with GFRP reinforcement in concrete structures (pile or retaining wall) successfully solves the problems mentioned above. This is because GFRP displays brittle behavior when loading reaches and surpasses failure conditions, and the TBM can directly cut the retaining structures. This technique is widely accepted in China, and numerous projects have applied GFRP to fully or partially replace traditional reinforcement. Some projects with GFRP geotechnical application (retaining walls) are the Shanmei subway station R_2 in Dongguan city (Ming 2011), the Shenzhen Subway R_5 in Shenzhen city (Zhang et al. 2011) and the Wuhan Changjiang tunnel in Wuhan city (Jiao 2007). Note that most of the project designs are based on ACI 440.1R-06s (2006) and GB 50010–2002 (2002). A recent review indicated that there is limited research on the deflection behavior of FRP bar–reinforced concrete piles considering supports installation and soil excavation; and no investigation referring to the FRP confinement on concrete piles (Zhou et al. 2016).

2.3.5 *Other Piles*

As a result of their small range of use, piles with unusual configurations and piles that are a combination of a traditional pile and other elements can be considered 'other piles'. Such piles include ribbed piles, T-shaped piles, H-steel-reinforced concrete pipe piles and stiffened DCM piles.

To achieve the requirement of piles with high capacity and low settlement, geotechnical engineers have designed ribbed piles, which can increase shaft resistance. Field trials indicated that these ribbed piles could increase capacity by 30 to 40% (Ground Engineering 2003). Recent research explored the influence of ribs on the ultimate bearing capacity of bored piles. As shown in Fig. 2.37, three different types of ribbed piles were tested and compared with a plain pile. It was found that the ribbed piles were more environmentally friendly. Further, a plastic failure envelope for the pile toe was proposed, and an ultimate design solution was presented (McNamara and Gorasia 2016).

In some countries where pretensioned spun High-strength Concrete (PHC) piles are commonly used, there have been new types of piles created to overcome the shortcomings of these PHC piles. These new piles are called 'extended end' piles (or 'Ext piles'), as shown in Fig. 2.38. Some research has focused on these piles'

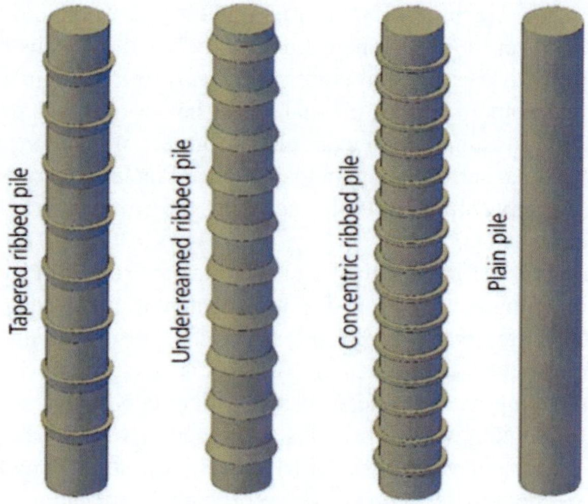

Fig. 2.37 Ribbed bored piles (McNamara and Gorasia 2016)

Fig. 2.38 Extended end concrete pile (Kim et al. 2017)

material properties and bearing capacity, based on laboratory trail tests (Lee and Song 2010; Shin et al. 2014). Recently, a further study investigated Ext piles with respect to time, cost and workability. Based on the field test, it was found that the bearing capacity of Ext piles is better than PHC piles by about 35–50%, and the use

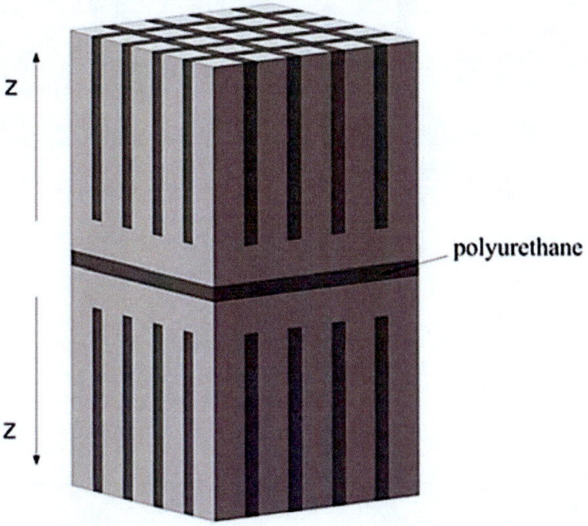

Fig. 2.39 Structure of the two-layered composite pile (Li et al. 2015)

of Ext piles will decrease the work duration and project cost by about 25% and 14%, respectively (Kim et al. 2017).

Composite piles that use monolithic piezoelectric materials have some shortcomings because this material exhibits brittle behavior and is easy to crack. In this context, some researches have commenced referring to the 1–3 composite materials that possess good flexibility. Previous work focused on the properties of the composite material and the manufacture procedures (Manh et al. 2013; Park et al. 2003; Zhou et al. 2012a). A recent study investigated the 1–3 composite piles, as shown in Fig. 2.39. In this study, PZT-5 was used as the piezoelectric material, while polyurethane with low stiffness was selected as the passive polymer matrix (Li et al. 2015). The limitation of this research was that there were no field tests conducted because the application of these piles is very rare.

For soft piles, the failure mode depends on the relative strength between piles and soils. The strength of DCM piles (soft piles) is relatively low; thus, the pile foundation system mostly fails because of failure of the pile head. To solve this problem, previous studies have been undertaken referring to the cement content, which can increase the strength of the pile by adding extra cement. However, these studies found that, with cement content increasing, the cement-treated clay did not linearly increase, and the efficiency was low (Jongpradist et al. 2010; Uddinet al. 1997). Further, in the manufacturing process of DCM piles of mixing cement with soil underneath, it is uneconomic to add extra cement content along the length of the DCM pile because only the pile head requires strength increase.

One method to increase the capacity of DCM piles is inserting a stiffer core into the piles, as shown in Fig. 2.40. The field tests indicated that DCM piles with an inserted core are an effective and economic approach for settlement control and capacity improvement (Dong et al. 2004). The configuration of the core can vary, though the

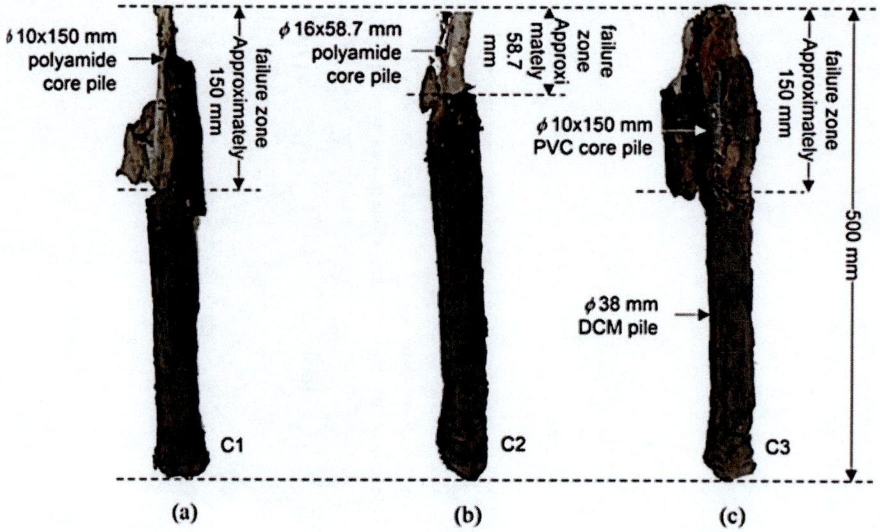

Fig. 2.40 Failure patterns of stiffened DCM piles (Wonglert and Jongpradist 2015)

common cores are circular or rectangular concrete columns, and sometimes H-steels (Jamsawang et al. 2011; Werasak and Meng 2013). Recent research focused on the shape of the core, such as the ratio between the length of the core and pile, as well as simulation (Voottipruex et al. 2011; Wonglert and Jongpradist 2015). Another method to increase the capacity of DCM piles is changing their configurations. As shown in Fig. 2.41, field tests have been conducted on the T-shaped deep mixed (TDM) soil piles. These tests found that, compared with the normal DCM piles, the T-shaped piles could save cement and time (Liu et al. 2011).

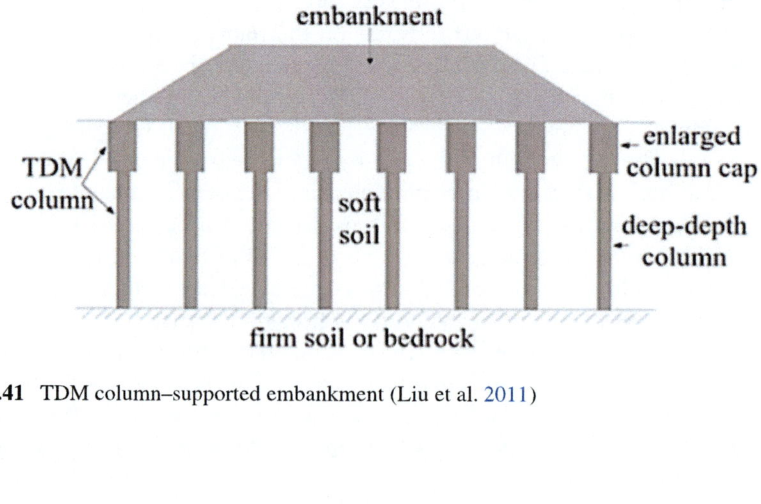

Fig. 2.41 TDM column–supported embankment (Liu et al. 2011)

2.4 Analytical Method for Bearing Capacity

2.4.1 *Meyerhof Method*

It seems that Meyerhof (1956) was the first person to develop a method of pile capacity estimation based on the N-value obtained from standard penetration resistance (Alansari 1999). The empirical correlations proposed by the Meyerhof were based on analysis of numerous pile load tests in a variety of cohesionless soil deposits. The report published by the FHWA highlighted that this method is quick and easy to use; however, it should only be used for preliminary estimates and not for final design because the SPT test data can be influenced by various factors (Hannigan et al. 2006).

Meyerhof (1976) reported empirical equations for the average unit shaft resistance (f_s) of displacement and non-displacement driven piles using the average corrected SPT resistance value $\overline{N}$ (blows per 300 mm) as shown in Eqs. 2.23 and 2.24, respectively. The recommended equation of base resistance (q_t) for driven piles in sands and gravels is provided in Eq. 2.25. Note that for piles driven in a uniform cohesionless stratum, the base resistance can be determined in Eq. 2.26. By correlating the SPT N-value into the average SPT N-value, the shaft and base resistance can be obtained. By multiplying the shaft and base areas respectively, the shaft and base capacity can be calculated, and then, the total ultimate bearing capacity can be evaluated.

$$f_s = 2\overline{N} \leq 100kPa \tag{2.23}$$

$$f_s = \overline{N} \leq 100kPa \tag{2.24}$$

$$q_t = 400\overline{N}_O + \frac{(40\overline{N}_B - 40\overline{N}_O)D_B}{b} \leq 400\overline{N}_B \tag{2.25}$$

$$q_t = \frac{40\overline{N}_B D_B}{b} \leq 400\overline{N}_B \tag{2.26}$$

where:

$\overline{N}_O$ = average corrected SPT N-value for the stratum overlying the bearing stratum.

$\overline{N}_B$ = average corrected SPT N-value of the bearing stratum.

D_B = pile embedment depth into the bearing stratum in meters.

b = pile diameter in meters.

2.4.2 Brown Method

Instead of using the average corrected N-value, the Brown (2001) method uses the value of N_{60} to calculate the shaft and base resistances. The Brown method is a simple empirical method that is based on capacity correlations from 71 SLTs. These tests are preformed from Caltrans projects, and there is a wide variety of soil types (Hannigan et al. 2006).

The recommended shaft resistance for impact piles is provided in Eq. 2.27, and base resistance in Eq. 2.28. Note that the F_{VS} is a reduction factor for vibratory installed piles, and parameters of A_b and B_b are obtained from the regression analyses. The detailed values are provided in Table 2.4. After the shaft and tip resistances are determined, they are multiplied with the shaft and toe area, and then the total capacity of a single pile can be determined. It should be noted that the shaft area of a H-pile is the 'box' area, and, for the shaft area of an open-ended pile, the external surface area is recommended.

$$f_s = F_{VS}(A_b + B_b N_{60}) \tag{2.27}$$

$$q_t = 0.17 N_{60} \tag{2.28}$$

Table 2.4 Input factors for Brown's method (Hannigan et al. 2006)

Loading condition	Installation method	Soil type	F_{vs}	A_b		B_b	
				kPa	(ksf)	kPa	(ksf)
Compression	Impact	Clay to sand	1.00	26.60	0.555	1.92	0.04
		Gravelly sand to boulders	1.00	42.60	0.888	42.60	0.888
		Rock	1.00	138.00	2.89	138.00	2.89
Tension	Impact	Clay to sand	1.00	25.00	0.522	1.80	0.0376
		Gravelly sand to boulders	1.00	40.00	0.835	0.00	0.000
		Rock	1.00	130.00	2.71	0.00	0.000
	Vibratory	Clay to sand	0.68	25.00	0.522	1.80	0.0376
		Gravelly sand to boulders	0.68	40.00	0.835	0.00	0.00
		Rock	0.68	130.00	2.71	0.00	0.00

2.4.3 Nordlund Method

Based on field observations involving considerations of pile taper and displacement of soil during several load test programs in cohesionless soils, Nordlund developed this method in 1963 and updated it in 1979 (Eq. 2.29), as shown in Fig. 2.42. Nordlund (1979) suggested that the shaft resistance is a function of the variables of the friction angle of the soil, friction angle on the sliding surface, taper of the pile, effective unit weight of the soil, pile length, minimum pile perimeter and volume of soil displaced. The Nordlund method tends to over-predict pile capacity for piles with widths larger than 600 mm. In addition, the effective overburden pressure, which is used to compute the pile base resistance, is limited to 150 kPa. For a pile with embedded length *d*, if it has a uniform cross-section, the angle of the pile taper from vertical (ω) is zero, and the Nordlund equation becomes Eq. 2.30. Diagrams and tables of the relationships between δ/φ and pile soil displacement (*V*), evaluation of K_δ with a friction angle ranging from 25 to 40°, α_t and $N_q^{'}$ can be found in FHWA NHI-16–009.

$$Q_u = \sum_{d=0}^{d=D} K_\delta C_F P_d \frac{\sin(\delta + \omega)}{\cos \omega} C_d \Delta d + \alpha_t N_q^{'} A_t p_t \tag{2.29}$$

$$Q_u = (K_\delta C_F P_d \sin\delta C_d d) + \alpha_t N_q^{'} A_t p_t \tag{2.30}$$

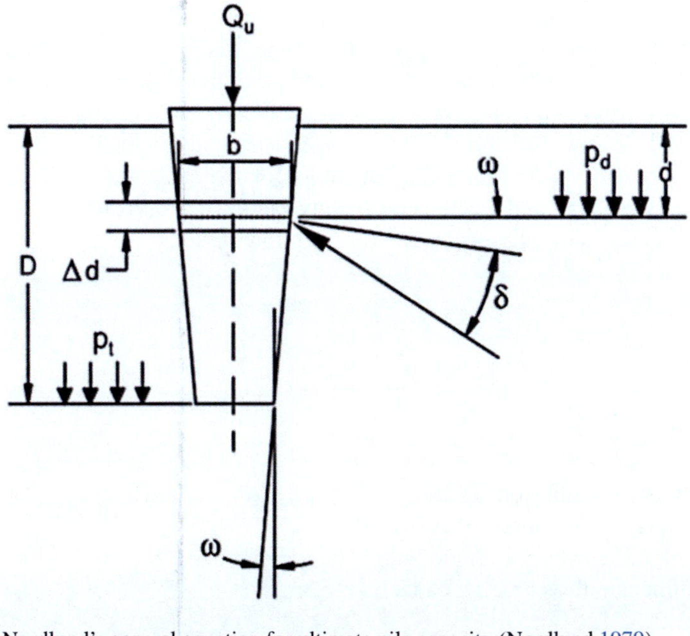

Fig. 2.42 Nordlund's general equation for ultimate pile capacity (Nordlund 1979)

where:

D = depth of embedment.
K_δ = lateral earth pressure coefficient at depth of d.
C_F = correction factor for K_δ.
P_d= effective overburden pressure.
δ = friction angle between soil and pile.
C_d= pile diameter at depth of d.
Δd = pile segment length.
α_t = depth–width relationship factor (dimensionless).
$N_q^{'}$= capacity factor.
A_t= pile tip area.
p_t= effective overburden pressure at pile tip.

2.4.4 Tomlinson Method

The static analysis for piles in cohesive soils includes the Tomlinson method (α method), effective stress method (β method) and a method based on CPT. The Tomlinson method is used to calculate the short-term load capacity of piles in cohesive soils, and the effective stress method can be used to calculate the short- and long-term load capacity of piles in both cohesive and cohesionless soils (Helwany and Wiley 2007).

The Tomlinson method—also known as α method—is a total stress method that uses the parameters acquired from undrained soil shear tests to calculate the capacity of the pile in cohesive soil. This approach assumes that the shaft resistance is independent of the effective overburden pressure and that the unit shaft resistance can be expressed in terms of an empirical adhesion factor times the undrained shear strength (Samtani and Nowatzki 2006b). Two factors are used to calculate the shaft resistance and toe resistance via this method. Equations for the shaft and toe resistance are provided in Eqs. 2.31 and 2.32. The capacity factor, N_c, used to calculate the toe resistance in clay is usually taken as 9 for deep foundations:

$$f_s = c_a = \alpha c_{u1} \tag{2.31}$$

$$q_t = N_c c_{u2} \tag{2.32}$$

where:

α = empirical (adhesion) factor.
c_{u1} = average undrained shear strength.
c_{u2} = undrained shear strength below the toe of a pile.
N_c = dimensionless bearing capacity factor.

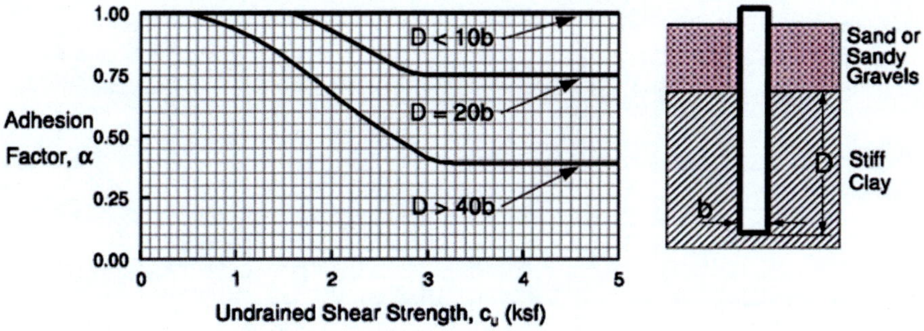

Fig. 2.43 Adhesion factor determination scenario A (Samtani and Nowatzki 2006b)

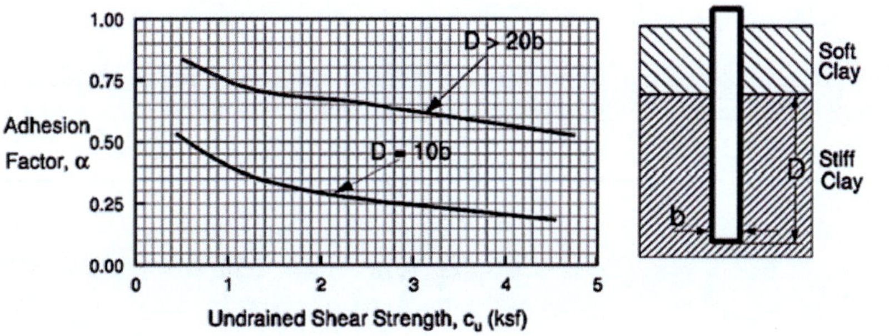

Fig. 2.44 Adhesion factor determination scenario B (Samtani and Nowatzki 2006b)

The shaft resistance is equal to the adhesion between the pile and soil in failure conditions, and the adhesion can be determined by using the undrained shear strength, c_u of clay along the pile multiplied with the empirical factor. This coefficient depends on the clay strength, load magnitude, pile dimension, installation method and time effect. Further, it is a function of the pile embedment and soil stratigraphy.

The adhesion factor, α, can be obtained from Figs. 2.43, 2.44 and 2.45 for piles driven through a sand or gravel layer and into an underlying stiff clay stratum; through soft clay and overlying stiff clay; and through stiff clays without any different overlying strata, respectively.

2.4.5 *Effective Method*

Besides the Tomlinson method, the effective stress method is an alternative method used to calculate static pile capacity in cohesive soil. Compared with the α method, the β method uses drained soil strength parameters for capacity determination, instead of parameters obtained from undrained soil tests. As such, the effective friction angle of soil, φ', should be used. Thus, the β method can be used to calculate the capacity

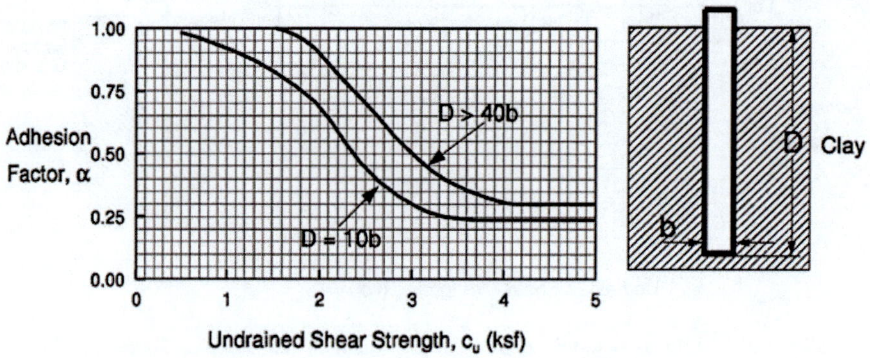

Fig. 2.45 Adhesion factor determination scenario C (Samtani and Nowatzki 2006b)

in both cohesive and cohesionless soil. The formulas of the unit shaft resistance and toe resistance are provided in Eqs. 2.33 and 2.34:

$$f_s = \beta p_o \tag{2.33}$$

$$q_t = N_t p_t \tag{2.34}$$

where:

β = Bjerrum-Burland beta coefficient = $K_s \tan \delta$.

p_o = average effective overburden pressure along the pile shaft.

K_s = earth pressure coefficient.

δ = interface friction angle between pile and soil.

N_t = toe bearing capacity coefficient.

p_t = effective overburden pressure at the pile toe.

To determine the designed capacity of a pile using β method, the effective overburden pressure should be calculated in each layer. Moreover, the effective friction angle, φ', should be determined from laboratory or in-situ tests in the soil profile. The φ' can also be determined via the SPT N-value or corrected N-value if laboratory tests are unavailable.

Figure 2.46 is used to determine the parameter of β to compute the unit shaft resistance of each soil layer. Table 2.5 and Fig. 2.47 can be used to determine the toe bearing capacity factor N_t. The total shaft resistance can be determined by summing up the values obtained from multiplying the shaft stress with the contact area (between each layer and pile surface) based on in Eq. 2.35. The ultimate bearing capacity of a pile based on the effective method is equal to the sum of the total shaft and end resistance in kN, and the allowable designed load is governed by Eq. 2.36. Based on the construction methods, the safety factor can be obtained, as illustrated in Table 2.6.

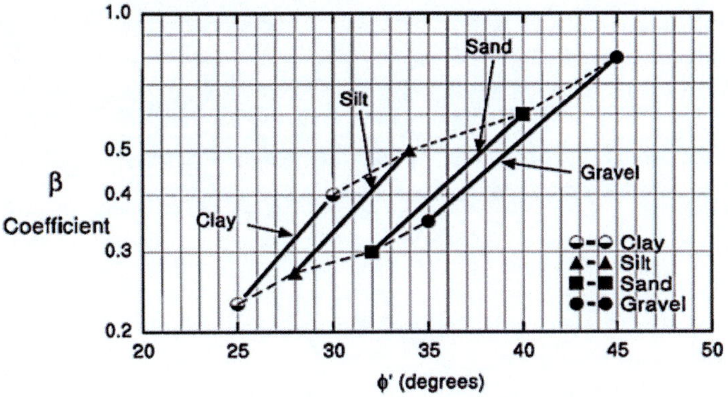

Fig. 2.46 Chart for estimating β coefficient as a function of soil type φ' (Fellenius 1991)

Table 2.5 Approximate range of β and Nt coefficients (Fellenius 1991)

Soil type	φ'	β	N_t
Clay	25–30	0.23–0.40	3–30
Silt	28–34	0.27–0.50	20–40
Sand	32–40	0.30–0.60	30–150
Gravel	35–45	0.35–0.80	60–300

Fig. 2.47 Chart for N_t coefficients as a function of soil type φ' angle (Fellenius 1991)

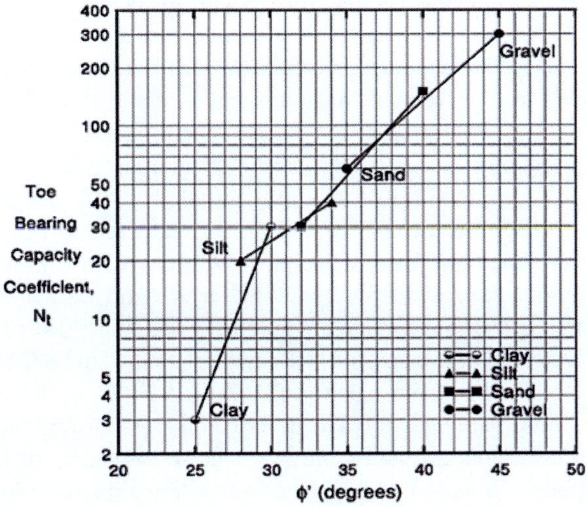

$$R_s = \sum f_s A_s \tag{2.35}$$

$$Q_a = \frac{Q_u}{Factor\ of\ Safety} \tag{2.36}$$

Table 2.6 Recommended safety factor (Hannigan et al. 2006)

Construction control method	Factor of safety
SLT with wave equation analysis	2.00
Dynamic testing with wave equation analysis	2.25
Indicator piles with wave equation analysis	2.50
Wave equation analysis	2.75
Gates dynamic formula	3.50

where:

R_s = total shaft resistance.
f_s = shaft resistance between soil layers with concrete.
A_s = shaft area.
Q_u = ultimate bearing capacity of pile foundation.
Q_a = allowable pile capacity.

2.5 Field Tests

2.5.1 Static Load Tests

2.5.1.1 Compressive Static Load Tests

The basic scheduled mechanism is provided in Fig. 2.48 for a vertically loaded pile test. Increment loads are applied from the pile head, and the settlements are monitored (Kyfor et al. 1992). A load–movement curve of the pile head can be plotted. In addition, the settlement and transferred loads anywhere along the pile can be determined if instrumenting telltales or stain gauges inside the pile.

ASTM-D-1143/D-07 (1994) provides recommendations for several alternative systems to perform SLTs under compressive loading. Commonly, the compressive loads are provided by hydraulic jacks against a reaction beam (as demonstrated in Fig. 2.49) that can be anchored by reaction piles or ground anchors or against a weighted platform. The settlement of a pile can be recorded through installation of a monitoring system, such as LVDTs or dial gauges. It should be noted that determining the ultimate capacity of a pile based on gross movements is not recommended because there is no consideration of elastic deformation of the pile shaft, which leads to underestimation of long piles and overestimation of short piles. The presentations and interpretation of typical SLTs are provided at the end of the Sect. 2.5.

There are seven types of methods to perform SLTs: quick tests, maintained tests, loading in excess of maintained tests, constant time interval tests, constant rate of penetration tests, constant movement increment test and cyclic loading tests. In this book, all compressive SLTs performed were based on maintained tests. This test

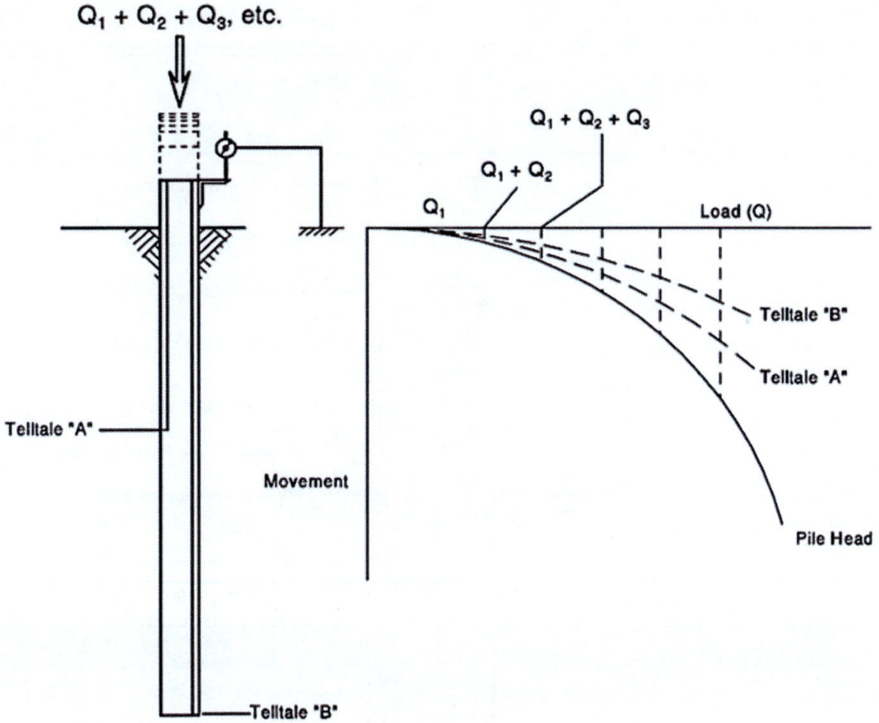

Fig. 2.48 Basic mechanism of a compression pile load test (Kyfor et al. 1992)

maintains loads up to twice the value of the original designed loads (ASTM International 1994; Standards Australia 2009). Some national codes or local standards may require maximum loading that varies from 1.5 to four times the design load. The test loads in increments of 10 to 25% of the design load, and each of the loads should be maintained until the recorded rate of axial movement does not exceed 0.1 to 0.25 mm per hour. During each applied load, the settlement needs to be recorded with incrementing intervals of five, 15, 20 and 45 min. One hour may also be required if the rate of axial movement does not meet the standard requirement. Similarly, the increment time intervals may also vary with different codes. It is recommended that local standards should be used with consideration of the geotechnical situation, instead of selecting one code.

2.5.1.2 Uplift Static Load Tests

The pile foundation can not only be used to resist the downwards loads from the upper structure, but can also be used as a structure to resist uplift loads—such as for the piles under a wind turbine. Compared with compressive SLTs, the basic mechanics of this test are similar, except for the loading direction, which moves upwards. The

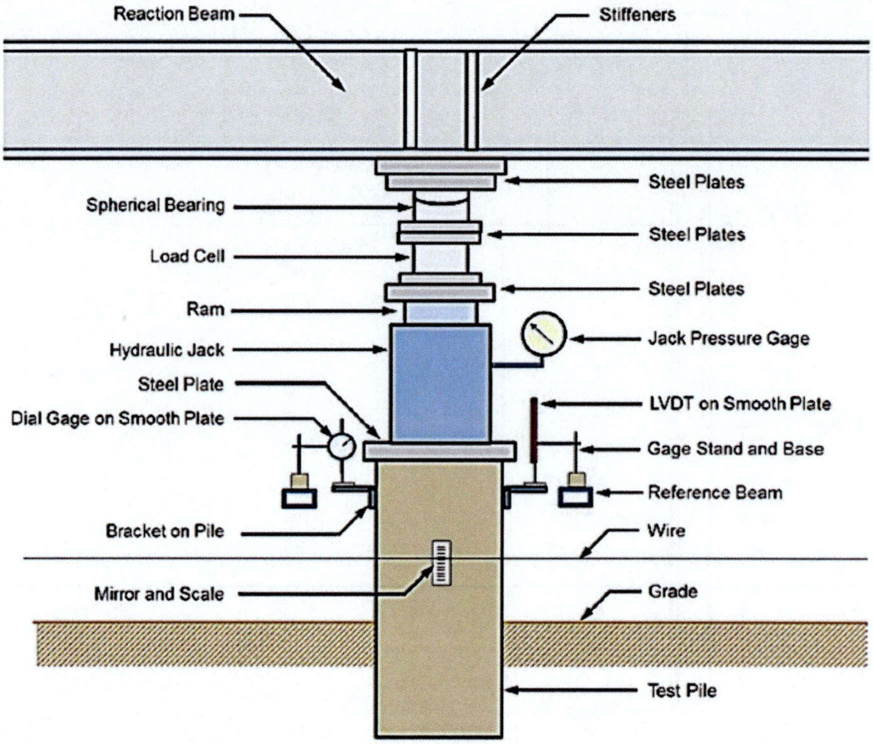

Fig. 2.49 Schematic of anchored reaction frame (Hannigan et al. 2016)

typical setup of uplift SLTs is provided in Fig. 2.50, which also shows the hydraulic jacks used for load provision that are located at the top of the test beam. If reaction piles are cast and connects with reaction beam which is also connected with pile reinforcement, the hydraulic jacks can be placed on the leveled ground and pushing reaction beam upward. The reason to this set-up is that more than two jacks can be used, which will increase the testing capacity. Detailed information can be found in Chap. 3. For the data acquisition, LVDTs and dial gauges are the commonly used equipment to monitor displacement of the pile head.

The test procedures consist of quick tests, maintained tests, loading in excess of maintained tests, constant time interval tests, constant rate of uplift tests and cyclic loading tests (ASTM International 1995a). For determining driven piles, the quick loading procedure is recommended (Hannigan et al. 2006). Increments of 10 to 15% of the design load are required for this test, with a constant time interval of 2.5 min to perform the uplift SLTs. Further, before the additional load increment is applied, the time, load value and total upwards displacement should be recorded immediately. After the final load is applied, unloading is required in five load decrements, and all five loads should be held for five minutes. Compared with other methods, the maintained test is comparatively time consuming, yet allows the soil to be stable.

Fig. 2.50 Typical setup for tensile load test (Courtesy of WKG[2])

This procedure requires maximum loads of up to 200% of the anticipated design load, and each load increment can only be applied when the rate of displacement of the pile head does not exceed 0.25 mm per hour.

2.5.1.3 Lateral Static Load Tests

In recent years, pile foundations have been used to resist lateral load, particularly with regard to special design event such as seismic and vessel effects. For example, structures that suffer from large wind loads, such as large overhead signs, will result in lateral loads on the foundation. The primary purpose of lateral SLTs is to obtain the p-y curves and capacity of a single pile. The instrumentation and observation equipment of lateral SLTs are similar to compressive and uplift SLTs. The reaction system used can involve reaction piles, weighted platforms or a deadman. In some projects, reaction systems may be different. For the convenience of performing lateral SLTs, sometimes, two piles are tested simultaneously because each can be considered the reaction pile to the other as shown in Fig. 2.51.

Lateral SLTs contain standard loading, excess loading, cyclic loading, surge loading, reverse loading, reciprocal loading, specified lateral movement and combined loading tests (ASTM International 1995b). The most common lateral SLTs are standard load testing. These tests apply a maximum load of 200% of the design load, similar to the compressive and uplift maintained procedure; however, the time intervals are different. At times, for cast-in-situ piles, a PVC tube may be installed inside the pile foundation, and the deflection data along the pile shaft can be determined by an inclinometer during this test for lateral behavior investigation.

Fig. 2.51 Setup for lateral static load test (Courtesy of WKG[2])

2.5.1.4 Extrapolations of Field Tests

The most reliable method to determine the actual compressive, uplift and lateral capacities of piles involves conducting SLTs until failure of pile or yield of soil. These methods refer to physical loading on piles at required time intervals, and then monitoring the vertical or horizontal displacement of piles. Under consideration of cost and time, loading until the pile experiences excessive displacement (failure of soil-pile system) is conducted infrequently. In practice, the proof tests are the most common tests instead of failure tests, which means that the maximum applied load does not represent the ultimate bearing capacity of pile. The interpretation of these field tests' results is required to determine the ultimate bearing capacity.

For compressive SLTs, the typical presentation is plotting vertical settlement versus loads, as illustrated in Fig. 2.52. To facilitate the interpretation, an implemented scale considering elastic deformation of single pile was proposed, and the elastic compression line was kept inclined at an angle of 20° (Vesic 1977). The elastic deformation is governed by Eq. 2.37. In some Asian countries, the standards also require diagrams of settlement-lg (time) curves (s-lgt) and settlement-lg (load) curves (s-lgQ). As shown in Fig. 2.53, s-lgt curves can provide information of the pile under failure load; however, for the proof test, the failure load curve cannot be easily determined. Note that the s-lgQ curves are similar to DeBeer's (1970) method.

$$\Delta = \frac{QL}{EA} \tag{2.37}$$

where:

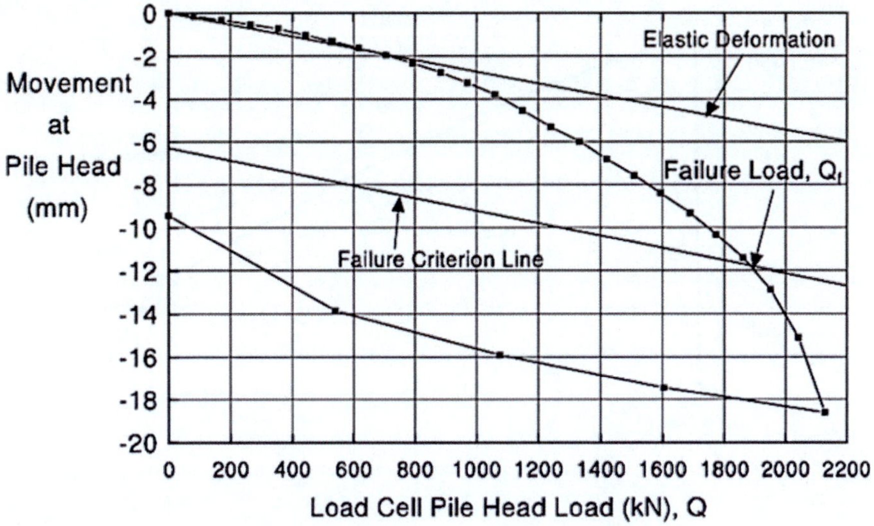

Fig. 2.52 Typical results of compressive SLT (Hannigan et al. 2006)

Fig. 2.53 s-lgt result of compressive SLT

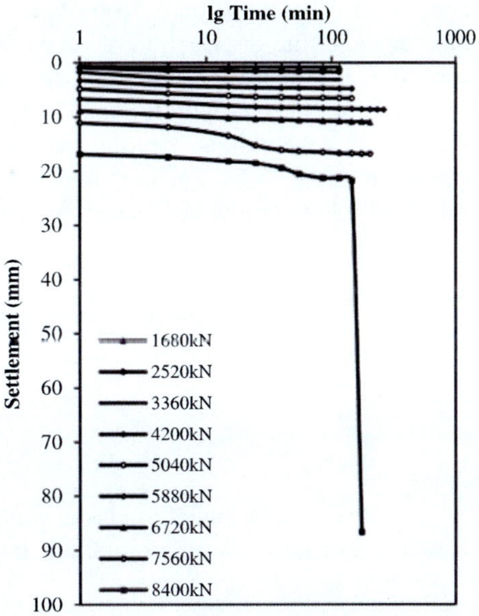

Δ = elastic deformation in mm.
Q = test loads in kN.
L = pile length in mm.
E = elastic modulus of pile in kPa.
A = pile cross-section area in m^2.

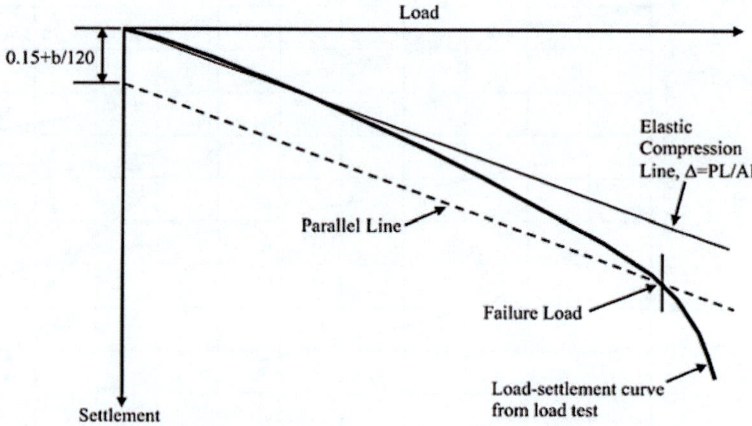

Fig. 2.54 Interpretation based on Davisson's method (US unit) (Samtani and Nowatzki 2006b)

Davisson (1972) proposed a method by offsetting the elastic compression line to determine the failure load and the corresponding settlement. As shown in Fig. 2.54, the offset settlement value, s_f, is equal to 0.15 inches plus the pile diameters in inches over 120. Equation 2.38 provides the s_f in SI units. Based on Davisson's offset method, the intersection of offset line with load–settlement curve (obtained from SLTs) provides the ultimate capacity of a single pile. At times, this offset line does not intersect with the load–settlement curve from loading phases, which illustrates that the ultimate bearing capacity of this pile exceeds the maximum loads applied from the SLTs. For large diameter piles, which are usually over 610 mm, additional pile toe movement must be considered, and the offset value needs to be replaced, as given in Eq. 2.39:

$$s_f = \Delta + (4.0 + 0.008b) \tag{2.38}$$

$$s_f = \Delta + (b/30) \tag{2.39}$$

where:

s_f = offset displacement of elastic compression line at failure condition.
Δ = elastic deformation of total pile length.
b = pile diameter or width.

Compared with Davisson's criteria, the double tangent method is more commonly used to determine cast-in-situ piles. As shown in Fig. 2.55, two tangent lines were first determined, and the intersection of these two lines represented the failure load of the pile with corresponding maximum failure settlement. Mind that when proof test performed or the applied loads being relatively small, the second tangent line will be very difficult to obtain. Under this condition, result handling based on experience and local data required.

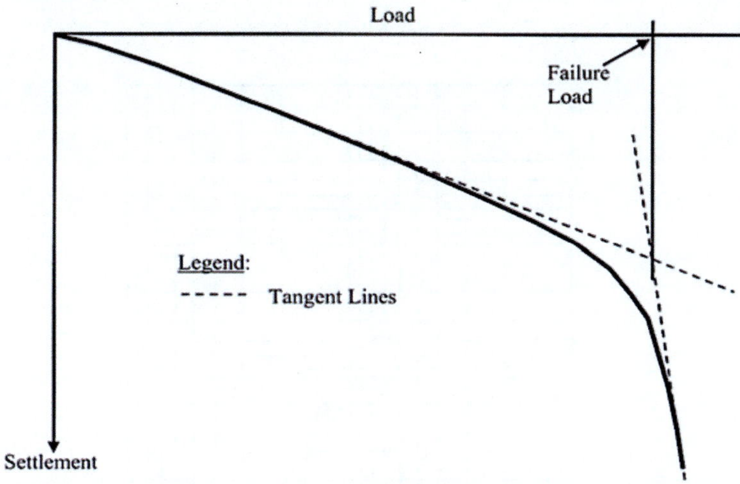

Fig. 2.55 Interpretation based on double tangent method (Samtani and Nowatzki 2006b)

DeBeer's method—also known as DeBeer's log–log method—is one of the other extrapolations of SLTs' results. As shown in Fig. 2.56, by plotting the load–settlement data into logarithmic scales, DeBeer (1970) defined the failure load as the load corresponding to the intersection of two distinct slopes. For plugging failure piles, these two slopes can be visible; for example, as shown in the Fig. 2.56, the 'turning point' of 7560 kN is discovered as the ultimate bearing capacity of the tested pile. However, for piles that experience local failure, the capacity may be a range of values (Paikowsky and Tolosko 1999).

Chin's method operates under the assumption that the load–settlement relationship is hyperbolic as illustrated in Eq. 2.40, and that the inverse slope, C_1, obtained from the plotted the settlement/load versus settlement gives the capacity of a single pile as provided in Eq. 2.41 (Chin 1970). An example is provided in Fig. 2.57, where the slope for pile with label P80 is determined as 0.0002 from the plotted settlement/load versus settlement diagram. The ultimate capacity of the pile is: $Q_u = 1/0.0002 = 5{,}000$ kN. Given that this method is based on a mathematical relationship, the capacity value of a single pile can pass beyond the maximum applied load from the SLTs. This interpretation may provide overestimated values.

$$\frac{s}{Q} = c_1 s + c_2 \tag{2.40}$$

$$Q_u = \frac{1}{c_1} \tag{2.41}$$

where:

s = settlements from SLTs.

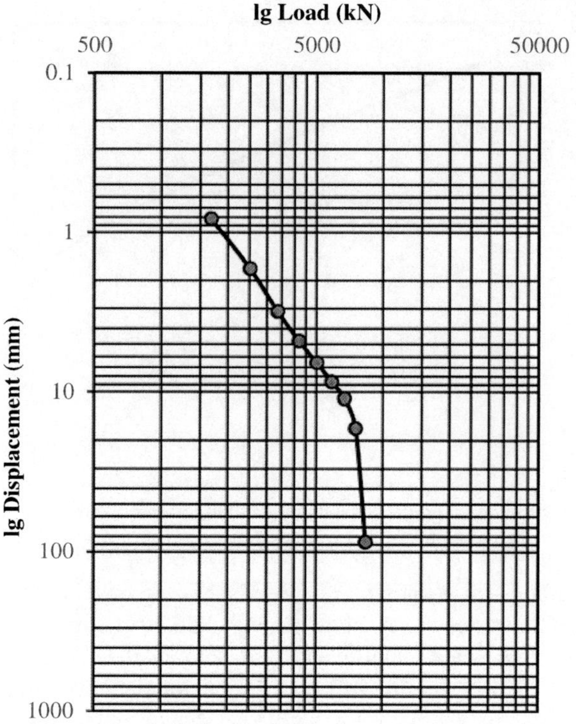

Fig. 2.56 Interpretation based on DeBeer's method

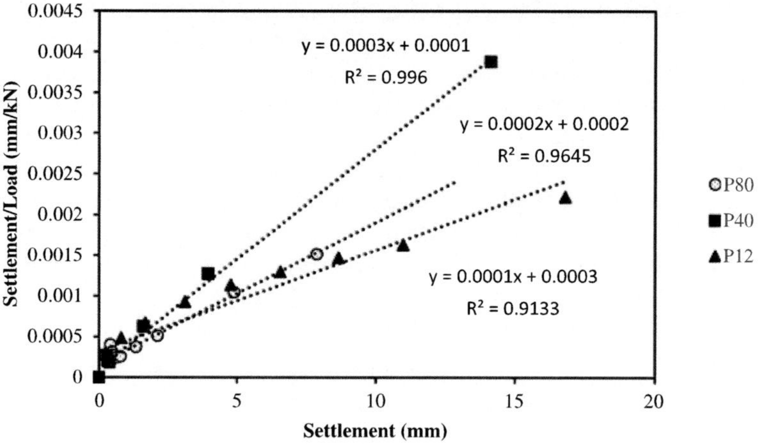

Fig. 2.57 Interpretation based on Chin's method

Q = loads applied from SLTs.
Q_u = ultimate bearing capacity of single pile.

Yang and Xiao (2011) pointed out that, theoretically, Chin's method illustrates two lines, as depicted in Fig. 2.58. The reciprocal of these first gradient is the shaft capacity before the end bearing capacity is mobilized, and the reciprocal of second gradient represents the total bearing capacity.

For the results of uplift SLTs, the typical presentation is similar to compressive SLTs, as shown in Fig. 2.59. However, a widely accepted method for ultimate capacity in uplift load testing has not been published. Fuller (1983) pointed out that the criteria for uplift capacity is related to elastic lengthening, as illustrated in Fig. 2.60. The offset line of this elastic lengthening will intersect with the displacement-load curve,

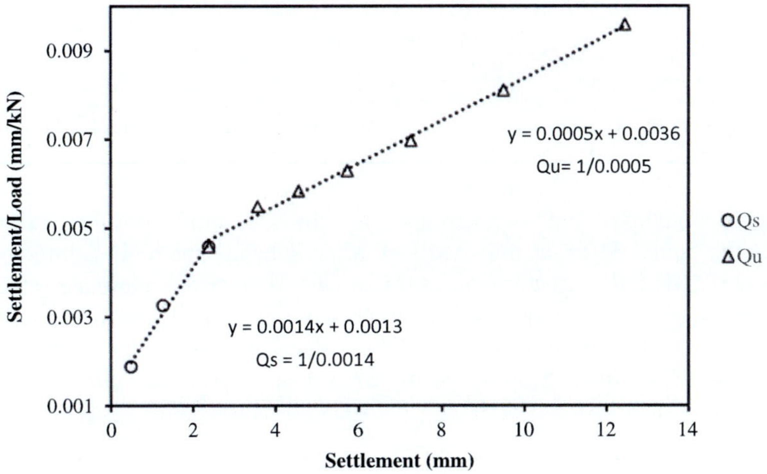

Fig. 2.58 Determination of shaft and end resistance based on Chin's method

Fig. 2.59 Typical results of uplift SLT

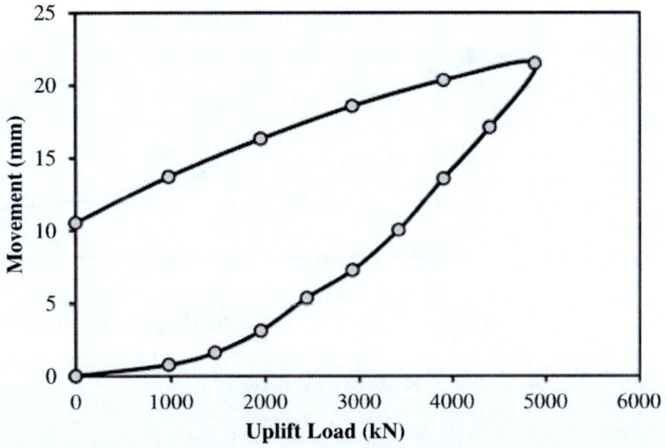

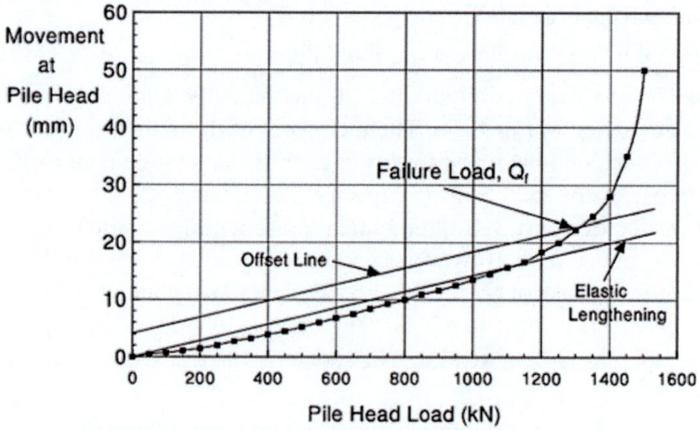

Fig. 2.60 Typical tension load test for load–movement curve (Hannigan et al. 2006)

which represents the failure load and corresponding failure settlement. FHWA NHI-16–009 (2006) recommended the offset value is to be 4.0 mm, and the uplift design load can be equal to half to two-thirds of the failure load.

Another method used to determine the ultimate uplift bearing capacity of non-failure uplift SLTs is the modified Mazurkiewicz method (Chim-oye and Marumdee 2013). The assumption in this method is that the nominate settlement (load/settlement) value is equal to 0 when the loading is small, and the settlement is very high. Therefore, as shown in Fig. 2.61, a line goes through the y-axis, and the intersection of two lines illustrates the uplift ultimate bearing capacity of the pile.

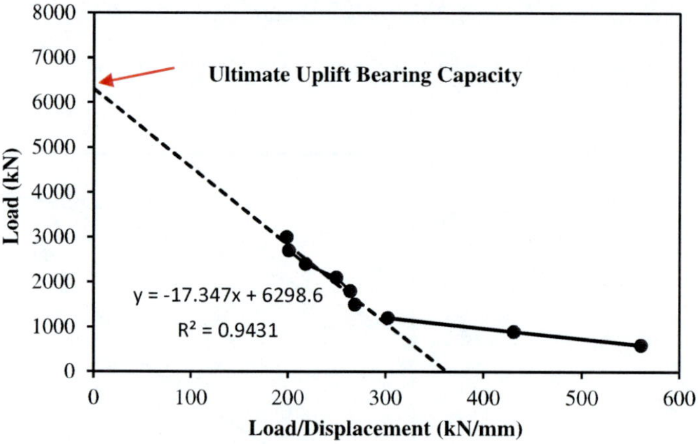

Fig. 2.61 Modified Mazurkiewicz method

Referring to Reese (1984) and Wang and Reese (1993), a report for lateral SLTs should contain a presentation such as pile head's deflection-load curve (Fig. 2.62) and deflection along the pile (Fig. 2.63). Compared with the compressive and uplift SLTs, the interpretation and analysis of horizontal SLTs are much more complicated. The ordinary results presentation (Fig. 2.62) can illustrate the lateral movement under increment horizontal loads; however, it is better to get the ultimate bearing capacity when cyclic loads being applied. Further, when researching the deflection behavior of pile foundation, the lateral loads are commonly distributed along the pile, results in Fig. 2.63 provides limited information because the applied load is a kind of point load

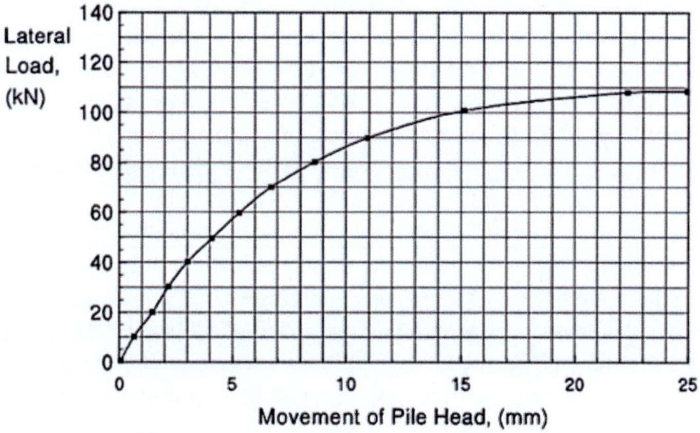

Fig. 2.62 Typical lateral load test for pile head load–deflection curve (Hannigan et al. 2006)

Fig. 2.63 Deflection behaviour versus depth (Kyfor et al. 1992)

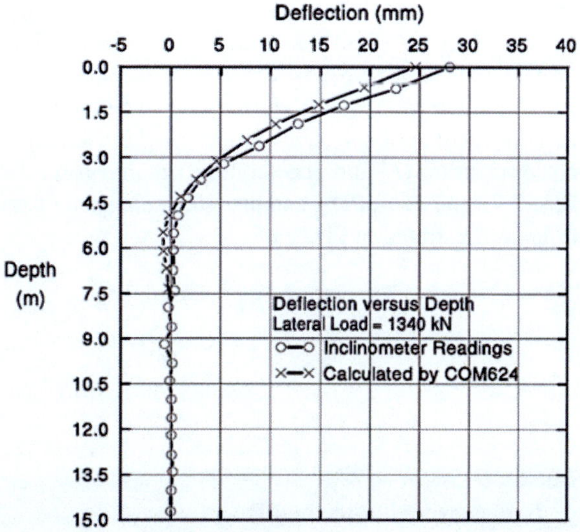

acting on pile head. Note that the case study referring to cyclic horizontal load test is presented in Chap. 4, the result presentation and interpretation is also included. Another case study referring to the deflection behavior of pile by performing the inclinometer monitoring is provided in the Chap. 5.

2.5.2 Dynamic Load Test

SLTs provide the most actual results for pile capacity and are accepted worldwide; however, these tests cannot be used to determine the long-term settlement and down-drag from consolidation and setting of soils. Further, these tests are costly and time consuming. For example, if performing a compressive SLT, it is necessary to transfer a reference beam or weighted platform (may exceed 900 tons) and cast reaction piles (two to four piles). In addition, the duration of maintained load tests may exceed to 24 h during some loading stages. In contrast, dynamic testing requires much less time and money (Cheney and Chassie 2000).

The development of dynamic load test techniques started with a thesis project at Case Institute Technology—known as the 'case method'. During 12 years of research, the dynamic load analysis was improved, including field solutions and a numerical method called the 'case pile wave analysis program' (CAPWAP). These previous projects and results can be found in Goble and Rausche (1970) and Goble et al. (1975). Recent updated research information can be found in Rausche et al. (2004).

The application of dynamic testing facilitated the determination of pile soil behavior, such as determining the soil resistance distribution by CAPWAP, evaluating the performance of pile foundations during the driving process, obtaining the installation stresses in piles and determining the pile integrity. As shown in Fig. 2.64, at least two strain transducers and two accelerometers need to be installed onto the surface of the tested pile, and need to be bolted to diametrically opposite sides of the pile (Hannigan et al. 2016). These transducers and accelerometers are used to record the strain and acceleration during pile dynamic tests. The data are then transferred to a data acquisition system, called the Pile Driving Analyzer (PDA), which finally converts strain (ε) and acceleration (a) into force (F) and velocity (V). Based on Eqs. 2.42 and 2.43, PDA can provide a diagram of force and velocity curves versus time, as illustrated in Fig. 2.65.

$$F_t = EA\varepsilon_t \tag{2.42}$$

$$V_t = \int a_t dt \tag{2.43}$$

where:

F_t = converted force by PDA.
E = elastic modulus of pile.

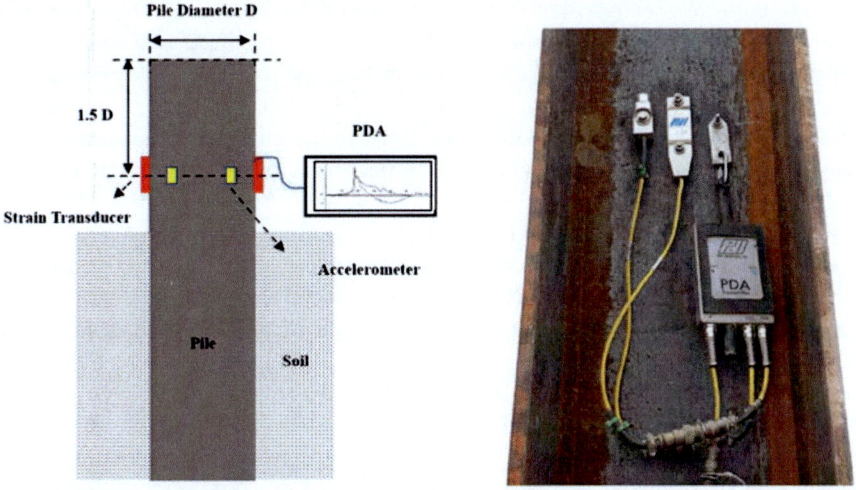

(a) (b)

Fig. 2.64 **a** Schematic diagram of apparatus for dynamic test; **b** Strain transducer and accelerometer installed on pile surface (Hannigan et al. 2016)

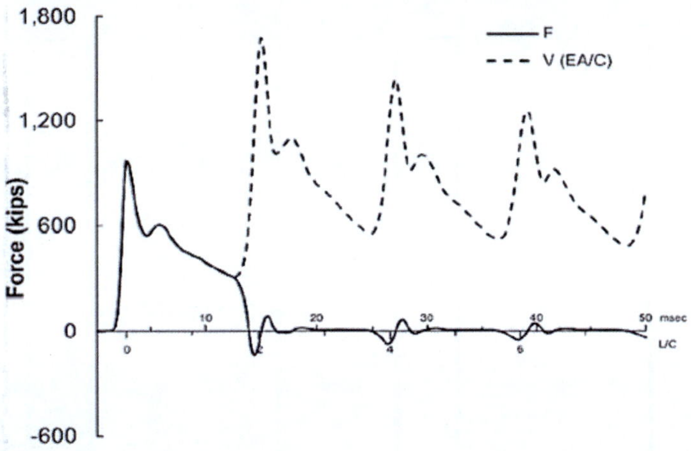

Fig. 2.65 Typical force and velocity traces (Hannigan et al. 2016)

A = cross-section of pile.
ε_t = recorded strain by strain transducers.
V_t = converted velocity by PDA.
a_t = recorded acceleration by accelerometer.

The dynamic load test was developed based on the theory of wave mechanics. A wave speed, C, can be created by a striking from a mass onto a uniform elastic rod, in which the cross-section area is A and the elastic modulus is E. The force, F, is generated at the surface of the rod after the strike, and this force will compress the other part of the rod with an acceleration, a, where the particles of this rod attain a velocity of V. If there is no resistance effect on this rod, the force wave (equal to EA/C) and material wave (C) will travel down to the end of rod at a time of L/C (length of rod is L), and these waves will reflect back to the rod head at a time of $2L/C$.

If there is no resistance at the end of the rod, as illustrated in Fig. 2.66a, tensile wave refection occurs, the force becomes zero and velocity doubles. If there is a fixed end, compression wave reflection occurs, the velocity becomes zero and the force magnitude doubles, as shown in Fig. 2.66b. Considering the rod as a pile with small and large soil resistances at a depth of A and B (Fig. 2.66c), where the strike is impacted on the pile head, the recorded force and velocity data versus time are provided in Fig. 2.67.

Before time $2A/C$ (A is the depth in meters, not the cross-section area), the force and velocity are proportional since there is no resistance. After $2A/C$, the force wave will increase and the velocity wave will decrease slightly because of the small soil resistance. Similarly, the reflection wave will be affected by large soil resistance at the time $2B/C$, the force wave will increase, and the velocity wave will decrease

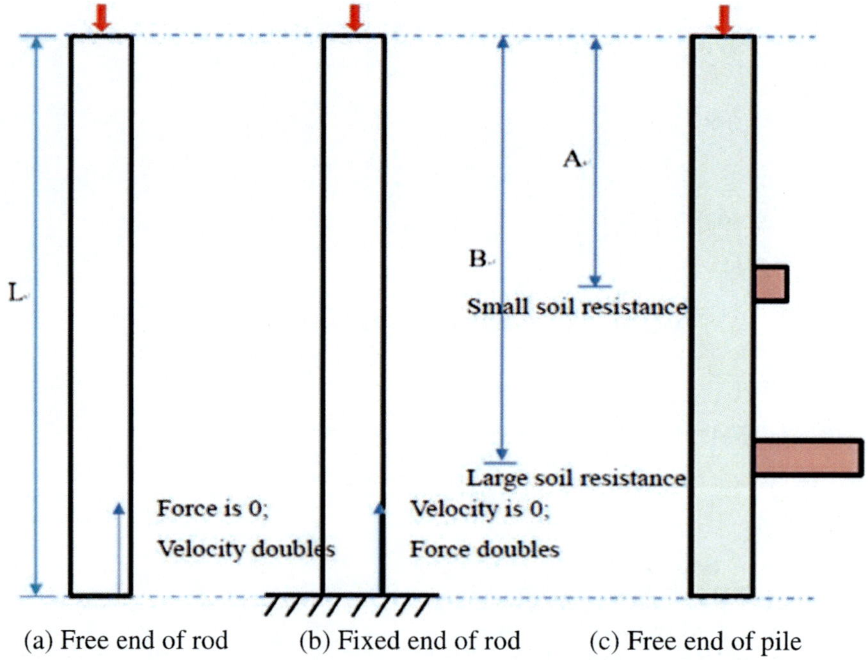

Fig. 2.66 Wave mechanics of rod and pile

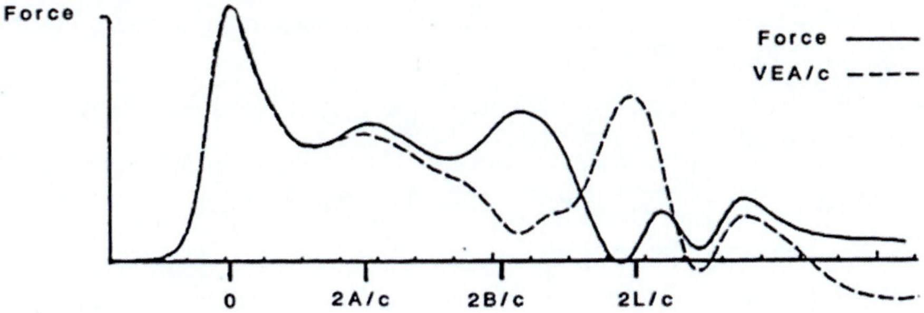

Fig. 2.67 Soil resistance effects on force and velocity records (Hannigan 1990)

by a large magnitude. When the wave passes to the pile end, which is a free end condition, the force wave will decrease and the velocity wave will increase (tensile wave reflection).

The CASE method investigation was conducted at Case Western Research University. Based on one-dimensional wave propagation, the assumption of closed form solution and pile material is linear elastic. The total static and dynamic resistance during driving, *RTL*, is governed by Eq. 2.44. Based on the findings of the dynamic resistance component function proposed by Goble et al. (1975), the static pile capacity, *RSP*, can be determined based on Eq. 2.45:

$$RTL = \frac{1}{2[F_{t_1} + F_{t_2}]} + \frac{EA}{2[V_{t_1} + V_{t_2}]}/C \tag{2.44}$$

$$RSP = RTL - \mathrm{J}\left[\frac{V_{t_1}\mathrm{EA}}{\mathrm{C}} + F_{t_1} - \mathrm{RTL}\right] \tag{2.45}$$

where:

F = force obtained from gauge location.
V = velocity obtained from gauge location.
E = elastic modulus of pile.
A = cross-section area.
C = wave speed of pile material.
L = length of pile below gauge location.
t_1 = time of initial impact.
t_2 = time of reflection of initial impact from pile toe.
J = dimensionless damping factor referring to soil type near pile toe.

The CAPWAP method can provide more rigorous static load capacity of piles and soil resistance along the pile as shown in Fig. 2.68. Similar to the CASE method, measured force versus time should first be plotted. A computed force is then used to match the obtained signal with the assumptions that: (*i*) the pile is simulated by a series of pile segments, and (*ii*) soil resistance is simulated by elasto-plastic springs

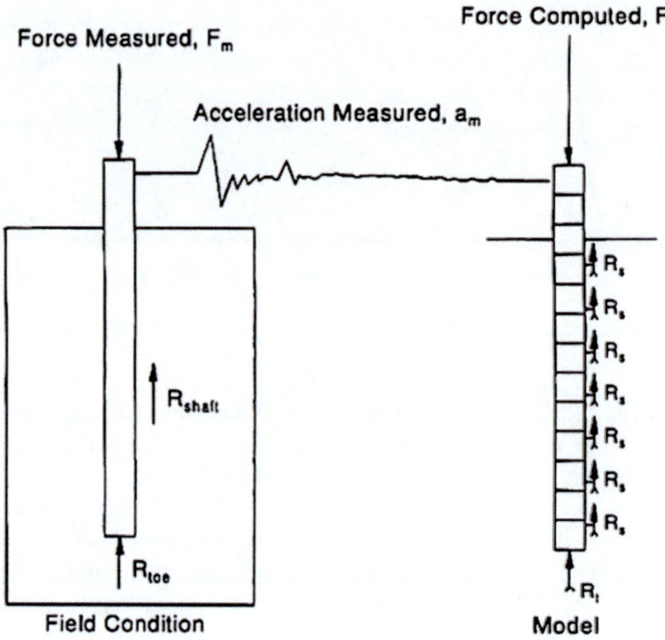

Fig. 2.68 Schematic of CAPWAP analysis method (Hannigan et al. 2016)

and dynamic resistance. Adjustments and a repeat process are required for the soil model, and this model is refined until no further agreement can be determined. Finally, this process can be terminated and the soil model can be used to evaluate the capacity of the tested pile. A detailed case study referring to dynamic load tests as well as CAPWAP method will be provided in Chap. 3.

2.5.3 Osterberg Cell Test

The Osterberg cell test, also known as the O-cell test, is a proprietary method to determine the capacity of driven piles and cast-in-place piles. Different to static and dynamic load tests, O-cell tests do not need a reaction frame and anchor system. The axial loading of these tests is provided by an expendable jack and load cell cast inside the shaft. As shown in Fig. 2.69, the O-cell, which is welded with bearing plates, is used to apply loading upwards and downwards. The vertical displacement of the tested pile can be recorded by displacement transducers. The example referring to the typical presentation (displacement versus loading) is provided in Figs. 2.70 and 2.71.

As shown in Fig. 2.70, after the maximum loading of 2,400 kN being applied, the settlement of the pile end is 0.5 mm and the vertical upward movement of the

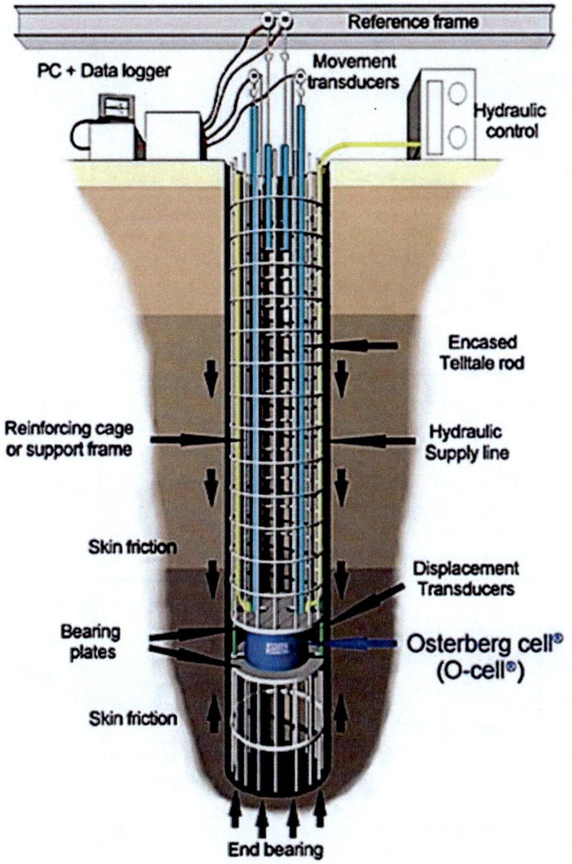

Fig. 2.69 Schematic of test pile by O-cell testing. *Source* www.loadtest.com

pile shaft is 2.7 mm. The shaft resistance of soil layers is fully developed. As shown in Fig. 2.71, the ultimate end bearing is reached because the shaft displacement is small and end displacement is large.

This test has many advantages compared with the traditional tests. First, O-cell equipment can apply large loading of up to 27 MN (Samtani and Nowatzki 2006b). Second, this test saves time because the testing can be conducted once the concrete has reached suitable strength. Third, this test is economic and can provide accurate data. Other advantages include convenient performance for offshore projects, high safety (no need to use heavy reaction system).

There are also some disadvantages for O-cell test performance. First, the O-cell device must be installed prior to construction or driving. Second, once the shaft has reached the ultimate condition, the ultimate end resistance cannot be obtained, and vice versa. Third, if the capacity of the O-cell is inadequate, the test fails because the

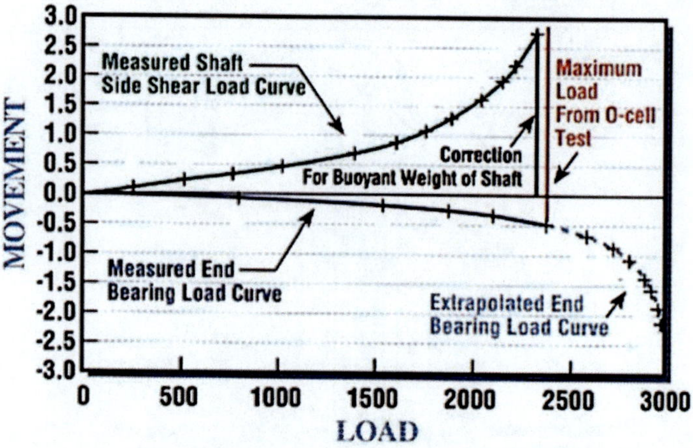

Fig. 2.70 Typical O-cell results—shaft failure. *Source* www.loadtest.com

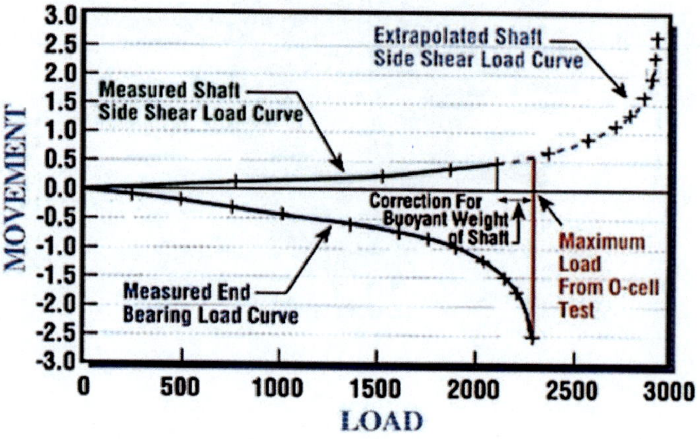

Fig. 2.71 Typical O-cell results—end failure. *Source* www.loadtest.com

device has already been cast inside the pile. Fourth, this device is not reusable after the test is finished. Finally, the tests on H-piles and sheet piles are not applicable.

2.5.4 Statnamic Test

The statnamic test method is also a proprietary method that was developed by the Berminghammer Foundation Corporation. This test contains four phases, as shown in Fig. 2.72. Phase I involves the test setup. The reaction mass is placed onto the test shaft. A small volume of propellant is placed inside the pressure chamber. After the

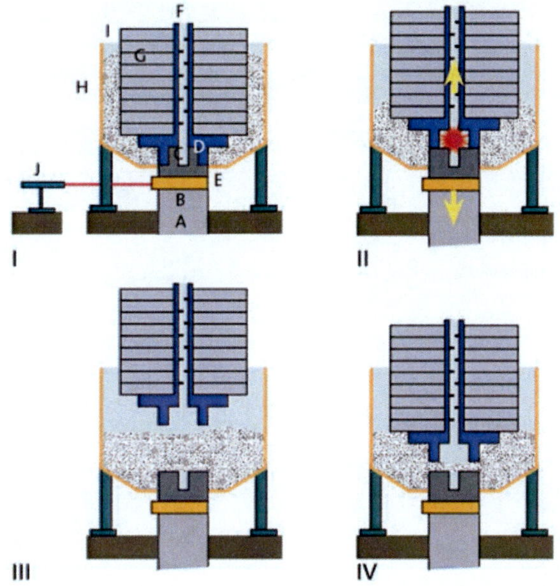

Fig. 2.72 Schematic of test pile by statnamic testing. *Source* www.profound.nl

propellant is ignited, the generated high-pressure gas accelerates the reaction mass upwards, and the downwards reaction force occurs in Phase II. In Phase III, the mass is forced upwards and the gravel fills the gravel container. Finally, in Phase IV, the gravel flows over the pile head as a layer. This layer then catches the reaction mass and transfers the impact forces. During Phases I to IV, the laser beam records the movement of the pile head, and a diagram of load versus time and displacement can be obtained, as shown in Fig. 2.73a. The typical result of load versus displacement can be determined by Profound's Foundation Pile Diagnostic System, as illustrated in Fig. 2.73b.

The theory of this statnamic test is based on Newton's three laws of motion. The measured statnamic force, F_{stn}, is the sum of the inertia force, F_{a}, and soil resistance, F_{soil}, which is governed by Eq. 2.46. The soil resistance is the sum of the static soil resistance, F_{u}; damping force from soil, F_{v}; and pore water pressure resistance, F_{p}, so the measured statnamic force is governed by Eq. 2.47. Under the assumption of pore water pressure resistance is included as part of the damping force, and pore water pressure can be ignored (less than 5%). The static soil resistance is then governed by Eq. 2.48:

$$F_{stn}(\mathrm{t}) = F_{soil}(t) + F_a(t) \tag{2.46}$$

$$F_{stn}(\mathrm{t}) = F_u(t) + F_v(t) + F_p(t) + F_a(t) \tag{2.47}$$

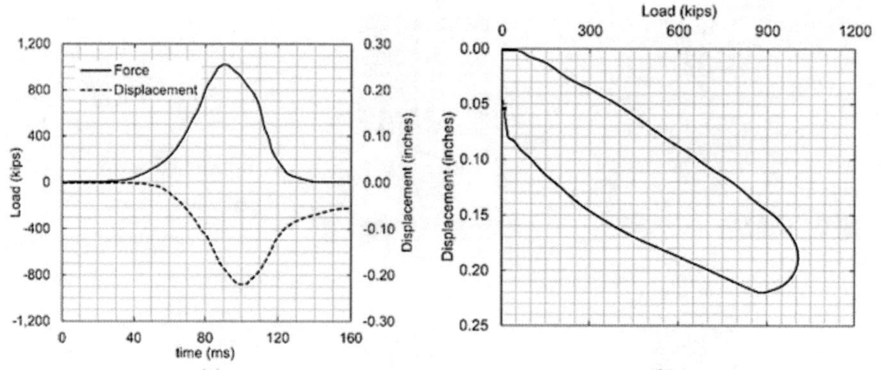

Fig. 2.73 Typical result presentation by statnamic test (Courtesy of Berminghammer Foundation Equipment)

$$F_u(t) = F_{stn}(\mathrm{t}) - k \times u(t) - c \times v(t) \tag{2.48}$$

where:

F_{stn} = measured statnamic force.
F_a = inertia force.
F_{soil} = soil resistance.
F_u = static soil resistance.
F_v = damping force from soil.
F_p = pore water pressure resistance.
k = spring stiffness in N/m.
$u(t)$ = measured displacement in m.
c = damping factor in Ns/m.
$v(t)$ = velocity du/dt in m/s.

Compared with SLTs, which are time consuming and cumbersome, the statnamic test can provide the same loading using a mass that is equal to only 5 to 10% of an equivalent SLT. Compared with O-cell testing, the statnamic load test does not need to cast the device inside the pile, and the vertical load is applied for a duration of 120 ms. This testing method can determine the behavior of high-capacity piles with capacity over 5.0 MN. The commercial test equipment nowadays is capable of producing loads of up to 40 MN. Different to the dynamic load test, piles tested by the statnamic load test are not dominated by stress wave propagation, and the created acceleration is relatively low. Moreover, this statnamic test can directly provide measurements, instead of processing the data by experienced engineers, as does the dynamic load test.

2.6 Concluding Remarks

This chapter reviewed subsurface explorations firstly, which included laboratory and in-situ tests. It presented the test setup, test procedures and case studies for performing the tests to provide a more comprehensive geotechnical understanding. This was followed by a discussion of the methods used to determine the soil parameters that were applied in previous projects' designs, as well as the literature review of the different types of piles, considering configuration, construction methods and investigations undertaken in the past. This chapter also reviewed the analytical methods used to determine pile capacity based on the parameters acquired from laboratory and in-situ tests.

Additionally, this chapter reviewed the SLTs associated with compressive, uplift and lateral loads, including test setup, procedures based on the codes' requirements, presentation of typical results and extrapolations of SLT results. Finally, this chapter reviewed the theories and methods used to determine the ultimate bearing capacity of a single pile based on dynamic load tests, Osterberg cell test, and statnamic test.

References

Abu-Farsakh, M. Y., Haque, M. N., & Tsai, C. (2017). A full-scale field study for performance evaluation of axially loaded large-diameter cylinder piles with pipe piles and PSC piles. *Acta Geotechnica, 12*(4), 753–772.

Alansari, O. M. A. (1999). *Capacity and behavior of steel pipe piles in dry sand* (Doctoral dissertation).

Alleyne, D., & Cawley, P. (1995). The long range detection of corrosion in pipes using Lamb waves. *IEE Colloquium (Digest), 240,* 6.

American Concrete Institute. (2006). *Guide for the design and construction of structural concrete reinforced with FRP bars* (ACI 440.1R-06s).

Ashford, S. A., & Jakrapiyanun, W. (2001). Drivability of glass FRP composite piling. *Journal of Composites for Construction, 5*(1), 58–60.

ASTM International. (1994). *Standard test methods for deep foundation under static load compressive load* (ASTM-D-1143/D-07) (pp. 1–15).

ASTM International. (1995a). *Standard test methods for deep foundations under static axial tensile load* (ASTM-D-3689/D-07) (pp. 1–13).

ASTM International. (1995b). *Standard test methods for deep foundations under lateral load* (ASTM-D-3966/D-07) (pp. 1–18).

ASTM International. (2007a). *Standard test methods for direct shear test of soils under consolidated drained conditions* (ASTM-D-3080–98) (pp. 1–6).

ASTM International. (2007b). *Standard test methods for one-dimensional consolidation properties of soils* (ASTM-D-2435–06) (pp. 1–10).

Balasubramaniam, A., Oh, E., & Phienwej, N. (2009). Bored and driven pile testing in Bangkok sub-soils. *Journal of Lowland Technology International, 11*(1), 29–36.

Bergado, D., Ruenkrairergsa, T., Taesiri, Y., & Balasubramaniam, A. (1999). Deep soil mixing used to reduce embankment settlement. *Ground Improvement, 3,* 145–162.

Boathong, P., Jamsawang, P., & Mairaing, W. (2014). Lateral movement of slope stabilized with DCM column rows. *Electron Journal Geotechnical Engineering, 19*(H), 1647–1664.

Buathong, P., & Mairaing, W. (2010). Failure behavior of large drainage canal reinforced by DCM piles. In *Proceedings of the EIT-JSCE Joint International Symposium*. Bangkok, Thailand.

Bowles, J. E. (1977). *Foundation analysis and design* (3rd ed.). New York, NY: McGraw-Hill Book Company.

Brinch-Hansen, J. (1963). Hyperbolic stress-strain response: Cohesive soils discussion. *American Society of Civil Engineers Journal of Soil Mechanics and Foundation Division, 89*(SM4), 241–242.

Brown, D. A., O'Neill, M., Hoit, M., McVay, M., El Naggar, M., & Chakraborty, S. (2001). *Static and dynamic lateral loading of pile groups*. In National Cooperative Highway Research Program Report (pp. 1–57).

Bustamante, M., & Gianeselli, L. (1982). *Pile bearing capacity prediction by means of static penetrometer CPT*. Paper presented at the Proceedings of the 2nd European Symposium on Penetration Testing.

Butler, H. D., & Hoy, H. E. (1976). *The Texas quick-load method for foundation load testing, user's manual*, NASA STI/Recon Technical Report N. (p. 77).

Casagrande, A. (1936). *The determination of the pre-consolidation load and its practical*. Paper presented at the 1st International Soil Mechanics and Foundation Engineering Conference, Cambridge, MA.

Chen, J.-J., Wang, J.-H., Ke, X., & Jeng, D.-S. (2011). Behavior of large-diameter rock-socketed piles under lateral loads. *International Journal of Offshore and Polar Engineering, 21*(04).

Cheney, R. S., & Chassie, R. G. (2000). *Soils and foundations workshop manual*. Washington, DC: Federal Highway Administration, National Highway Institute.

Cheng, A., & Cheng, A. (1999). Characterization of layered cylindrical structures using cylindrical waves. *In Review of progress in quantitative nondestructive evaluation* (pp. 223–230). Springer.

Chim-oye, W., & Marumdee, N. (2013). Estimation of uplift pile capacity in the sand layers. *International Transaction Journal of Engineering, Management, & Applied Sciences & Technologies, 4*(1), 57–65.

Chin, F. K. (1970). Estimation of the ultimate load of piles not carried to failure. In *Proceedings of the* 2nd *Southeast Asian Conference on Soil Engineering* (pp. 81–91).

Coulomb, C. A. (1776). Essai sur une application des regles des maximis et minimis a quelquels problemesde statique relatifs, a la architecture. *Mem. Acad. Roy. Div., 7*, 343–387.

Craig, R. F. (1983). *Soil mechanics* (3rd ed.). Department of Civil Engineering, University of Dundee: Springer Science & Business Media LLC.

Dapp, S., Muchard, M., & Brown, D. (2006). Experiences with base grouted drilled shafts in the southeastern United States. In *Proceedings of the 10*th *International Conference on Piling and Deep Foundations* (pp. 1553–1562). Amsterdam, the Netherlands: Deep Foundations Institute.

Davisson, M. T. (1972). *High capacity piles*. Proceedings of the Soil Mechanics Lecture Series on Innovations in Foundation Construction, American Society of Civil Engineers, IIlinois Section, Chicago, 81–112.

De Kuiter, J., & Beringen, F. (1979). Pile foundations for large North Sea structures. *Marine Georesources & Geotechnology, 3*(3), 267–314.

DeBeer, E. E. (1970). Experimental determination of the shape factors and the bearing capacity factors of sand. *Geotechnique, 20*(4), 387–411.

Dong, P., Qin, R., & Chen, Z. (2004). Bearing capacity and settlement of concrete-cored DCM pile in soft ground. *Geotechnical and Geological Engineering, 22*(1), 105–119.

Fam, A. Z. (2000). *Concrete-filled fibre-reinforced polymer tubes for axial and flexural structural members* (Doctoral thesis).

Fellenius, B. H. and Samson, L. (1976). Testing of drivability of concrete piles and disturbance to sensitive clay. *Canadian Geotechnical Journal, 13*(2), (pp. 139–160).

Fellenius, B. H. (1991). Pile foundations. In H. S. Fang (Ed.), *Foundation engineering handbook* (pp. 511–536). New York, NY: Van Nostrand Reinhold Publisher.

Fellenius, B. H. (2002). *Determining the True Distributions of Load in Instrumented Piles* (p. 116). Paper presented at the ASCE International Deep Foundation Congress: Geotechnical Special Publications.

Fuller, F. M. (1983). *Engineering of pile installations*. McGraw-Hill Companies.

Goble, G., & Rausche, F. (1970). *Pile load test by impact driving*. Washington, DC: Paper presented at the Highway Research Board Annual Meeting.

Goble, G. G., Likins Jr, G., & Rausche, F. (1975). *Bearing capacity of piles from dynamic measurements* (No. OHIO-DOT-05–75 Final Rpt.). United States Department of Transportation.

Gregersen, O. S., Aas, G., & Dibiagio, E. (1975). Load tests on friction piles in loose sand. *International Journal of Rock Mechanics and Mining Sciences and Geomechanics Abstracts, 12*(7), 98–98.

Ground Engineering. (2003). Getting to grips with friction. *Ground Engineering, Magazine of the British Geotechnical Association, 26*, 20–21.

Guades, E., Aravinthan, T., Islam, M., & Manalo, A. (2012). A review on the driving performance of FRP composite piles. *Composite Structures, 94*(6), 1932–1942.

Hannigan, P. J. (1990). *Dynamic monitoring and analysis of pile foundation installations—Continuing education short course text*. Deep Foundations Institute.

Hannigan, P. J., Goble, G. G., Likins, G. E., & Rausche, F. (2006). *Design and construction of driven pile foundations(FHWA-NHI-05-043) (FHWA-NHI-05-043)*. Federal Highway Administration: United States.

Hannigan, P. J., Goble, G. G., Likins, G. E., & Becker, M. L. (2016). *Design and construction of driven pile foundations(FHWA-NHI-16-009) (FHWA-NHI-16-009)*. Federal Highway Administration: United States.

Hassan, M., & Iskander, M. G. (1998). State of the Practice Review in FRP Composite Piling. *Journal of Composites for Construction, 2*(3), 116–120.

Helwany, S., & Wiley, B. (2007). *Applied soil mechanics: With ABAQUS applications*. Hoboken, NJ: John Wiley & Sons.

Ho, C. E. (2003). Base grouted bored pile on weak granite. In *Proceedings of the Third International Conference on Grouting and Ground Treatment* (pp. 716–727).

Hsu, S.-T. (2014). Behaviors of large-scale driven PC piles. *Journal of Marine Science and Technology, 22*(4), 487–497.

Huo, S., Chao, Y., Dai, G., & Gong, W. (2015). Field test research of inclined large-scale steel pipe pile foundation for offshore wind farms. *Journal of Coastal Research, SI*(73), 132–138.

Hussein, M. H., Woerner, I., Wayne, A., Sharp, M., & Hwang, C. (2006). Pile driveability and bearing capacity in high-rebound soils. In *GeoCongress 2006: Geotechnical Engineering in the Information Technology Age* (pp. 1–4).

Jamiolkowski, M., Lo Presti, D., & Manassero, M. (2003). Evaluation of relative density and shear strength of sands from CPT and DMT. *Soil behavior and soft ground construction* (pp. 201–238).

Jamsawang, P., Bergado, D. T., & Voottipruex, P. (2011). Field behaviour of stiffened deep cement mixing piles. *Proceedings of the Institution of Civil Engineers—Ground Improvement*, 164(1), 33–49.

Jiao, Q. Z. (2007). Shaft wall design for a shield to directly cut through. *Modern Tunneling Technology, 44*(4), 20–23.

Jongpradist, P., Youwai, S., & Jaturapitakkul, C. (2010). Effective void ratio for assessing the mechanical properties of cement-clay admixtures at high water content. *Journal of Geotechnical and Geoenvironmental Engineering, 137*(6), 621–627.

Juran, I., & Komornik, U. (2006). *Behavior of fiber-reinforced polymer composite piles under vertical loads* (No. FHWA-HRT-04–107). United States Department of Transportation.

Khaleghi, B., Lehman, D., & Roeder, C. (2016). *Concrete Filled Steel Tube Bridge Pier Connections-An ABC Solution*. Produced by Accelerated Bridge Construction: Center & Florida International University.

Kim, S., Whang, S.-W., Kim, S., & Hyung, W. G. (2017). Application of extended end composite pile design in pile foundation work. *Proceedings of the Institution of Civil Engineers-Geotechnical Engineering, 170*(5), 455–465.

Knappett, J., & Craig, R. F. (2012). *Craig's soil mechanics* (8th ed.). Abingdon, Oxon, New York: Spon Press.

Kumar, S., Alarcon, C., & Hosin, A. (2004). O-cell testing of reinforced concrete driven piles. *In International Conference on Case Histories in Geotechnical Engineering* (pp. 1–7).

Kundu, T., & Ryu, Y.-S. (2002). Underwater inspection of concrete-filled steel pipes using guided waves. *KSCE Journal of Civil Engineering, 6*(1), 25–31.

Kyfor, Z. G., Schnore, A. R., Carlo, T. A., & Baily, P. F. (1992). *Static testing of deep foundations* (No. FHWA-NHI-16–009). US Department of Transportation.

Lai, P., Mullins, G., & Dapp, S. D. (2000). *Pressure-grouting drilled shaft tips in sand*. Paper presented at the New Technological and Design Developments in Deep Foundations.

Lai, Y., Bergado, D., Lorenzo, G., & Duangchan, T. (2006). Full-scale reinforced embankment on deep jet mixing improved ground. *Proceedings of the Institution of Civil Engineers-Ground Improvement, 10*(4), 153–164.

Lao, W. K., Zhou, L. Y., & Wang, Z. (2004). Field test and theoretical analysis on flexible large-diameter rock-socketed steel pipe piles under lateral load [J]. *Chinese Journal of Rock Mechanics and Engineering, 10,* 1–35.

Lee, J., & Song, K. (2010). Material properties and bearing capacities of extended PHC pile with enlarged pile thickness. *Architectural Institute of Korea, 30*(1), 207–208.

Lehane, B. M., Williams, R., & Li, Y. (2013). Shaft capacity of displacement piles in clay using the cone penetration tests. *Journal of Geotechincal and Geoenvironmental Engineering, 139*(2), 253–266.

Li, W. J., Qin, L., Zhong, C., & Wang, L. K. (2015). Study of vibration characteristics about two-layered composite pile. *Integrated Ferroelectrics, 167*(1), 41–51.

Li, X., Xie, K., Zeng, G., & Hou, X. (2000). Research of bored pile slurry effect created during construction. *Structural Construction, 30*(5), 21–23.

Liao, S. S., & Whitman, R. V. (1986). Overburden correction factors for SPT in sand. *Journal of geotechnical engineering, 112*(3), 373–377.

Liew, S., Ng, H., & Lee, K. (2004). *Comparison of HSDPT and SLT results of driven piles in Malaysian residual soils*. Paper presented at the Malaysian Geotechnical Conference.

Liu, J., Yuan, H. J., Li, J. F., Zhou, H., & Sun, H. Y. (2014). Current state of research and application of GFRP in shield engineering. *Urban Rapid Rail Transit, 27*(1), 81–86.

Liu, S.-Y., Du, Y.-J., Yi, Y.-L., & Puppala, A. J. (2011). Field investigations on performance of T-shaped deep mixed soil cement column–supported embankments over soft ground. *Journal of Geotechnical and Geoenvironmental Engineering, 138*(6), 718–727.

Liu, X. J. (2014). Comparative study on expansive soil steep slope FRP materials bolt support. *Applied Mechanics and Materials, 454,* 250–254.

Lunne, T., Robertson, P., & Powell, J. (1997). *Cone penetration testing*. Geotechnical Practice.

Luo, L. (2014). Development and application of FRP materials in the structural in China. In *Recent advances in material, analysis, monitoring, and evaluation in foundation and bridge engineering* (pp. 126–132).

Madhyannapu, R. S., & Puppala, A. J. (2015). Design and construction guidelines for deep soil mixing to stabilize expansive soils. Journal of Geotechnical and Geoenvironmental Engineering, *141*(9).

Malik, A. A., Kuwnao, J., Tachibana, S., & Maejima, T. (2016). Interpretation of screw pile load test data using extrapolation method in dense sand. *International Journal of GEOMATE, 10*(1), 1567–1574.

Manh, T., Jensen, G. U., Johansen, T. F., & Hoff, L. (2013). Microfabricated 1–3 composite acoustic matching layers for 15MHz transducers. *Ultrasonics, 53*(6), 1141–1149.

Mansur, C. I., & Hunter, A. H. (1970). Pile tests-Arkansas river project. *Journal of Soil Mechanics & Foundations Division.*

Marcos, M. C. M., Chen, Y.-J., & Kulhawy, F. H. (2013). Evaluation of compression load test interpretation criteria for driven precast concrete pile capacity. *KSCE Journal of Civil Engineering, 17*(5), 1008–1022.

Mayne, P. W., Christopher, B., Berg, R., & DeJong, J. (2002). *Subsurface Investigations(Geotechnical Site Characterization), (FHWA NHI-01-031) (Geotechnical Site Characterization), (FHWA NHI-01-031).* National Highway Institute, Federal Highway Administration, Washington, D.C.: U.S. Dept. of Transportation.

Mayne, P. (2007). *Cone penetration testing—A synthesis of highway practice.* Washington DC: Transportation Research Board, National Academies Press.

McNamara, A., & Gorasia, R. J. (2016). High-capacity ribbed pile foundations. *Proceedings of the Institution of Civil Engineers: Geotechnical Engineering, 169*(3), 264–275.

Meyerhof, G. G. (1956). Penetration tests and bearing capacity of cohesionless soils. *Journal of the Soil Mechanics and Foundations Division, 82*(1), 1–19.

Meyerhof, G. G. (1976). Bearing capacity and settlement of pile foundations International *Journal of Rock Mechanics and Mining Sciences & Geomechanics Abstracts, 13*(6), A67–A67.

Ming, Z. (2011). The design and application of GFRP concrete structure in Dongwan subway tunnel project. *China Academic Journal Electronic Publishing House, 6,* 40–42.

Ministry of Construction of the People's Republic of China. (2002). *Code for design of concrete structures (GB 50010–2002).* Beijing, China: National Standard of the People's Republic of China.

Mirmiran, A., & Shahawy, M. (1996). A new concrete-filled hollow FRP composite column. *Composites Part B-Engineering, 27*(3–4), 263–268.

Miura, N., Horpibulsuk, S., & Nagaraj, T. (2001). Engineering behavior of cement stabilized clay at high water content. *Soils and Foundations, 41*(5), 33–45.

Mokhlesur, M., Rahman, P., & Rabbi, M. Z. (2011). Strength and deformation characteristics of cement treated soft Bangladesh clays. *IEM Journal, 72*(4), 21–31.

Naesgaard, E. (1992). Lateral load tests to examine large-strain (seismic) behaviour of piles. *Canadian Geotechnical Journal, 29*(2), 245–252.

Nakamura, S. (1998). Design strategy to make steel bridges more economical. *Journal of Constructional Steel Research, 1*(46), 58.

Nguyen, V. L., Nie, L., & Zhang, M. (2012). Method Cement Post-grouting to Increase the Load Capacity for Bored Pile. *Research Journal of Applied Sciences, Engineering and Technology, 5*(19), 4727–4732.

Nordlund, R. L. (1963). Bearing capacity of piles in cohesionless soils. *American Society of Civil Engineers Journal of the Soil Mechanics and Foundations Division, SM3,* 1–35.

Nordlund, R.L. (1979). Point Bearing and Shaft Friction of Piles in Sand. *Missouri-Rolla* 5th *Annual Short Course on the Fundamentals of Deep Foundation Design.*

Paikowsky, S. G., & Tolosko, T. A. (1999). *Extrapolation of pile capacity from non-failed load tests* (No. FHWA-RD-99–170). US Department of Transportation.

Pando, M. A. (2003). *A laboratory and field study of composite piles for bridge substructures* (Doctoral dissertation).

Pando, M. A., Ealy, C. D., Filz, G. M., Lesko, J. J., & Hoppe, E. J. (2006). *A laboratory and field study of composite piles for bridge substructures* (No. FHWA-HRT-04–043). US Department of Transportation.

Pando, M. A., Hoppe, E. J., Filz, G. M., & Dove, J. E. (2002). Interface shear tests on FRP composite piles. *In Deep Foundations 2002: An International Perspective on Theory, Design, Construction, and Performance* (pp. 1486–1500).

Park, J.-S., Lee, S.-H., Park, S.-S., Cho, J.-W., Jung, S.-W., Han, J.-H., & Kang, S.-G. (2003). Acoustic and electromechanical properties of 1–3 PZT composites for ultrasonic transducer arrays fabricated by sacrificial micro PMMA mold. *Sensors and Actuators a: Physical, 108*(1), 206–211.

Patel, D., Glover, S., Chew, J., & Austin, J. (2015). The Pinnacle-design and construction of large diameter deep base grouted piles in London. *Ground Engineering*, 24–31.

Peck, R. B., Hanson, W. E., & Thornburn, T. H. (1974). *Foundation engineering* (Vol. 10). New York, NY: Wiley.

Perko, H. A. (2009). *Helical piles: A practical guide to design and installation.* John Wiley & Sons.

Peterson, T. (1999). *Structural properties of steel-encased concrete piles* (master's dissertation).

Rausche, F., Moses, F., & Goble, G. G. (2004). Soil resistance predictions from pile dynamics Current Practices and Future Trends in Deep Foundations (pp. 418–440).

Reese, L. C. (1984). *Handbook on design of piles and drilled shafts under lateral load* (No. FHWA-NHI-16–009). US Department of Transportation.

Robertson, P. (1990). Soil classification using the cone penetration test. *Canadian Geotechnical Journal, 27*(1), 151–158.

Robinson, B., & Iskander, M. (2008). Static and dynamic load tests on driven polymeric piles. In *GeoCongress 2008: Geosustainability and Geohazard Mitigation* (pp. 939–946).

Rose, J. L., Cho, Y., & Ditri, J. J. (1994). Cylindrical guided wave leakage due to liquid loading. *Review of Progress in Quantitative Nondestructive Evaluation, 13,* 259–259.

Samtani, N. C., & Nowatzki, E. A. (2006a). *Soils and foundations—Volume I* (No. FHWA-NHI-16–009). US Department of Transportation.

Samtani, N. C., & Nowatzki, E. A. (2006b). *Soils and foundations—Volume II* (No. FHWA-NHI-16–009). US Department of Transportation.

Schmertmann, J. H. (1978). *Guidelines for cone penetration test (performance and design)* (Final Report No. FHWA-TS-78–209). US Department of Transportation.

Shi, C. (2005). The application of the pile-end mud-jacking technique in the construction of bored caisson pile. *Sci/Tech Information Development & Economy, 15*(23), 293–296.

Shin, Y., Kim, M., Ko, J., & Jeong, S. (2014). Proposed design chart of mechanical joints on steel-PHC composite piles. *Materials and Structures, 47*(7), 1221–1238.

Sinnreich, J., & Simpson, R. C. (2013). Base Grouting Case Studies Including Full Scale Comparative Load Testing. *Seventh International Conference on Case Histories in Geotechnical Engineering.* No. 2.16, 1–8.

Skempton, A. W. (1986). Standard penetration test procedures and the effects in sands of overburden pressure, relative density, particle size, ageing and overconsolidation. *Geotechnique, 36*(3), 425–447.

Standards Australia. (1998a). *Methods of testing soils for engineering purposes—Method 6.2.2: Soil strength and consolidation tests—Direct shear test using a shear box* (AS 1289.6.2.2) (pp. 1–25).

Standards Australia. (1998b). *Methods of testing soils for engineering purposes—Method 6.6.1: Soil strength and consolidation tests—Determination of the one-dimensional consolidation properties of a SOI* (AS 1289.6.6.1) (pp. 1–12).

Standards Australia. (2009). *Piling—Design and installation* (AS 2159) (pp. 1–97).

Standards Australia. (2016a). Methods of testing soils for engineering purposes—Method 6.4.1: *Determination of compressive strength of a soil—Compressive strength of a specimen tested in undrained triaxial compression without measurement of pore water pressure (AS 1289.6.4.1)* (pp. 1–9).

Standards Australia. (2016b). *Methods of testing soils for engineering purposes—Method 6.4.2: Determination of compressive strength of a soil—Compressive strength of a saturated specimen tested in undrained triaxial compression with measurement of pore water pressure* (AS 1289.6.4.2) (pp. 1–17).

Taesiri, Y., & Chantaranimi, P. (2001). Slope stabilizations of highway embankments adjacent to irrigation/drainage canal. In *Proceedings of Soft Ground Improvement and Geosynthetics Applications* (pp. 211–227). Thailand.

Taylor, D. W. (1948). *Fundamentals of soil mechanics.* New York, NY: John Wiley & Sons.

Terzaghi, K. (1944). *Theoretical soil mechanics.* New York: Chapman and Hali, Limited John Wiley and Sons.

Tomlinson, M. J. (1980). *Foundation design and construction* (5th ed.). Harlow, England: Longman Scientific & Technical.

Tomlinson, M. J. (2001). Foundation design and construction. *Pitman Publishing Limited, 1963,* 1–583.

Tomlinson, M. J., & Boorman, R. (2001). *Foundation design and construction* (7th ed.). New York; Harlow, England: Prentice Hall.

Uddin, K., Balasubramaniam, A., & Bergado, D. (1997). Engineering behavior of cement-treated Bangkok soft clay. *Geotechnical Engineering, 28,* 89–119.

Vesic, A. S. (1977). *Design of pile foundations. NCHRP synthesis of highway practice* (42).

Voottipruex, P., Suksawat, T., Bergado, D. T., & Jamsawang, P. (2011). Numerical simulations and parametric study of SDCM and DCM piles under full scale axial and lateral loads. *Computers and Geotechnics, 38*(3), 318–329.

Wang, S. T., & Reese, L. C. (1993). *Laterally loaded pile analysis program for the microcomputer* (No. FHWA-SA-91-048). US Department of Transportation.

Werasak, R., & Meng, J. (2013). Field testing of stiffened deep cement mixing piles under lateral cyclic loading. *Earthquake Engineering and Engineering Vibration, 12*(2), 261–265.

Wilkins, E., & Castelli, R. J. (2004). Osterberg load cell test results on base grouted bored piles in Bangladesh. In *GeoSupport 2004: Drilled Shaft, Microiling, Deep Mixing, Remedial Methods, and Specialty Foundation Systems* (pp. 587–602).

Wonglert, A., & Jongpradist, P. (2015). Impact of reinforced core on performance and failure behavior of stiffened deep cement mixing piles. *Computers and Geotechnics, 69,* 93–104.

Xu, G. (2009). *Effects of frozen soils on site response and lateral behavior of concrete-filled steel pipe pile* (dissertation). ProQuest Dissertations Publishing.

Yang, H., & Xiao, D. (2011). Back analysis of static pile load test for SPT-based pile design: A Singapore experience. In *Advances in Pile Foundations, Geosynthetics, Geoinvestigations, and Foundation Failure Analysis and Repairs* (pp. 144–152).

Yang, J., Wang, F., Lu, S., & Wang, C. (2014). Application of compactness detection to complicated concrete-filled steel tube by ultrasonic method. *Transactions of Tianjin University, 20*(2), 126–132.

Yang, P., Hu, H.-S., & Xu, J.-F. (2012). Settlement characteristics of pile composite foundation under staged loading. *Procedia Environmental Sciences, 12,* 1055–1062.

Yttrup, P., & Abramsson, G. (2003). Ultimate strength of steel screw piles in sand. *Australian Geomechanics: Journal and News of the Australian Geomechanics Society, 38*(1), 17.

Yu, F., & Yang, J. (2012). Base capacity of open-ended steel pipe piles in sand. *Journal of Geotechnical and Geoenvironmental Engineering, 138*(9), 1116–1128.

Zhang, H., Chen, S., Zhao, Y. B., & Li, M. W. (2011). The application of GFRP in shield tunnel construction. *Railway Standard Design, 3*(24), 73–76.

Zhang, H. W., Smith, S. T., & Kim, S. J. (2012). Optimisation of carbon and glass FRP anchor design. *Construction and Building Materials, 32,* 1–12.

Zhou, J., Oh, E., Zhang, X., Jiang, H., Bolton, M., & Wang, P. (2017a). Compressive and Uplift Static Load Tests of Shaft and Base Grouted Concrete Bored Piles. *27*th *International Ocean and Polar Engineering Conference* (pp. 685–692).

Zhou, J. L., Zhang, X., Jiang, H. S., Bolton, M., & Oh, E. (2016). A review of geotechnical application of fibre reinforced polymer materials. In *8*th *International Conference on Fibre-Reinforced Polymer Composite in Civil Engineering* (pp. 856–862).

Zhou, D., Lam, K. H., Chen, Y., Zhang, Q., Chiu, Y. C., Luo, H., & Chan, H. L. W. (2012). Lead-free piezoelectric single crystal based 1–3 composites for ultrasonic transducer applications. *Sensors and Actuators a: Physical, 182,* 95–100.

Chapter 3
Field Tests of Post Grouted Concrete Piles

3.1 General Introduction

With the soaring requirement of building space in metropolises, high rise building development is indispensable and thus deep piles are imperative. Pile length can range from 20 to 60 m due to large loads from upper structures needing to be transferred into soils. Apart from designing long pile foundations, which can provide more friction resistance, grouting technology is also a good way to increase the ultimate bearing capacity. Sometimes this technology is also applied in construction of large diameter deep piles. One example of this is the two bored piles with depths of 65 and 91 m tested in western Bangladesh. The tests results showing increments of end bearing and shaft bearing capacity and decrements of base settlement (Castelli and Wilkins 2004). Another example is a skyscraper project in London, where the diameter of piles was 2.4 m with a length of 63 m (Patel et al. 2015).

Compared to bored piles, precast concrete piles are frequently more costly (Tomlinson and Woodward 2007). However, during the process of pile foundation construction, some soil deposits can remain at the base area after drilling the hole. The remaining deposit, which is accumulated by collapse during drilling, will lead to a decrease in the end resistance capacity of the pile, and an increase of pile settlement. This potential for reduced pile capacity was affirmed during construction and performance of static load tests in Taiyuan City, China where capacities were much lower than the designed requirement (Shi 2005). Furthermore, the drilling operation can loosen the soil underneath the base of the bored hole, which can also lead to excessive working load settlement. Engineers nowadays use admixtures like polymeric slurry or bentonite as a support to avoid the collapse of soils during drilling construction; this, however, creates other issues. The use of these materials may lead to capacity decrease of pile shafts because these admixtures create a layer between soil and pile when combined with soil and water, which consequently leads to a decrease of friction resistance of the pile. Li et al. (2000) reported that this admixture layer or composite layer could decrease the pile bearing capacity by 30–40%.

J. Zhou and E. Oh, *Full-Scale Field Tests of Different Types of Piles*,
Advanced Topics in Science and Technology in China 62,
https://doi.org/10.1007/978-981-33-6183-6_3

Such limitations, however, can be addressed with a base and shaft grouting technique (or post grouting technique). This construction technology of cast-in situ piles started in the mid of 1970s. It can generally be categorized into a flat jack system which consists of grout delivery pipes connected to a steel plate with a rubber membrane, and a sleeve-port system that consists of 2–4 U-tubes installed at the bottom of the pile. This U-tube is covered by rubber and can be arranged in various configurations (Dapp et al. 2006). Cement admixture which is blended into various ratios is the popular material for grouting. It is injected to the pile toe through U-tubes under high pressure, consequently restoring the original density of the base soil and reducing the settlement of the pile when loading transfers from the upper structure (Tomlinson and Woodward 2007). It is worth noting that due to high pressure, partial cement admixture will force the admixture layers upwards, which leads to a recovering and improvement of shaft resistance around the pile as shown in Fig. 3.1.

Field geotechnical testing generally includes dynamic load tests, compressive and uplift static load tests and O-Cell tests (Zhussupbekov and Omarov 2016). Several compressive static load tests of base grouted piles have been conducted. Due to grouting being more popular in granular soils, some researchers have focused on the soil type effect and load transfer mechanism (Ho 2003; Rollins et al. 2008; Thasnanipan et al. 1998). Other researchers have concentrated on the construction process and capacity improvements (Wang et al. 2006, 2015).

Recent tests and numerical simulations have shown that the post grouted concrete piles can increase twice the ultimate capacity of defected piles, and increase about 20% compared to normal piles (Nguyen et al. 2012). Compared to grouted piles and

Fig. 3.1 Grouting concrete pile with cement layer (Zhou et al. 2017b)

non-grouted piles, results have shown grouting could increase 18–19% of shaft resistance and 63% base resistance (Wang and Zhang 2013). Through tests on seven piles, Zhang et al. (2012c) have pointed out that post technology has had an enhancing effect on shaft and base resistance but the increase in pile capacity was affected by slenderness ratio. For the research of grouting techniques on precast piles, a recent experimental and FEM study has proposed the capacity estimation and method of grout pressure (Thiyyakkandi et al. 2013). Several other projects with application of base grouting techniques have been provided by Sinnreich and Simpson (2013); however, these results are ambivalent because some projects illustrate increased capacity of pile by grouting and some projects do not.

A large number of investigations have been carried out to obtain the behaviors of piles under uplift loading. Some researchers have focused on the behaviors of different types of piles such as precast piles, cast-in situ bored piles and steel pipe piles etc. (Huo et al. 2013; Kusakabe et al. 1994; Shelke and Patra 2011). Other researchers have concentrated on the uplift behavior of piles in cohesionless soil (Gaaver 2013; Thiyyakkandi et al. 2014; Verma and Joshi 2010). Through experimental tests, a study revealed that pile behavior under uplift force depends mainly on pile embedment depth-to-diameter ratio and soil properties (Gaaver 2013). A further study investigated pile behavior under combined uplift and lateral loading (Madhusudan Reddy and Ayothiraman 2015). Recent research on belled and multi-belled piles in dry loose sand has also been conducted, with uplift resistance being found to increase up to 60% in comparison with straight piles (Moayedi and Mosallanezhad 2017). The investigations of base-and-shaft grouted uplift field tests, capacity improvement, as well as load transfer mechanism, however, remain limited.

To determine the load transfer behavior of piles under compressive and uplift static load tests, wire vibration strain reinforced bars are commonly used. The disadvantage of these strain gauge applications is that labor is required and it is time consuming. In order to avoid the damage of strain gauges, the transfer of the steel cages is always slow, which may lead to project being delayed. With technology developing, dynamic load test appears which can provide good results with load transfer information. Compared to static load tests, Pile Driven Analyzer (PDA) tests are relatively cheaper and time saving (Budi et al. 2015). A popular program is the Case Pile Wave Analysis Program or CAPWAP. The bearing capacity and the stress distribution along the pile shaft and toe as well as simulated static load tests can be determined by matching signals obtained from the dynamic load tests (Likins et al. 1996). The strain sensors and accelerometers are used to measure the force and velocity inside the pile after applying load provided by a hammer. The load transfer mechanism can be obtained when the matched wave approaches the acquired wave onsite. Investigations of driven polymeric piles using dynamic load tests were conducted in Elizabeth, New Jersey, providing load testing installation and comparison between static and dynamic load tests. That paper also focused on the possible application of plastic piles under axial loading (Robinson and Iskander 2008).

As mentioned above, there is limited research on compressive and uplift loaded grouted piles. This Chapter, therefore, aims to investigate the ultimate capacity of post grouted concrete piles under compressive and uplift static load and dynamic load tests. By comparing the compressive capacity among piles with shaft-and-base grouting, base grouting only and no grouting, the increment of pile capacity by grouting technique is determined. The increment of uplift capacity by grouting is also observed by uplift static load tests. In addition, the interpretation of the static load test result including the double tangent, DeBeer's, Chin's and Mazurkiewicz's methods are provided. This Chapter also provides dynamic load tests for load transfer mechanism analysis. It should be noted that this Chapter investigated the piles behavior and shaft mechanisms under different grouting techniques, which has not been reported based on results' comparison between static load tests and dynamic load tests.

3.2 Site Conditions

The examined project aimed to build a 22-level office building at a height of 82.95 m, with a construction area of 50.8 × 42.2 m^2. The construction site was located in Jinan city, China. The subsurface exploration was determined through laboratory and in situ tests. In situ SPTs and CPTs and laboratory consolidation tests, direct shear tests and triaxial tests were conducted based on the Chinese Standard for Soil Test Method (GB/T 50123-1999 1999) and Code for Investigation of Geotechnical Engineering (GB 50021-2001 2001), respectively.

A hammer with a weight of 63.5 kg was selected for the SPTs, and counts of 300 mm penetration were recorded as the N-value or $N_{63.5}$-value. The soil classification was determined based on these N-values and soil sampling: Clause 3.3 of Soil Classification (GB 50021-2001 2001). Laboratory tests including consolidation test, shear test and triaxle tests (CU and UU) were conducted for soil characteristics of specific gravity, relative density, porosity, void ratio, saturation ratio, water content, plastic limit, liquid limit, plasticity index, liquidity index, cohesion and friction angle. These tests were conducted based on Clauses 5, 8, 9, 13, 14, 16 and 18 of Soil Test Method respectively (GB/T 50123 1999). Based on the borehole logs, the Simplified soil layers were then discovered as follows, with the properties of soil layers illustrated in Table 3.1.

1. miscellaneous fill: loose, containing gravel and rubble, diameter ranging from 2.0 to 7.0 mm, average thickness of layer of 2.61 m; 2. silty clay: medium stiff, yellowish with high plasticity, average thickness of 3.0 m; 3. gravel: medium dense, average thickness of 3.5 m; 4. silty clay: medium stiff, low plasticity, average depth of 3.5 m; 5. gravel: medium dense with dark clay, diameter of gravel ranging from 5.5 m; 6. residual soil: dense, layer containing iron and manganese oxides, average thickness of 5.0 m; 7. weathered diorite: medium to dense, average depth of 4.0 m; 8. highly weathered diorite: medium to dense, average thickness of 1.7 m; 9. bearing stratum: very dense, rock-quality designation = 40, average strength of 10.65 MPa.

This project required underground carparks, so the deep excavation construction is required (excavation depth of 9 m). After the soil was removed, the cast-in situ piles were cast and surrounded by four layers. The compressive and uplift loaded piles with labels of P51, P121, P126, P15 and P16 were adjacent with each other, so

Table 3.1 Simplified soil layers

Soil layers	Depth (m)	Average N-value	γ (kN/m^3)	C (kPa)	φ' (^{o})	F_{ak} (kPa)
Miscellaneous fill	2.61	5	19.0	5.0	10.0	
Silty clay	3	7	18.9	16.6	19.7	150
Gravel	3.5	21	20.0	5.0	30.0	260
Silty clay	3.5	21	19.3	18.2	16.8	200
Gravel	5.5	22	20.0	8.0	30.0	280
Residual soil	5	23	18.0	5.0	28.0	200
Weathered diorite	4	41	18.5	10.0	30.0	240
Highly weathered diorite	1.7	43	25.5	12.0	48.0	450
Bearing stratum	N/A	>50	28.8	12.0	54.0	2000

the soil condition was similar. The soil layers were residual soil with a thickness of 5 m, weathered diorite with an average thickness of 4 m, highly weathered diorite with an average thickness of 1.7 m and bearing stratum.

3.3 Pile Description

Three and two concrete bored piles with labels of P51, P121, P126, P15 and P16 were cast for conducting compressive and uplift SLTs, respectively. All these piles were made of the same concrete and reinforcement, with the same diameter and length. Detailed information is provided in Tables 3.2 and 3.3. However, the used grouting technology was different. In contrast to the base grouting, the base-and-shaft grouting system contained a circular tube (circular diameter $R = 800$ mm) with tube diameter of 5 mm and an additional shaft grouting pipe, as shown in Fig. 3.2, which was welded to steel cages. This grouting tube was drilled with small holes with spacing of 150 mm, and covered by a membrane that was used to avoid the fine sand blocking the grouting.

As depicted in Table 3.3, during the construction process, P126 was cast using base-and-shaft grouting technology and P121 was cast using base grouting only. To

Table 3.2 Reinforcement properties of tested piles

Steel bars	Diameter (mm)	Type	Spacing (mm)	f_y (MPa)	Es (MPa)	v
Longitudinal	28	HRB400	N/A	360	200,000	0.3
Stirrups top 4 m	16	HRB400	100	360	200,000	0.3
Stirrups of other part	16	HRB400	200	360	200,000	0.3
Stiffening ring	14	HRB400	2000	360	200,000	0.3

Table 3.3 Description of test piles

Pile label	Technology used	Types of tests	Length (m)	Diameter (mm)	Concrete type	Compressive strength (kPa)
P51	No grouting	Compressive SLTs	19.5	800	C50	50
P121	Base grouting	Compressive SLTs Dynamic load test	19.5	800	C50	50
P126	Base-and-shaft grouting	Compressive SLTs Dynamic load test	19.5	800	C50	50
P15	Base-and-shaft grouting	Uplift SLTs	19.5	800	C50	50
P16	No grouting	Uplift SLTs	19.5	800	C50	50

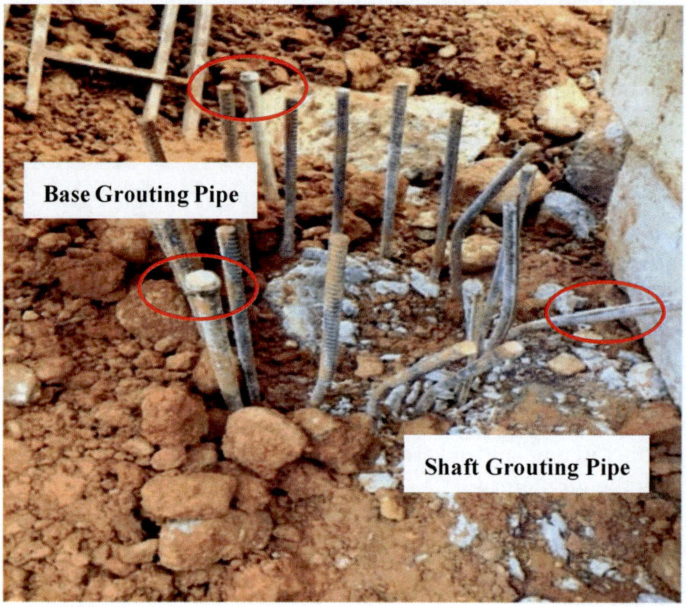

Fig. 3.2 Grouting pipe system of concrete pile

obtain results of a trial comparison, P51 was cast without any grouting, these three piles were designed to apply the compressive force. Previously, all these three piles were designed to be uplift loaded after compressive SLTs being completed, but with consideration of loading influence by compressive loads, the team determined to cast other two piles with labels of P15 and P16 for the uplift loading tests. The difference

Fig. 3.3 Mud slurry support

between these two piles was that P15 was applied with base-and-shaft grouting technology. Further, the dynamic load tests were conducted on P121 and P126 to determine load transfer behaviors. All piles were cast with application of mud slurry supporting, as shown in Fig. 3.3. Note that the mud slurry replaced the admixture slurry during drilling process, which was only for economic considerations.

3.4 Static Load Test Setup

SLTs were conducted based on the Chinese Technical Code for Testing of Building Foundation Piles (JGJ 106-2014 2014). The maintained load SLT procedure was used. According to the code, the maximum applied loads should be twice the designed loads, with increment value of 1/10 Q_{Max}. i.e., for compressive SLTs: $Q_{Max} = 2 \times$ 4,500 kN; increment loading = 900 kN; for uplift SLTs: $Q_{Max} = 2 \times 1{,}500$ kN; increment loading = 300 kN.

As shown in Figs. 3.4 and 3.5, four and two hydraulic jacks (QF630T-20) were used to provide loads to piles for compressive and uplift SLTs, respectively. For the compressive SLTs (P51, P121, P126), the loading started at 1,800 kN, with increments of 900 kN, and released with decrements of 1,800 kN back to 0 kN. For the uplift load tests (P15, P16), the loading started at 600 kN, with increments of 300 kN, and released with decrements of 600 kN back to 0 kN.

Following Chinese codes GB 50202-2002 (2002) and JGJ 106-2014 (2014), for the compressive SLTs, the settlement was recorded with intervals of five, 15, 30, 45 and 60 min (and 1 h later if required) under the maintained loads. The followed loading could only be applied when the variable quantity of the recorded movement

Fig. 3.4 Compressive SLTs setup

Fig. 3.5 Uplift SLTs setup

was less than 0.1 mm. The test needed to stop when the recorded settlement was five-times of the previous settlement or the recorded settlement was observed over 40 mm. Same with the compressive SLTs, for the uplift SLTs, low-speed maintenance tests were conducted. The applied load was maintained until the rate of axial movement did not exceed to 0.1 mm. After each load was applied, the vertical movement of the piles was recorded with intervals of five, 10 and 15 min, and 30 min. If the accumulated time exceeded 1 h, record movement of pile with interval of 30 min. The loading could be terminated in the case of the following conditions:

1. The applied loading was equal to the value of 0.9 times the ultimate strength of reinforcement;
2. Quintuple movement change between two loading stages was discovered;
3. The pile head movement increased up to 100 mm.

3.5 Dynamic Load Test Setup

The setup of the dynamic load tests is illustrated in Fig. 3.6. For this test, the dynamic penetration analyzer (PDA) (RS-1616 K[S]) was used for signal analysis. Two accelerators (SY-2) and two strain transducers (CYB-YB-FIKA) were installed symmetrically onto the pile surface with a distance equal to or over the value of 1.5 D (D—diameter of pile) from the pile head. The wave signal was recorded by accelerators and strain transducers after a heavy hammer dropped onto the pile head, and the PDA matched the signal based on the CAPWAP.

Fig. 3.6 Dynamic load tests setup

3.6 Static Load Tests Results

3.6.1 Compressive Static Load Tests

3.6.1.1 Load Settlement Curves

Figure 3.7 provides the load-settlement (Q-s) curves of three compressive loaded bored piles. After the maximum loading of 9,000 kN was applied, the maximum pile head settlement of the non-grouted (P51), based-and-shaft grouted (P126), and base grouted (P121) piles were 11.68, 13.37 and 11.74 mm, respectively. From the curves, it represents approximately linear from the loading stage. The final settlements of P51 and P121 were discovered to be 8.11 and 7.36 mm, respectively. This was likely because the bearing stratum was diorite, and the base grouting does not effectively improve the settlement. Also, as shown in this Figure, P126 showed a smaller permanent settlement of 2.8 mm. This demonstrated that base-and-shaft grouting will decrease the final settlement of a pile foundation. Also, this illustrates that the base-and-shaft grouting has changed characteristic of the pile-soil system from semi-plastic state to a relatively elastic rigid body, and the maximum settlement is primarily caused by the pile-soil elastic compression.

Fig. 3.7 Results of Q-s curvatures of compressive loaded piles

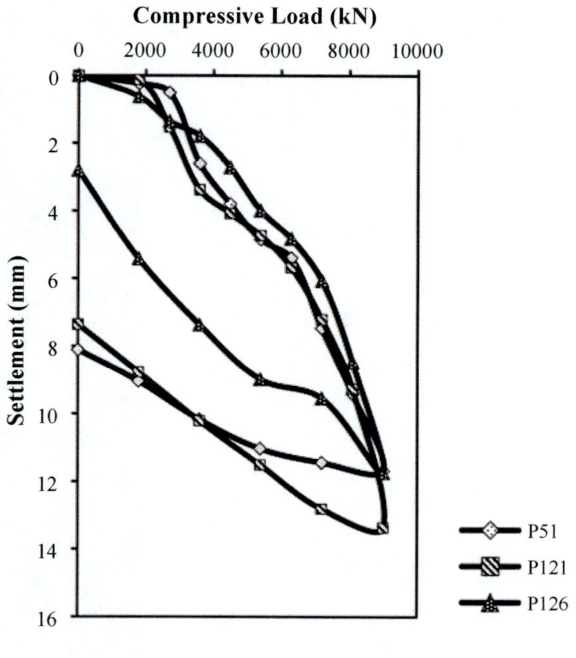

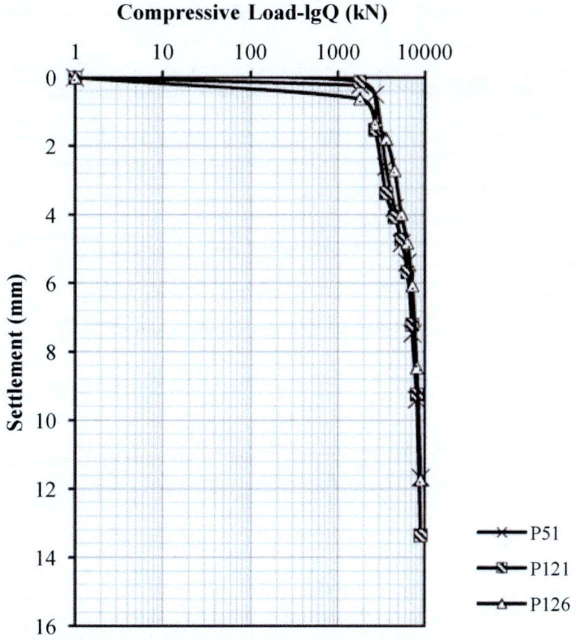

Fig. 3.8 s-lgQ curves of tested piles

3.6.1.2 Settlement lg (Load) Curves

Through the settlement-lg load (s-lgQ) curvatures shown in Fig. 3.8, the ultimate compressive bearing capacities of the piles were determined, albeit with difficulty, to be 6,800, 7,100 and 8,000 kN, respectively. This was because the tests were 'proof tests', which aimed to ensure the pile settlement being acceptable under the maximum loading (equal to twice the design load). The non-failure settlements led to difficulty in determining the capacity from the s-lgQ curves. The determination of ultimate pile capacity needed more comprehensive analysis.

3.6.1.3 Settlement lg (Time) Curves

The settlement-lg time (s-lgt) curves of P51, P121 and P126 are provided in Figs. 3.9, 3.10 and 3.11, respectively. There were no conditions indicating an extreme settlement decreasing trend under a specific loading stage, and the change of settlements between the adjacent two loading stages was relatively close. Further, it was discovered that, in each loading, the settlement did not change much as time went by, and this demonstrated that all three piles were stable after the loading was applied. The ultimate bearing capacity of the pile was unknown via analyzing these diagrams; thus, further comprehensive analysis was required.

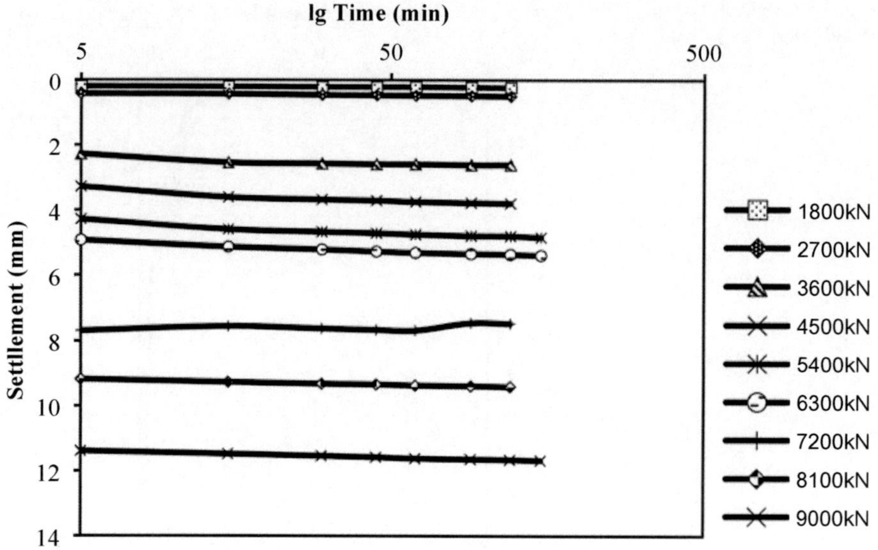

Fig. 3.9 s-lgt curves of P51

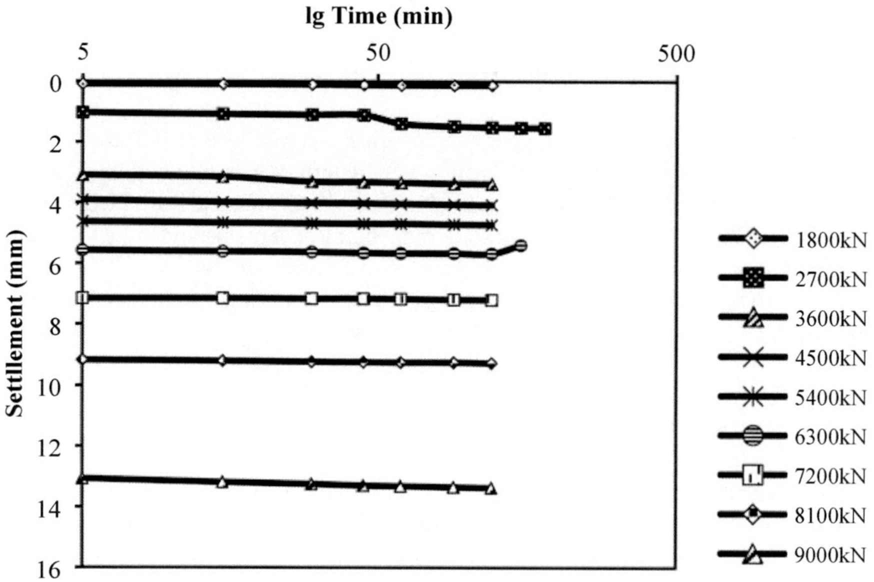

Fig. 3.10 s-lgt curves of P121

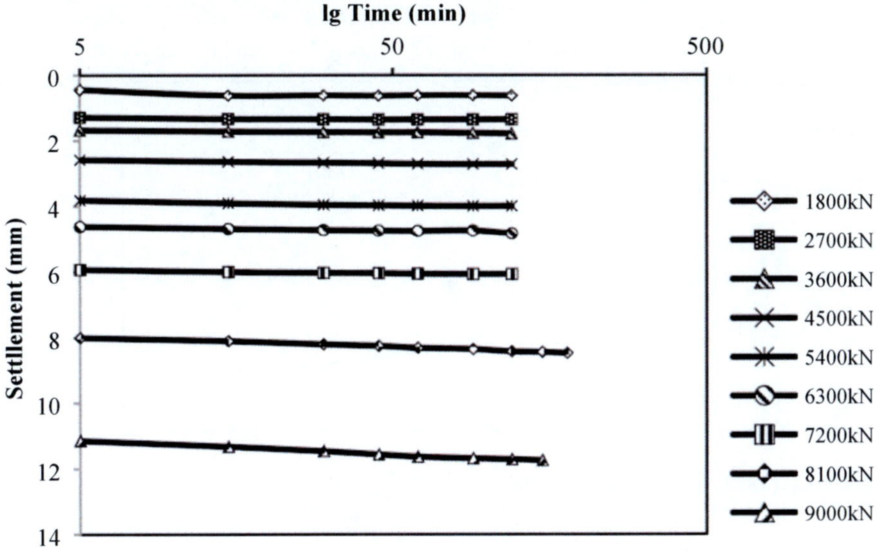

Fig. 3.11 s-lgt curves of P126

3.6.1.4 Interpretations of Static Load Tests

Double Tangent Method

As illustrated in the General Principles and Practice in Chap. 2, the double tangent method determines two tangent lines from the load–settlement curves, and finds the interaction between these two tangent lines. The point shows the corresponding ultimate capacity and ultimate settlement. In this study, the capacities of the three compressive loaded piles with labels of P51, P121 and P126 were determined to be 7,560, 7,700 and 8,250 kN, with corresponding ultimate settlements of 6.0, 6.2 and 8.0 mm, respectively, as shown in Figs. 3.12, 3.13 and 3.14. It should be noted that these values are conservative because of these tests are the proof test instead of failure test. Under this condition, the double tangent method can be appropriately adjusted based on the local experience.

DeBeer's Method

The curves based on DeBeer's method are provided in Fig. 3.15. As shown in the lg settlement-lg load diagram, finding the intersection of two distinct slopes of each pile is difficult. Similar to the settlement to lg load (s-lgQ) results, the ultimate bearing capacities of those piles were determined with difficulty to be 7,600, 7,900 and 8,300 kN, respectively.

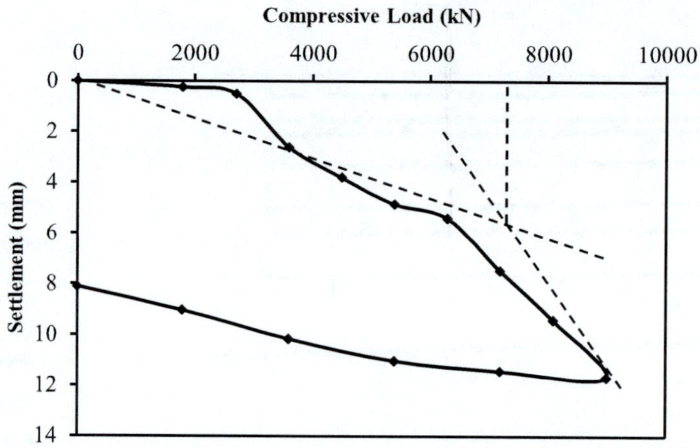

Fig. 3.12 Double tangent method of P51

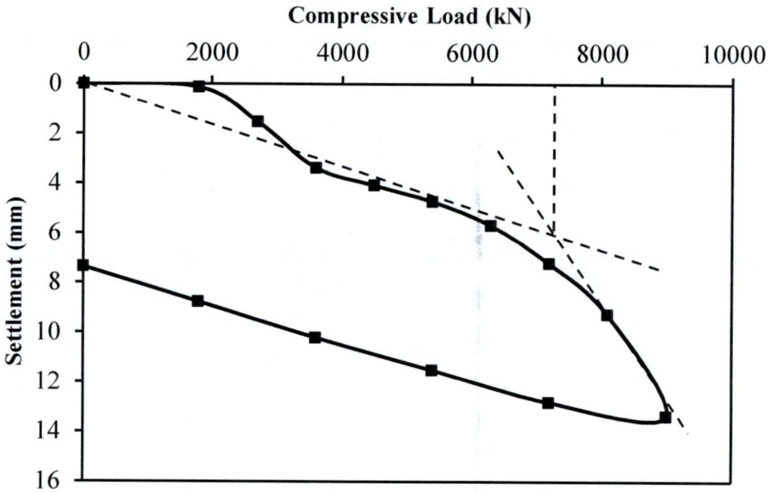

Fig. 3.13 Double tangent method of P121

Davisson's Offset Method

By calculation of the QL/EA (Eq. 2.37), the elastic deformation line can be obtained. As shown in Fig. 3.16, the elastic deformation line and the parallel line based on Davisson's offset formula (Eq. 2.39) is plotted in the Q-s curves. It can be seen that there is no intersection between the Q-s curves and the offset line. This is because the total settlement of these piles was small under maximum applied loads. Hence, this method did not fit this case to determine the ultimate bearing capacity. However, this provided information that the ultimate bearing capacity of the pile was greater than the maximum applied loads from the SLTs. This result also illustrates that for

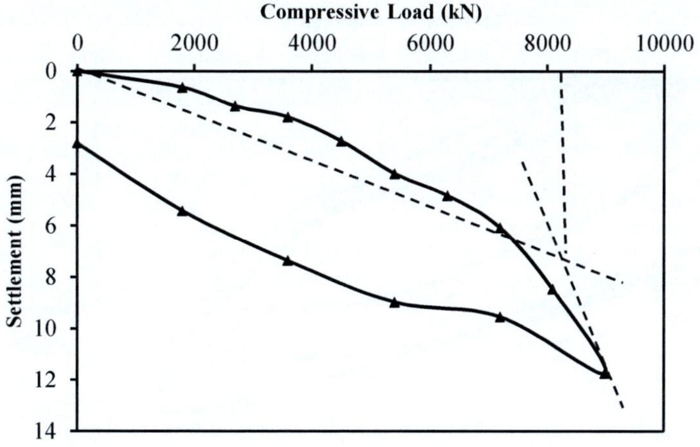

Fig. 3.14 Double tangent method of P126

Fig. 3.15 DeBeer's method of tested piles

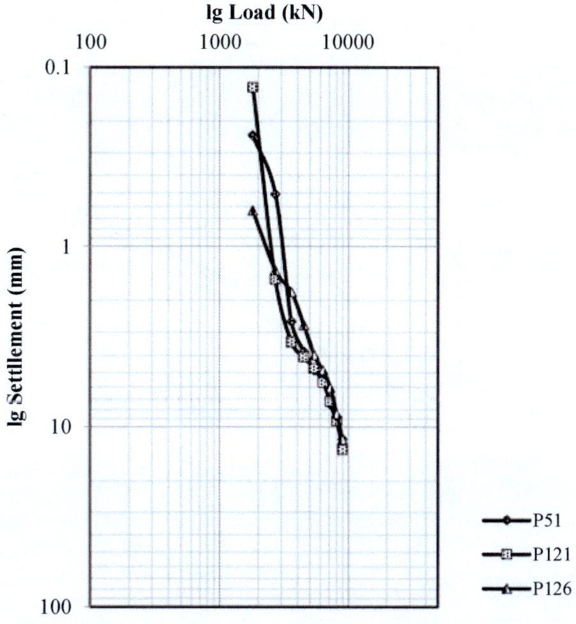

the proof test result, the Davisson's offset method may not be used for interpretation to determine the ultimate bearing capacity.

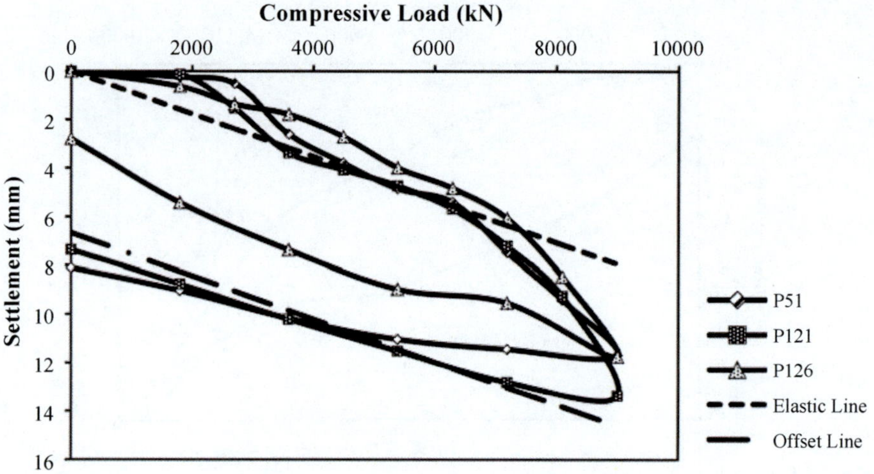

Fig. 3.16 Davisson's offset method of tested piles

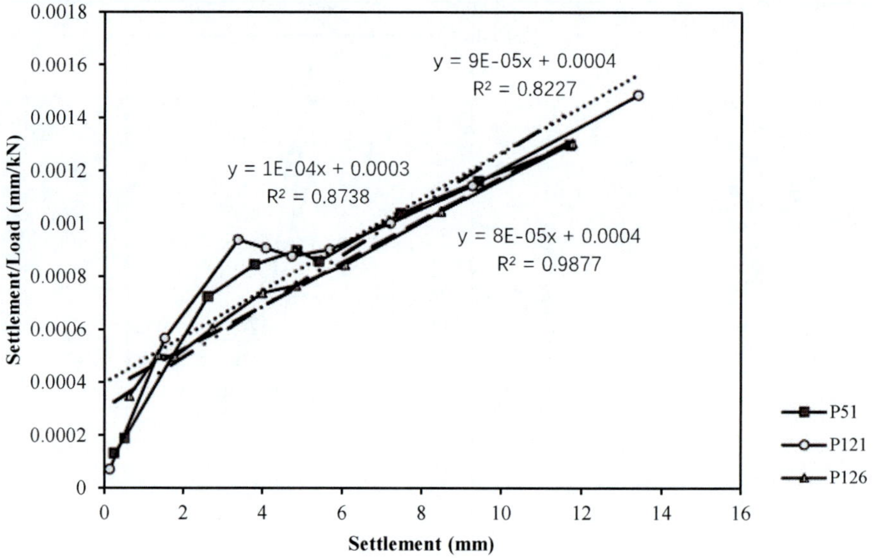

Fig. 3.17 Chin's method of tests piles

Chin's Method

As discussed in the literature review in Chap. 2, the inverse slopes of Δ/P results in the failure value. In this study, the results of the three piles are provided in Fig. 3.17. It was found that the ultimate bearing capacities of P51, P121 and P126 were 10,000,

Table 3.4 Results of piles under different interpretation methods

Pile Label	Double tangent		DeBeer's		Chin's	
	Load (kN)	Settlement (mm)	Load (kN)	Settlement (mm)	Load (kN)	Settlement (mm)
P51	7,560	6.0	7,600	7.9	10,000	N/A
P121	7,700	6.2	7,900	8.3	11,111	N/A
P126	8,250	8.0	8,300	8.2	12,500	N/A

11,111 and 12,500 kN, respectively. Given that the determined ultimate loads were beyond the maximum applied loading, the corresponding ultimate settlements were unknown.

Based on the different interpretations, the ultimate bearing capacities of piles suffering from compressive loading were determined. The ultimate bearing capacity with corresponding ultimate settlement of these compressive loaded piles are summarized in Table 3.4.

3.6.2 Uplift Load Static Load Tests

3.6.2.1 Load Settlement Curves

To determine if the piles achieved the design requirement when suffering from uplift force and to determine the improvement of the base-and-shaft grouting technique, two piles were selected for uplift SLTs—P15 and P16. As shown in Fig. 3.18, when the maximum loading of 3,000 kN was applied, the maximum vertical movements of

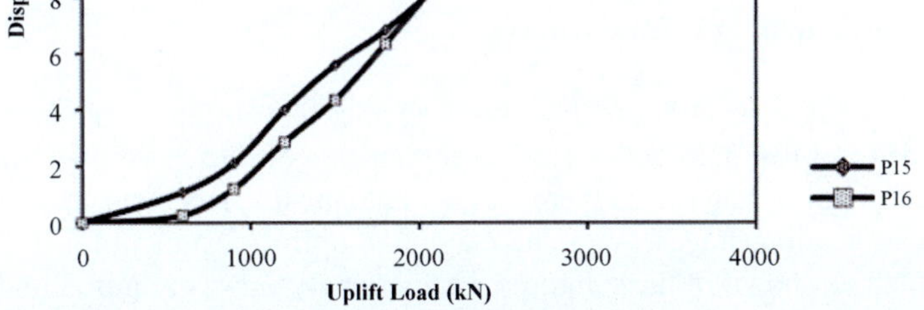

Fig. 3.18 Results of Q-s curvatures of uplift loaded piles

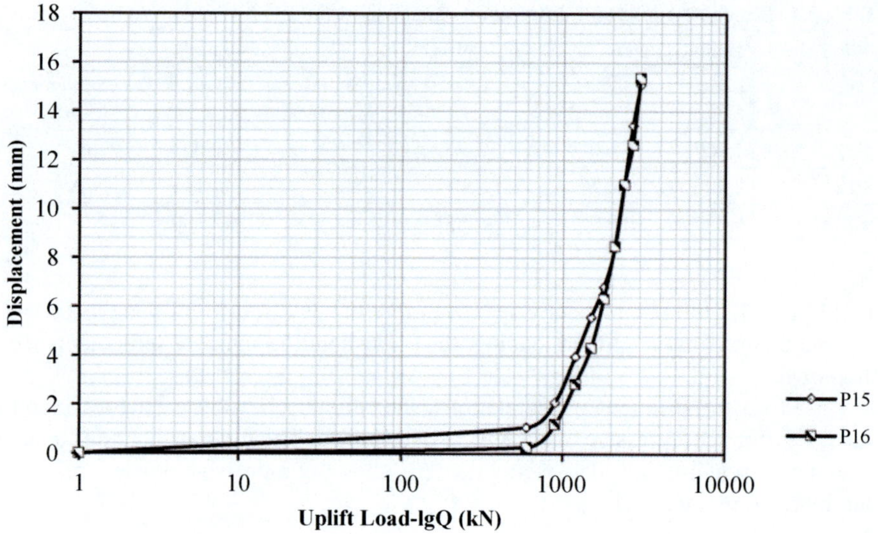

Fig. 3.19 s-lgQ curves of tested piles

P15 and P16 were 15.1 and 15.38 mm, respectively. These two tests also illustrated that, when loading from 0 to 3,000 kN, the load line was approximately linear. This result illustrated that these two piles could resist more loading. Further analysis was required to determine the ultimate uplift capacity of the pile.

3.6.2.2 Settlement lg (Load) Curves

The analysis of the s-lgQ curves of these two uplifted piles indicated no plunging points (the movement increase upward dramatically), as shown in Fig. 3.19. Also, it can be found that the maximum upward displacements of both piles were very small (less than 16 mm). This illustrated that the pile achieved the designed load, but the maximum loading was not the ultimate bearing loading. Further research was required to determine the ultimate bearing.

3.6.2.3 Settlement lg (Time) Curves

The settlement-lg time curves of the two tested piles that suffered from uplift loading are provided in Figs. 3.20 and 3.21. The settlements of these two piles were stable during each loading stages, and there was no dramatic displacement change between two adjacent loading stages. Also, the maximum uplift movement for P15 and P15 is very small as shown in these figures. Further analysis was required to determine the ultimate bearing capacity of these piles.

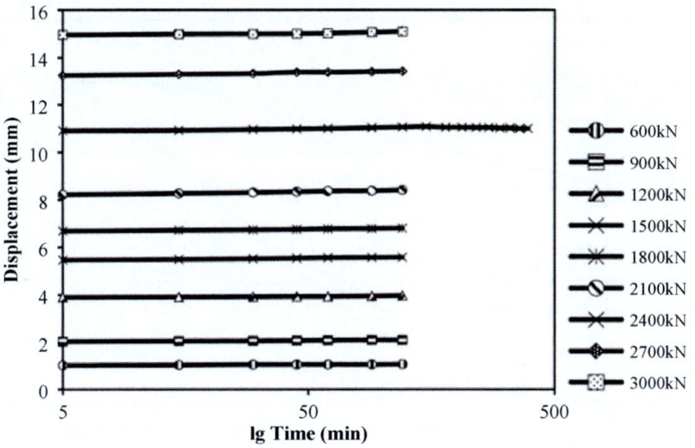

Fig. 3.20 s-lgt curves of P15

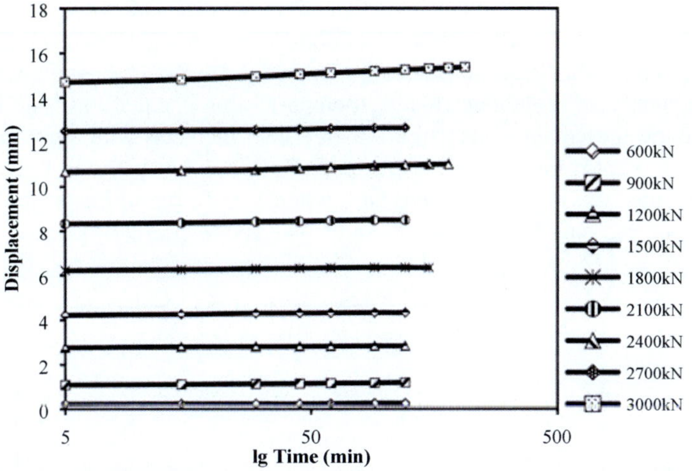

Fig. 3.21 s-lgt curves of P16

3.6.2.4 Interpretations of Static Load Tests

As illustrated in the literature review in Chap. 2, for uplift SLTs, an offset method similar to Davisson's offset method can be employed. The offset value is recommended to be 4.0 mm, and the intersection of the uplift load–settlement curves and offset line represents the ultimate load. Based on this method, the ultimate bearing capacities of P15 and P16 were determined to be 1,500 and 1,700 kN, with corresponding vertical displacements of 5.5 and 5.6 mm, respectively, as shown in Fig. 3.22.

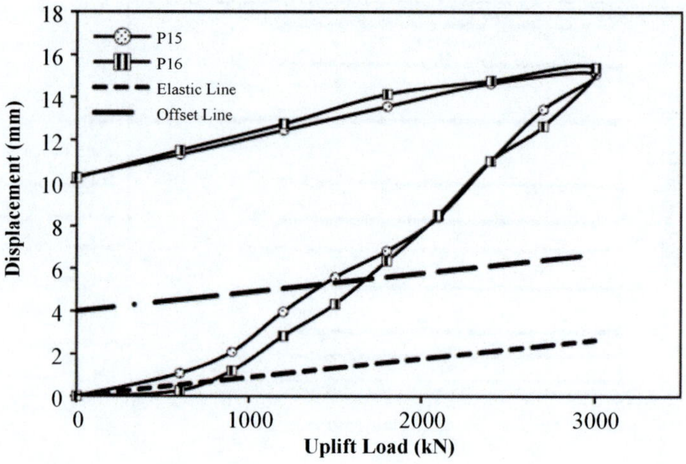

Fig. 3.22 Offset method of tested piles

The ultimate uplift bearing capacity of non-failure uplift SLTs can be determined by the modified Mazurkiewicz method (Thanadol 1998). The assumption is that, when the nominate settlement (load/settlement) value is equal to 0 (the loading is small and the settlement is very high), then a line intersects with the y-axis, which illustrates the uplift ultimate bearing capacity of pile. As shown in Figs. 3.23 and 3.24, two functions are determined, and, when x = 0, the ultimate uplift bearing capacities of P15 and P16 were 6,299 and 5,442 kN, respectively.

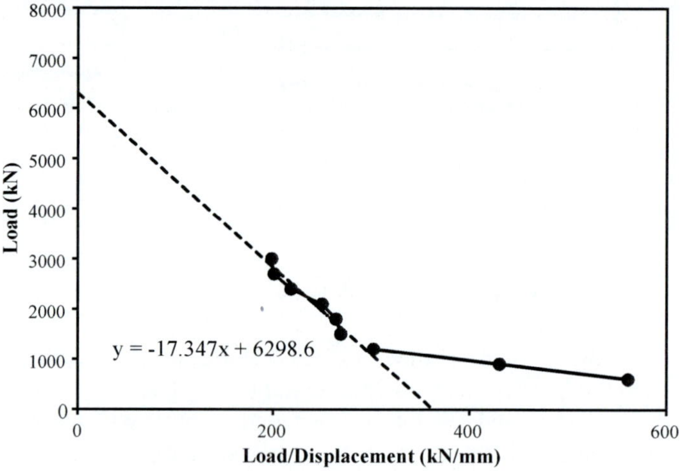

Fig. 3.23 Mazurkiewicz method of P15

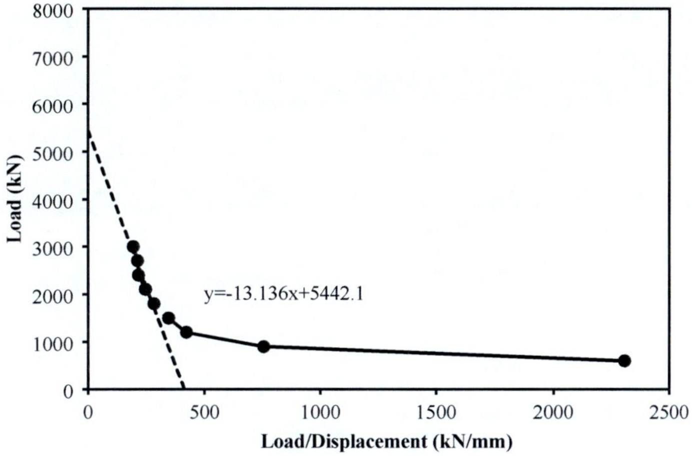

Fig. 3.24 Mazurkiewicz method of P16

3.7 Dynamic Load Tests Results

Two dynamic load tests were conducted on piles P121 (base grouting) and P126 (base-and-shaft grouting). Through CAPWAP, via many trials, the final matched signals are provided in Figs. 3.25 and 3.26. Based on these two matched signals, the simulated load–settlement curves, unit shaft resistance along the pile length and load transfer characteristics are provided in Figs. 3.27 and 3.28. As indicated by the dynamic load tests results, the ultimate bearing capacities of P121 and P126 were 9,071 and 9,851.5 kN, with corresponding maximum settlements of 26.37 and 18.93 mm, respectively.

As shown in Fig. 3.27, the shaft resistance of the base grouted pile was not uniformly distributed along the pile, and, compared with the results illustrated in Fig. 3.28, shaft grouting technology changed the shaft resistance and showed a relatively uniform distribution. This phenomenon occurred because the shaft grouting

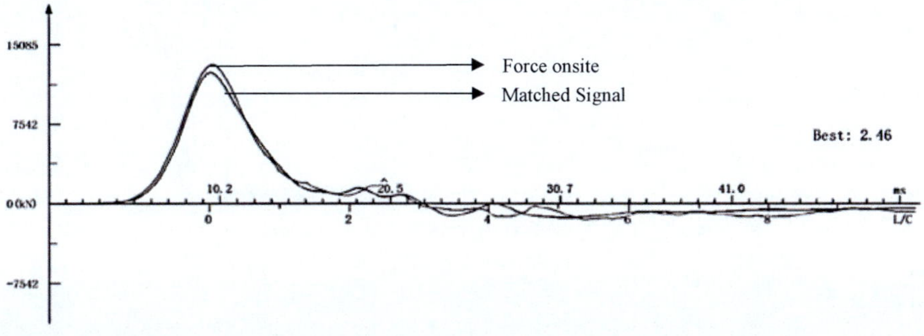

Fig. 3.25 CAPWAP results of P121

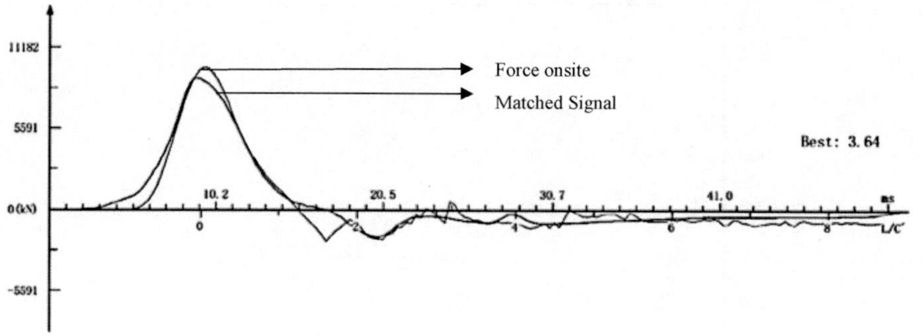

Fig. 3.26 CAPWAP results of P126

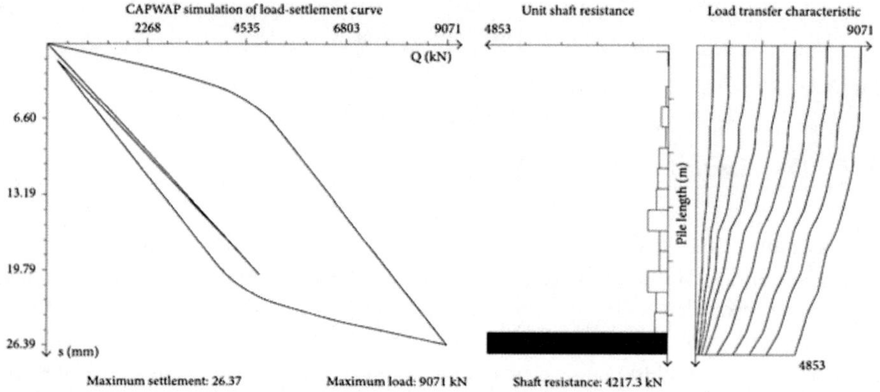

Fig. 3.27 Simulated load–settlement curves, unit shaft resistance and load transfer characteristic of P121

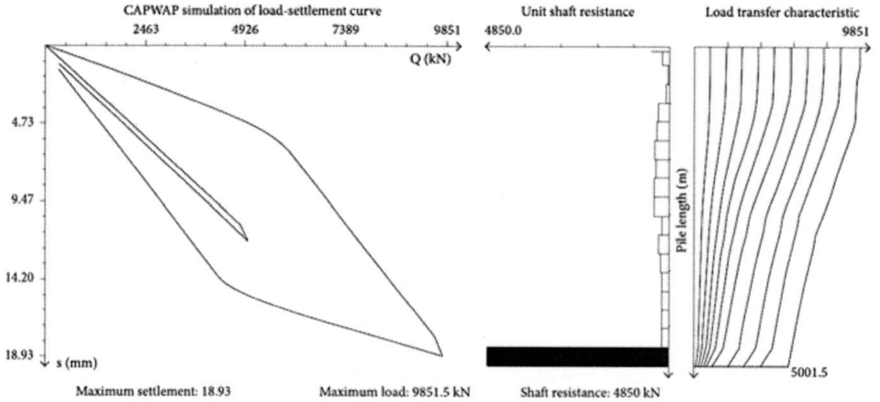

Fig. 3.28 Simulated load–settlement curves, unit shaft resistance and load transfer characteristic of P126

changed the property of the soil from the pile shaft. It can also be seen that the shaft resistances of P121 and P126 were 4,217.3 and 4,850.0 kN, respectively. From the load transfer characteristic diagrams, it can be determined that the load decreased along the pile length, yet increased with increasing applied loads. The comparison of the load transfer characteristic diagrams in Figs. 3.27 and 3.28 also indicated that the shaft grouting increased the shaft resistance of the pile.

All tested piles' results are summarized in Tables 3.5 and 3.6. As shown in Table 3.5, all these results obtained from different methods demonstrated that the base grouting and base-and-shaft grouting improved the ultimate bearing capacity of the piles. The ultimate load and settlement acquired from the double tangent method was close to the outcome obtained from DeBeer's method. Further, the results obtained from Chin's method and the dynamic method were close to each other.

For the friction resistance analysis of static uplift load tests and compressive dynamic load tests, Table 3.6 demonstrates that the base-and-shaft grouted pile possessed better capacity than did the base grouted pile and pile without grouting. It can also be seen that the grouting technique improved the shaft capacity and end bearing capacity of the piles. Further, compared with the shaft resistance results obtained from the dynamic load tests, Mazurkiewicz's method overestimated the ultimate capacity of the uplift loaded piles.

3.8 Concluding Remarks

This chapter provided the set-up of static and dynamic load tests, and various methodologies that were used to determine the ultimate compressive and uplift bearing capacity under non-plunging failure conditions. Three types of pile foundations were considered in this chapter, which were non-grouted pile, based grouted pile, and the base-and-shaft grouted pile. By test result analysis, it illustrated that the results from double tangent method and DeBeer's method were close to each other, and Chin's method and dynamic method provided a close result.

This chapter also provided the load transfer characteristic of piles obtained from dynamic load tests. Because the base grouting and shaft grouting techniques were the same, these tests were conducted without consideration of grouting materials and pressure. Further study should be initiated focusing on the influence of various materials, grouting pressure as well as proportion of the grouting material to determine the best way to increase pile capacity.

Table 3.5 Static and dynamic load tests of compressive loaded piles

Pile label	Technology used	Double tangent		DeBeer's		Chin's		Dynamic load test	
		Load (kN)	Settl. (mm)	Load (kN)	Settl. (mm)	Load (kN)	Settl. (mm)	Load (kN)	Settl. (mm)
P51	No grouting	7,560	6.0	7,600	7.9	10,000	N/A	N/A	N/A
P121	Base grouting	7,700	6.2	7,900	8.3	11,111	N/A	9,071	26.37
P126	Base and shaft	8,250	8.0	8,300	8.2	12,500	N/A	9,851.5	18.93

Table 3.6 Static and dynamic load tests of uplift loaded piles

Technology used	Dynamic load test (kN)		Mazurkiewicz's method (kN)	
	Ultimate bearing	Shaft resistance	Ultimate bearing	Shaft resistance
No grouting	N/A	N/A	5,442	5,442
Base grouting	9,071	4,217.3	N/A	N/A
Base and shaft	9,851.5	4,850.0	6,299	6,299

References

Budi, G. S., Kosasi, M., & Wijaya, D. H. (2015). Bearing capacity of pile foundations embedded in clays and sands layer predicted using PDA test and static load test. *Procedia Engineering, 125,* 406–410.

Castelli, R. J., & Wilkins, E. (2004). Osterberg load cell test results on base grouted bored piles in Bangladesh. In *GeoSupport 2004: Drilled shafts, Micropiling, deep mixing, remedial methods, and specialty foundation systems* (pp. 587–602).

Dapp, S., Muchard, M., & Brown, D. (2006). Experiences with base grouted drilled shafts in the southeastern United States. In *Proceedings of the 10th International Conference on Piling and Deep Foundations* (pp. 1553–1562). Amsterdam, the Netherlands: Deep Foundations Institute.

Gaaver, K. E. (2013). Uplift capacity of single piles and pile groups embedded in cohesionless soil. *Alexandria Engineering Journal, 52*(3), 365–372.

Ho, C. E. (2003). Base grouted bored pile on weak granite. In *Proceedings of the Third International Conference on Grouting and Ground Treatment* (pp. 716–727).

Huo, K. C., Qin, X., & Yue, H. H. (2013). Research on uplift static load test of large-diameter steel pipe pile based on mechanics. *Applied Mechanics and Materials*, 410–415.

Kusakabe, O., Kakurai, M., Ueno, K., & Kurachi, Y. (1994). Structural capacity of precast piles with grouted base. *Journal of Geotechnical Engineering, 120*(8), 1289–1306.

Li, X., Xie, K., Zeng, G., & Hou, X. (2000). Research of bored pile slurry effect created during construction. *Structural Construction, 30*(5), 21–23.

Likins, G., Rausche, F., Thendean, G., & Svinkin, M. (1996). CAPWAP correlation studies. In *Proceedings of the Fifth International Conference on the Application of Stress-Wave Theory to Piles*, Orlando, FL (pp. 447–455).

Madhusudan Reddy, K., & Ayothiraman, R. (2015). Experimental studies on behavior of single pile under combined uplift and lateral loading. *Journal of Geotechnical and Geoenvironmental Engineering, 141*(7).

Ministry of Construction of the People's Republic of China. (1999). *Standard for soil test method (GB/T 50123-1999).* Beijing, China: Ministry of Construction of People's Republic of China.

Ministry of Construction of the People's Republic of China. (2001). *Code for investigation of geotechnical engineering (GB 50021-2001).* Beijing, China: Ministry of Construction of People's Republic of China.

Ministry of Construction of the People's Republic of China. (2002). *Code for acceptance of construction quality of building foundation (GB 50202-2002).* Beijing, China: Ministry of Construction of People's Republic of China.

Ministry of Construction of the People's Republic of China. (2014). *Technical code for testing of building foundation piles (JGJ 106-2014).* Beijing, China: National Standard of the People's Republic of China.

Moayedi, H., & Mosallanezhad, M. (2017). Uplift resistance of belled and multi-belled piles in loose sand. *Measurement, 109*(2017), 346–353.

Nguyen, V. L., Nie, L., & Zhang, M. (2012). Method cement post-grouting to increase the load capacity for bored pile. *Research Journal of Applied Sciences, Engineering and Technology, 5*(19), 4727–4732.

Patel, D., Glover, S., Chew, J., & Austin, J. (2015). The Pinnacle-design and construction of large diameter deep base grouted piles in London. *Ground Engineering*, 24–31.

Robinson, B., & Iskander, M. (2008). Static and dynamic load tests on driven polymeric piles. In *GeoCongress 2008: Geosustainability and geohazard mitigation* (pp. 939–946).

Rollins, K. M., Kwon, K. H., & Gerber, T. M. (2008). Static and dynamic lateral load tests on a pile cap with partial gravel backfill. In *Geotechnical Earthquake Engineering and Soil Dynamics IV* (pp. 1–6).

Shelke, A., & Patra, N. (2011). Effect of compressive load on uplift capacity of cast-insitu bored piles. *Geotechnical and Geological Engineering, 29*(5), 927.

Shi, C. (2005). The application of the pile-end mud-jacking technique in the construction of bored caisson pile. *Sci/Tech Information Development & Economy, 15*(23), 293–296.

Sinnreich, J., & Simpson, R. C. (2013). Base grouting case studies including full scale comparative load testing. In *Seventh International Conference on Case Histories in Geotechnical Engineering*, no. 2.16, pp. 1–8.

Thanadol, K. (1998). *Study of Pile Capacity from Ultimate Pile Load Test* (M. Eng), Kasetsart University, Bangkok, Thailand.

Thiyyakkandi, S., McVay, M., Bloomquist, D., & Lai, P. (2013). Measured and predicted response of a new jetted and grouted precast pile with membranes in cohesionless soils. *Journal of Geotechnical and Geoenvironmental Engineering, 139*(8), 1334–1345.

Thiyyakkandi, S., McVay, M., Bloomquist, D., & Lai, P. (2014). Experimental study, numerical modeling of and axial prediction approach to base grouted drilled shafts in cohesionless soils. *Acta Geotechnica, 9*(3), 439–454.

Tomlinson, M., & Woodward, J. (2007). *Pile design and construction practice*. CRC Press.

Verma, A., & Joshi, R. K. (2010). Uplift load carrying capacity of piles in sand. In *Indian Geotechnical Conference*, India (pp. 858–860).

Wang, B., & Zhang, J. (2013). Mechanical mechanism of the post-grouting pile. *Applied Mechanics and Materials, 351–352,* 510–514.

Wang, W., Wu, J., & Di, G. (2006). Performance of base grouted bored piles in specially big excavation constructed using top-down method. In *Underground construction and ground movement* (pp. 393–400).

Wang, D. D., Wang, L. F., & Zhang, L. P. (2015). Experimental study on post grouting bearing capacity of large diameter bored piles. *MATEC Web of Conferences, 22*, 1–7.

Zhang, R. K., Shi, M. L., Zhang, H., & Wang, J. (2012c). The enhancement effect analysis of pile-base post-grouting piles. *Applied Mechanics and Materials*, 227–231.

Zhou, J. L., Zhang, X., Jiang, H. S., Lyu, C., & Oh, E. (2017). Static and dynamic load tests of shaft and base grouted concrete piles. *Advances in Civil Engineering, 2017,* 1–11.

Zhussupbekov, A., & Omarov, A. (2016). Modern advances in the field geotechnical testing investigations of pile foundations. *Procedia Engineering, 165,* 88–95.

Chapter 4
Field Tests of Precast Concrete Piles

4.1 General Introduction

Pile foundation is a long structural element that cooperates with soil to resist the loading transferred from the upper structure. It can be primarily categorized into cast-in-place and precast pile based on the piling construction condition. Precast pile can resist more loading comparing to shallow foundation, and sometimes cost less since it does not need to pre-drilling as the bored pile does. Mostly, the precast pile foundation is considered as displacement pile because the precast pile will compact soil layers during driving. Based on the configuration, it can be categorized into small or large displacement piles.

For the precast concrete square piles (large displacement piles), these sizes can be ranging from 200 to 700 mm in diameter and 12 to 25 m in length, and the working loads that can be resisted vary from 200 to 1,200 kN. These piles can be a normally reinforced structure or be a prestressed structure. Past studies were focused on the pile tests illustration (Mansur and Hunter 1970), tests in different soil profiles (Balasubramaniam et al. 2009; Gregersen et al. 1975), drivability (Ashford and Jakrapiyanun 2001; Fellenius and Samson 1976; Hussein et al. 2006), load transfer mechanism (Fellenius 2002; Hsu 2014) and configurations of piles and strength of concrete (Liew et al. 2004). Recently, there has been intensive investigations aimed to research the materials effect on concrete pile. For example, full-size tests were performed on concrete piles with Illinois PCC bottom ash, and these piles were compared to the traditional reinforced concrete pile with fly ash admixture (Kumar et al. 2004).

The behavior of open-ended piles is more complicated because the Soil Plug Effect (SPE) which created inside the pile should be considered when investigating the open-ended piles. Recently, a new CPT-based HKU method was proposed for base capacity estimation of the open-ended pipe piles with mechanical consideration of annulus resistance and plug resistance (Yu and Yang 2012). Detailed reviews of these precast piles were provided with consideration of soil profiles, methods of driven technique and configurations of precast piles (Marcos et al. 2013). Furthermore, another aspect

J. Zhou and E. Oh, *Full-Scale Field Tests of Different Types of Piles*,
Advanced Topics in Science and Technology in China 62,
https://doi.org/10.1007/978-981-33-6183-6_4

that associated with precast pile in clay is the 'setup' or 'freeze' of the foundation, which is a normal phenomenon of pile capacity increases with the increment of time. Recent research has commenced to determine the amount of setup time of precast piles by performing dynamic load tests, stanamic load tests and static load tests (Gwizdala and Wieclawski 2013; Abu-Farsakh et al. 2017).

Because of the process of driving pile into soil layer is relatively similar to some tests like standard penetration test, dynamic load tests etc., numerous studies have been focused on the relationships between bearing capacity with penetration tests' parameters (Lehane et al. 2013; Zhusupbekov et al. 2014). The real capacity is literally accepted after the field tests being conducted and hence there are a lot of field tests research performed. Practically, if static load tests are selected to be conducted, the Proof Test (PT) is preferred because it only apply twice the allowable load instead of failing it. Then, there are many extrapolation methods proposed to determine the ultimate bearing pile capacity. Marcos et al. (2013) reviewed 8 interpretation criteria based on the database of 72 sites with 152 field compression load tests. These tested piles included round and square shape, and the soil profile was categorized into drained and undrained condition. It is concluded the Davisson and slope tangent methods (or double tangent method) give a lower interpreted failure load by 10–15%, and Chin's method gives the upper bound solution.

The utilization of bored piles or driven piles is accepted worldwide in deep excavation engineering. Different to a Soil Mixing Wall (SMW) in which the inside H-steel pile can be pulled out after the underground construction is finished, the driven pile or cast-in-situ pile can be used as part of the building. The use of these passive piles as a retaining structure can improve the slope stability and hence allow much deeper excavation to be performed. These piles protect the buildings including underground structures near the construction site, and to some extent, they can also resist water permeation.

Compared to driven piles which need to be transferred from the manufacturing factory, cast-in-situ piles are used more commonly in excavation construction as these bored piles are cast once the holes are drilled and soils are removed, which improves the construction process. However, the disadvantage of this time-saving construction is that the welding of the reinforcement cage and blending the concrete at the construction site can cause severe environmental issues. Since 1995, the discharge of carbon dioxide in China has been 5.5 times higher than in America and 13.8 times higher than in Japan (Ma and Li 2006). It has been reported in the Chinese news that the ejection of industrial dust increased to 11.75 million tons in 2011 (Environtology 2011a) and that construction waste accounted for 30–40% of total municipal solid waste in 2011 (Environtology 2011b). This environmental problem can be controlled when substituting bored piles with driven piles. However, because of the proximity between passive piles in excavation construction and their immense size, driving these piles can squeeze the soils, consequently leading to a huge soil compaction effect. The best way to handle this issue is to excavate the soil first and then lower the driven pile into the drilled hole; it is worth noting that this method also diminishes noise pollution. The diameter of this hole cannot be larger than the diameter of the pile because once there is a gap between the concrete surface of the pile and soil, there will be no friction resistance which will result in decreased pile capacity. The construction method that removes an amount of soil first and later lowers the driven pile raises questions concerning the mechanism of friction resistance of the pile.

There is much research related to static load tests for determining the bearing capacity of the pile. By comparing grouted bored piles with traditional slurry stabilized bored piles under static load tests, a significant increase in skin friction resistance has been observed (Chu 2004). Most researchers have considered vertical compressive static load tests and interpretation of the test results for pile capacity determination; others have investigated the load transfer characteristic of piles. However, studies of the load transfer mechanism of the non-displacement precast pile under vertical uplift loads are very limited.

With technological developments, the statnamic method and dynamic methods which contain CASE and CAPWAP approach, are also used for capacity determination under vertical loading and are replacing the conventional static load test progressively. For the determination of ultimate lateral bearing capacity of a non-displacement driven pile which is used as a retaining structure, a lateral static load test still needs to be conducted. Different to the vertical static load test, the regular lateral load tests are conducted under five cyclic loads. In detail, after the first horizontal loads are applied, the loads will be released and later, the pile will be lateral loaded and released again. This process will be repeated five times, and then the second loading stage (increase the loads) can be performed.

This chapter examines the behaviors of small, large and non-displacement piles through performing compressive, uplift and lateral SLTs. Two case studies are provided. The first case study researched a total number of 40 piles tested by compressive SLTs to evaluate the pile behavior. The pile foundations included 25 pipe concrete piles (small displacement precast piles) and 15 rectangular concrete piles (large displacement precast piles). The first case study focused on researching piles' behavior under compressive loads. The piles lengths varied, and all reached the angular gravel which was the bearing stratum. It should be noted that the capacity comparison between interpretations' outcome as well as the designed bearing capacity, and the example with detailed calculation steps are provided in Chapter 8. The second case study consisted of three piles tested by uplift loading and three piles with horizontal loading (non-displacement precast concrete piles). This second study aimed to determine the pile behavior under uplift and lateral loads and the load transfer mechanism. Further, it included interpretation of non-displacement precast piles under uplift and lateral loads.

4.2 Small and Large Displacement Concrete Piles

4.2.1 Project Introduction

The first examined construction project aimed to construct a 28-level apartment with an underground garage. The driven location of the tested piles was in Shandong Province, China. The precast piles were driven in the construction site, which is shown in Fig. 4.1. Generally, the piles from three areas were tested by the compressive SLTs. Areas A and B were the locations of the tested concrete pipe piles, and Area C was the location of the tested solid rectangular piles. The solid rectangular pile driven in Area C was used to resist the loads transferred from the upper building.

4.2.2 Subsurface Conditions

The subsurface exploration was achieved through laboratory and in-situ tests. The in-situ SPTs and CPTs and laboratory tests of consolidation, direct shear and triaxial tests were conducted based on the Chinese Standard for Soil Test Method (GB/T 50123-1999 1999b), Standard for Test Methods of Engineering Rock Mass (GB/T 50266-2013 2013b) and Code for Investigation of Geotechnical Engineering (GB 50021-2001 2001), respectively. Based on the borehole logs, the soil layers were discovered as follows.

1. planting soil (Q^{4ml}): yellowish brown, loose and wet, contains clay, partially contains plant roots, average thickness of 0.75 m; a. miscellaneous fill (Q^{4ml}): loose, partially contains gravel and rubble, diameter ranging from 2.0 to 7.0 mm, average

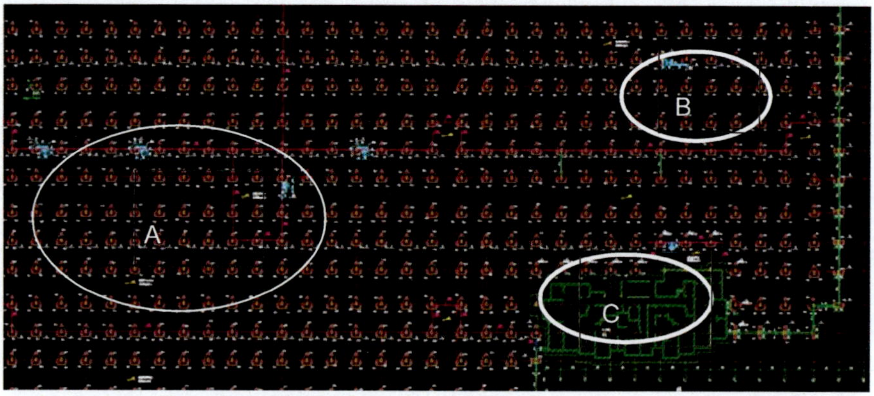

Fig. 4.1 Location of tested piles (not to scale)

thickness of 1.22 m; 2. silty clay (Q^{4al+pl}): medium stiff, yellowish with low plasticity, average thickness of 3.91 m; 3. silty clay (Q^{4m}): grey to ash black, liquid-plastic state, contains fine sand, partially contains clayey silt, average thickness of 4.92 m; a. fine sand (Q^{4m}): grey and cinereous, loose, saturated, average thickness of 2.01 m; 4. silty clay (Q^{4al+pl}): grey to yellowish grey, medium stiff, low plasticity, average depth of 2.57 m; 5. silty clay (Q^{4al+pl}): yellowish brown, plastic state, partially contains medium to dense sand, average thickness of 2.57 m; a. medium sand (Q^{4al+pl}): yellowish brown, medium dense, loose and saturated, average thickness of 4.08 m; b. coarse gravel sand (Q^{4al+pl}): yellowish brown, medium to dense, saturated, average thickness of 4.19 m; 6. angular gravel (Q^{4pl+dl}): yellowish brown, medium dense, saturated, particle size ranges from 2 to 50 mm, maximum to 100 mm, highly weathered, average thickness of 5.66 mm; 7. highly weathered mica schist (P_t): yellowish brown to grey, highly weathered, thickness of 4.52 m; 8. medium weathered mica schist (P_t): grey, medium hard.

Fig. 4.2 Driven small displacement pipe pile

4.2.3 Pile Preparation

For precast open-ended pipe piles, the diameter was 400 mm and the length varied from 20 to 26 m. In Area A, one 20 m pile, seven 21 m piles, seven 22 m piles, two 23 m piles and five 24 m piles were tested. In Area B, one 23 m pile and two 26 m piles were tested. For precast solid rectangular piles, the cross-section was 400 $\times$ 400 mm^2. All rectangular piles were tested in Area C. The tested piles consisted of two 22 m piles, two 25 m piles, one 26 m pile, four 27 m piles and six 28 m piles. The concrete strength of these tested piles was 30, 50 and 40 MPa from Areas A, B and C, respectively.

4.2.4 Test Setup

Different to the case study illustrated in Chap. 3, all the pile foundation was driven underground (Fig. 4.2), and the weighted platform was selected for SLT instead of the reaction beams and piles, as demonstrated in Fig. 4.3. As a result of the designed capacity of these piles being relatively small, compared with the bored piles depicted in Chap. 3, the reaction system only needed to provide small reaction loads. Further,

Fig. 4.3 Weighted platform of SLT

as numerous piles needed testing, it would have been very expensive to cast reaction piles. This weighted platform is more practical to select when a large number of tests is required because the transfer of the platform is much cheaper.

To perform compressive SLTs, two hydraulic jacks were used to provide loads. Four dial gauges were symmetrically installed on a reference bar to measure the vertical displacement of the pile head. For Area A, maintained loads with an increment of 130 kN were applied on the precast pipe pile until a maximum load of 1,300 kN being applied. For the pipe pile in Area B, an increment of 314 kN was applied until a maximum load of 3,140 kN. For the rectangular concrete pile, maintained loads with increments of 488 kN were applied on most piles (to maximum load of 4,880 kN), except for three piles that had increment loads of 360 kN (to maximum load of 3,240 kN) for data comparison. All the information about these tested piles is summarized in Tables 4.1, 4.2 and 4.3.

4.3 Non-displacement Square Concrete Piles

4.3.1 Project Introduction

This project was conducted in western Jinan city, China—coordinates of E116°54′–E117°02′, N36°35′–N36°40′ (Fig. 4.4). It aimed to construct a subway station with total construction area of $356.6 \times 19.7\ \text{m}^2$. The excavation depth was 16.8 m for this project and this project was based on the following local standards:

GB 50308-2008 (2008a) (Code for Urban Rail Transit Engineering Survey), GB/T 18314-2009 (2009a) (Specifications for Global Positioning System [GPS] Surveys), GB 50299-1999 (1999a) (Code for Construction and Acceptance of Metro Engineering), JGJ 8-2007 (2007) (Code for Deformation Measurement of Building and Structure), CJJ 76-2012 (2012) (Specification for Dynamic Observation of Groundwater in Urban Area), JGJ 120-2012 (2012) (Technical Specification for Retaining and Protection of Building Foundation Excavations), GB 50497-2009 (2009b) (Technical Code for Monitoring of Building Excavation Engineering), GB 50157-2013 (2013a) (Code for Design of Metro), GB 50111-2006 (2006) (Code for Seismic Design of Railway Engineering), QCR 9218-200-15 (2015) (Code for Tunneling Monitoring Technology Procedures) and GB 50911-2013 (2013b) (Code for Monitoring Measurement of Urban Rail Transit Engineering).

4.3.2 Subsurface Conditions

Before the projects started, boreholes were driven to depths ranging from 31.0 to 52.0 m to determine the subsurface. In total, 401 SPTs and dynamic penetration tests were conducted, with the latter reaching the bottom of gravel and medium sand. In addition, based on the Chinese code GB/T 50123-1999 (1999b), laboratory

Table 4.1 Load–settlement results of open-ended pipe pile (1)

No	1	2	3	4	5	6	7	8	9	10	11	12	13	14	15
Label	P1172	P1470	P976	P991	P995	P1163	P1165	P1175	P630	P1473	P643	P1509	P1485	P1475	P1580
Length (m)	20	21	21	21	21	21	21	21	22	22	22	22	22	22	22
Load (kN)	Settl. (mm)	Settl. (mm)	Settl. (mm)	Settl. (mm)	Settl. (mm)	Settl. (mm)	Settl. (mm)	Settl. (mm)	Settl. (mm)	Settl. (mm)	Settl. (mm)	Settl. (mm)	Settl. (mm)	Settl. (mm)	Settl. (mm)
0	0	0	0	0	0	0	0	0	0	0	0	0	0	0	0
260	0.56	0.6	1	0.62	0.67	0.93	0.62	0.55	0.73	0.54	0.61	0.72	0.7	0.75	0.89
390	1.74	1.61	2.43	1.97	1.86	2.23	1.79	1.9	1.66	1.84	1.26	1.64	1.62	1.94	1.87
520	4.04	3.04	4.28	3.74	2.8	3.62	2.97	3.54	2.72	3.24	2.33	3.01	3.41	2.95	3.17
650	5.45	4.51	5.77	5.97	4.5	5.81	4.72	5.78	4.13	4.84	3.47	4.63	5.27	4.34	5.08
780	7.43	6.45	8.32	7.84	5.98	8.19	6.12	8.24	5.78	6.51	4.83	6.77	7.2	5.56	7.29
910	9.54	8.47	11.18	10.58	8.33	11.12	8.24	10.83	7.32	8.68	6.45	8.67	9.21	7.46	10.41
1,040	12	10.9	13.4	13.36	11.91	14.78	10.35	13.21	9.11	11.58	7.92	11.33	11.39	9.4	13.35
1,170	15.14	13.26	17.22	16.47	16.33	18.5	13.67	16.24	11.37	13.99	9.34	14.9	14.63	11.63	16.84
1,300	18.59	15.8	21.8	19.58	20.97	22.64	17.48	19.56	14.08	16.55	11.48	19.19	18.01	14.53	21
1,040	17.13	14.57	20.74	18.29	19.91	21.45	16.48	18.18	13.34	15.19	10.78	17.99	16.44	13.73	19.82
780	15.2	13.24	19.03	16.71	18.34	19.85	14.98	16.67	12.42	13.81	10.06	16.56	14.84	12.5	18.36
520	13.01	11.39	16.85	14.54	16.26	17.69	12.58	14.53	10.91	11.93	8.75	14.88	12.93	10.83	16.47
260	10.81	9.52	14.08	12.15	13.5	15.06	10.37	12.39	8.74	9.75	6.88	12.53	10.89	8.83	14.19
0	8.34	7.19	10.41	9.26	10.53	11.8	8.09	9.66	6.4	7.57	5.36	9.74	8.67	6.66	11.58
R_{P-S}	55.14	54.49	52.25	52.71	49.79	47.88	53.72	50.61	54.55	54.26	53.31	49.24	51.86	54.16	44.86

Table 4.2 Load–settlement results of open-ended pipe pile (2)

No	16	17	18	19	20	21	22	No	23	24	25
Label	P1481	P639	P1275	P1287	P1299	P1453	P1465	Label	S73	S103	S126
Length (m)	23	23	24	24	24	24	24	Length (m)	26	26	23
Load (kN)	Settl. (mm)	Settl. (mm)	Settl. (mm)	Settl. (mm)	Settl. (mm)	Settl. (mm)	Settl. (mm)	Load (kN)	Settl. (mm)	Settl. (mm)	Settl. (mm)
0	0	0	0	0	0	0	0	0	0	0	0
260	0.89	0.66	0.91	0.96	0.7	0.47	0.66	628	0.8	0.91	0.93
390	1.6	1.14	2.06	2.06	1.67	1	1.53	942	1.38	1.56	1.72
520	3.53	2.28	3.5	3.36	3.34	1.84	2.64	1,256	2.71	2.41	3.1
650	5.27	3.11	5.05	5.18	4.85	3.1	3.95	1,570	4.65	3.69	5.01
780	7.67	4.51	7.02	7.92	6.63	4.41	5.12	1,884	7.17	5.29	7.38
910	9.6	5.79	9.41	10.72	8.77	6.2	6.92	2,198	9.79	7.48	10.06
1,040	12.7	7.56	11.8	14.3	10.93	7.84	8.8	2,512	13.15	9.9	13.55
1,170	15.6	9.26	14.58	18.66	13.22	10.46	11.13	2,826	16.58	13.09	17.65
1,300	20.1	12.08	18.2	23.58	15.63	13.2	13.73	3,140	20.95	16.6	22.34
1,040	18.87	11.11	16.99	22.33	14.89	12.32	12.77	2,512	19.99	15.44	21.17
780	17.54	9.87	15.53	20.51	13.43	11.03	11.48	1,884	18.16	14.03	19.54
520	15.47	9.07	13.86	18.12	11.79	9.37	10.13	1,256	15.87	12.25	17.22
260	13.47	7.59	11.51	15.14	9.4	7.57	8.2	628	13.37	9.93	14.02
0	10.73	5.51	8.69	11.56	7.08	5.85	6.07	0	10.05	6.98	10.07
R_{P-S}	46.62	54.39	52.25	50.98	54.70	55.68	55.79	R_{P-S}	52.02864	57.95181	54.9239

Table 4.3 Load–settlement results of solid rectangular concrete pile

No	1	2	3	4	5	6	7	8	9	10	11	12	No	13	14	15
Label	P1	P91	P238	P124	P137	P173	P147	P93	P85	P48	P16	P4	Label	P14	P41	P106
Length (m)	25	26	27	27	27	27	28	28	28	28	28	28	Length (m)	22	22	25
Load (kN)	Settl. (mm)	Settl. (mm)	Settl. (mm)	Settl. (mm)	Settl. (mm)	Settl. (mm)	Settl. (mm)	Settl. (mm)	Settl. (mm)	Settl. (mm)	Settl. (mm)	Settl. (mm)	Load (kN)	Settl. (mm)	Settl. (mm)	Settl. (mm)
0	0	0	0	0	0	0	0	0	0	0	0	0	0	0	0	0
976	0.77	0.89	0.72	0.56	1.29	0.78	0.75	0.71	0.96	0.67	0.94	0.72	720	0.7	0.81	0.8
1,464	1.59	1.89	1.58	1.2	2.56	2.33	1.97	1.35	2.09	2.15	1.7	2.27	1,080	1.59	1.63	1.89
1,952	3.09	3.34	2.76	2.4	4.35	4.05	3.82	2.6	3.95	3.41	3.39	4.5	1,440	2.46	2.59	3.3
2,440	5.34	5.22	4.27	4.36	7.28	6.96	6.39	4.3	5.72	6.09	5.09	6.75	1,800	3.74	3.73	5.02
2,928	7.25	7.4	6.06	6.67	10.05	9.59	9.77	6.61	8.09	8.46	7.11	9.9	2,160	5.08	5.24	6.93
3,416	10.03	9.66	8.42	8.92	13.54	12.74	13.98	9.1	10.42	11.22	9.35	12.8	2,520	6.57	7.31	9.19
3,904	13.59	13.06	11.19	12.23	17.78	16.19	17.85	12.22	13.92	14.11	12.4	16.36	2,880	9.18	10.14	11.83
4,392	17.11	16.57	14.61	15.93	22.35	19.79	22.17	15.91	18.17	17.13	15.59	20.13	3,240	11.85	13.63	14.97
4,880	21.53	21.72	18.77	21.27	27.3	24.5	26.35	20.27	23.82	21.82	19.35	24.5	3,600	15.24	17.77	18.58
3,904	20.35	20.45	17.63	20.46	25.99	23.33	25.16	19.08	22.2	20.91	18.19	23.23	2,880	14.48	16.8	17.54
2,928	18.57	18.88	16.26	18.9	24.07	21.46	23.27	17.57	20.39	19.72	16.82	21.34	2,160	13.59	15.46	16.11
1,952	16.32	16.78	14.28	16.72	21.25	19.07	20.49	15.61	18.2	18.03	14.64	18.86	1,440	12.06	13.83	14.33
976	13.72	13.35	11.98	14.31	17.62	16.36	17.26	13.12	15.51	15.92	12.45	16.08	720	10.41	11.72	12.12
0	10.54	9.91	9.27	11.03	13.45	12.45	13.43	10.35	13.43	12.75	9.51	12.52	0	7.77	8.89	9.46
R_{P-S}	51.05	54.37	50.61	48.14	50.73	49.18	49.03	48.94	43.62	41.57	50.85	48.90	R_{P-S}	49.02	49.97	49.09

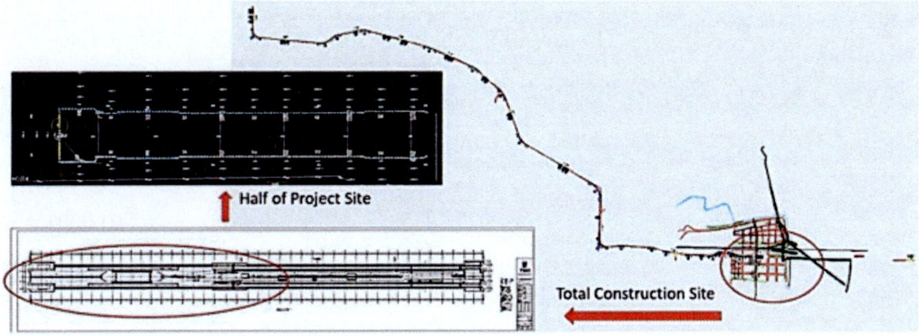

Fig. 4.4 Location of general engineering construction site (not to scale)

tests were conducted to determine the soil characteristics of specific gravity, relative density, porosity, void ratio, saturation ratio, water content, plastic limit, liquid limit, plasticity index, liquidity index, cohesion and friction angle (direct shear test and triaxle tests [CU and UU]). Through interpreting the borehole logs and parameters obtained from laboratory tests, the subsurface investigation found the following:

1. miscellaneous fill: contains gravel, concrete and brick stone, construction waste and plant roots, average depth of 2.5 m; 2. yellowish brown loess: plastic iron-manganese concretion, average depth of 2.9 m; 3. tawny silt: slightly wet, low tenacity, average depth of 2 m; 4. silty clay: yellow and brown, medium to high plasticity, contains iron and manganese oxides, average depth of 11.6 m; 5. fine sand: brown, medium dense and wet, contains quartz arkose, average depth of 2.1 m; 6. dark brown silty clay: contains iron and manganese oxides, average depth of 10 m; 7. silty clay: yellow, light brown, iron-manganese and calcareous concretion, contains little sand and gravel, average depth of 8.3 m; 8. coarse sand: brown, dense, saturated, average depth of 3.6 m; 9. Clay: brown, high viscous, average depth of 2 m. The soil parameters of these subsurface layers are summarized in Table 4.4.

Table 4.4 Soil parameters of subsurface layers

Soil layer	ω	G_s	ρ	Es	e	I_P	I_L	c	φ	N	q_{sik}	q_{pk}
	%	–	g/cm^3	MPa	–	–	–	kPa	o	–	–	–
Fill	N/A	N/A	1.8	N/A	N/A	N/A	N/A	N/A	N/A	N/A	N/A	N/A
Loess	24.5	2.7	1.93	9.3	0.75	11.8	0.55	29	24	10	30	N/A
Silt	23.2	2.69	1.96	15.8	0.69	8.7	0.51	19	20	15	35	N/A
Tawny clay	26.4	2.72	1.95	9.1	0.77	12.8	0.53	47	22	22	45	N/A
Fine sand	N/A	N/A	1.96	N/A	N/A	N/A	N/A	N/A	N/A	18	50	N/A
Silty clay	27.3	2.72	1.93	7.2	0.8	13.8	0.47	51	23	29	60	N/A
Silty clay	26.5	2.72	1.93	6.8	0.780	13.2	0.46	44	18	25	60	2,200
Coarse sand	N/A	N/A	1.98	20	N/A	N/A	N/A	N/A	N/A	39	60	6,800
Clay	26.8	2.74	1.91	7	0.766	19.2	0.27	39	17	N/A	62	2,400

Table 4.5 Strain gauge information

Depth (m)	Label of strain gauge	Initial value (Hz)	*K*-value
2.3	285,349	1,442.9	0.00006695
6.9	287,918	1,431.0	0.00006768
11	28,588	1,410.5	0.00006423
16	281,245	1,428.1	0.00007353
21	286,639	1,415.2	0.00006673
27	286,556	1,452.2	0.00006921

4.3.3 Pile Descriptions

All six precast piles contained two parts with bolt connections, as shown in Fig. 4.5. The cross-sections of these tested piles were all rectangular, with side lengths of 700 mm. The three lateral loaded piles and three uplift loaded piles were labelled P15, P163 and P152 and P138, P151 and P165, respectively, with the same concrete strength of C50 (characteristic strength of 50 MPa, cubic test sample dimension of $150 \times 150 \times 150\ mm^3$), and the same horizontal reinforcement of HRB400

Fig. 4.5 Tested non-displacement precast pile

(hot-rolled ribbed bar with yield strength of 400 MPa) and vertical reinforcement of HPB300 (hot-rolled plane bar with yield strength of 300 MPa).

The lengths of lateral loaded non-displacement piles P15, P163 and P152 were 24.7, 26.6 and 28 m, respectively. For the uplift loaded concrete piles P138, P151 and P165, the lengths were 24.7, 26.6 and 28 m, respectively. Specifically, six pairs of strain gauges were installed in P165 to determine the uplift load transfer characteristics of this non-displacement precast pile.

The locations, initial value and K-value of the strain gauges are summarized in Table 4.5. The stress of the strain gauge can be calculated by Eq. 4.1, using the parameters from Table 4.5 and the collected frequency data recorded onsite. The axial loading of the pile cross-section can then be determined through Eq. 4.2:

$$\bar{\sigma_s} = \frac{1}{n}\sum_{i=1}^{n}\left[K_i\left(f_{ini}{}^2 - f_i{}^2\right)/A_s\right] \tag{4.1}$$

$$N_P = \bar{\sigma_s}\left(\frac{E_c}{E_s}A_c + A_s\right) \tag{4.2}$$

where:

$\bar{\sigma_s}$	= average stress of vibration wire strain bar.
N_P	= force of the concrete pile.
E_c	= Young's modulus of concrete.
E_s	= Young's modulus of steel.
A_c	= cross-section area of concrete.
A_s	= cross-section area of steel.
K_i	= coefficient of vibration wire stress rebar.
f_{ini}	= initial recorded frequency.
f_i	= recorded frequency of vibration wire stress rebar.

4.3.4 Installation Process of Piles

These piles were cast in the factory, which can control the contamination caused by cement dust and exhaust gas released during welding, as shown in Fig. 4.6. The construction process of installing the non-displacement precast concrete piles was as follows:

1. The soil was removed from the required location with an auger drilling machine, as shown in Fig. 4.7. The drilled hole was 550 mm, which was less than the side length of the pile cross-section, so there was no gap between the pile and soil (Fig. 4.8); 2. After the first piece of the precast concrete pile being transferred to the location of the drilled hole (Fig. 4.9), the pile could automatically descend by gravity; 3. A squeezing machine clamped the first piece of the concrete pile, so it could be bolted with the second piece of pile, as shown in Fig. 4.10.

Fig. 4.6 Precast piles in factory

4.3.5 Test Setup

As shown in Figs. 4.11 and 4.12, two reaction beams were used for lateral and uplift SLTs. One hydraulic cylinder and two hydraulic jacks were used to provide loads for the lateral and uplift tests, respectively. Two dial gauges were located at the pile side for recording the horizontal movement of the pile head, and four dial gauges were symmetrically installed on the reference bar to measure the vertical uplift movement of the pile head.

Based on the Chinese codes JGJ 106-2014 (2014), JTJ 041-2000 (2000) and JGJ 94-2008 (2008b), horizontal cycling load tests were conducted. After the maintained loads were applied, a duration of four minutes was required before recording the horizontal movement. And after releasing the loads back to 0 kN, a duration of two minutes was required before recording the residual displacement of pile. During this test, five load circulations were required before the next loading was applied. The tests should be ceased if:

1. The monitored horizontal movement of the pile head exceeded 30 mm, or
2. A concrete crack emerged.

For the uplift SLTs, low-speed maintenance tests were conducted. The applied load was maintained until the rate of axial movement did not exceed 0.1 mm. After

Fig. 4.7 Auger drilling machine

each maintained load was applied, the vertical movement of the piles was recorded with intervals of five, 10 and 15 min, and 30 min. If the accumulated time exceeded one hour, record movement of pile with interval of 30 min. The loading could be terminated in the case of the following conditions:

1. The applied loading was equal to the value of 0.9 times the ultimate strength of reinforcement
2. Quintuple movement change of the previous loading was discovered
3. The pile head movement increased up to 100 mm.

4.4 Static Load Test Results of Precast Concrete Piles

4.4.1 Results of Small Displacement Precast Piles

As shown in Fig. 4.13, the load–settlement curves of the 21 m pipe pile showed similar behavior. The total settlement of the 21 m piles was found to be an average of 19 mm, and, after the loading released, the average permanent settlement was found to be 10 mm. It was also found that the curves after loading released were parallel with each other.

Fig. 4.8 Drilled hole for precast pile

The load–settlement curves of 22 m piles are presented in Fig. 4.14. It was found that the total settlement of the 22 m piles was an average of 15 mm, and, after the loading released, the average settlement was found to be 8 mm. Similar to the 21 m piles, the curves during unloading between the tested piles were parallel.

The load–settlement curves of the 24 m piles are provided in Fig. 4.15. The behavior of P1287 was relatively different to the other piles, which requires further discussion. The other 24 m piles showed similar behavior, and the total settlement was found to be 15 mm. After the loading was released, the average settlement was found to be 7 mm. The load–settlement curves of piles with different lengths are provided in Fig. 4.16. It was found that the total average settlement was 21 mm, and the permanent average displacement was 10 mm. The curves of the tested piles during the unloading stages were parallel with each other. The load–settlement curve of the tested pile located at Area B is provided in Fig. 4.17, the total average settlement was found to be 18 mm, and the average permanent settlement was found to be 8 mm. Similar to the pile behaviors tested at Area A, the curves of these tested piles during unloading were also found to be parallel, but the gradient (this is defined as the settlement change per load from the load release period) was different to the piles tested in Area A.

Fig. 4.9 Transferring precast piles

4.4.2 Results of Large Displacement Precast Piles

The load–settlement curves of the 27 m rectangular piles are presented in Fig. 4.18. The total average settlement was found to be 23 mm, and the average permanent settlement was found to be 12 mm. For the 28 m piles shown in Fig. 4.19, the load–settlement curves illustrated that the total average settlement was 22 mm, and the average permanent settlement was 12 mm.

As shown in Fig. 4.20, two piles with the same dimensions and soil condition (distance between piles of 5 m) were tested with a different load increment, and the same behavior was identified. Figure 4.21 presents the rectangular piles' behavior with various pile lengths. The results indicated that the total average settlement was 24 mm, and the average permanent settlement was 12 mm. Similar to the results obtained from the pipe piles, it was found that, during unloading stages, these curves were relatively parallel and the gradients were close to each other.

Fig. 4.10 After descending the first piece of pile

4.4.3 Results of Non-displacement Precast Piles

The horizontal loads, H, and movement, X, of three tested precast piles were recorded as shown in Figs. 4.22, 4.23 and 4.24. It can be seen that the critical load (H_{cr}) of P15 was 100 kN (to be conservative). The ultimate load appeared to be 400 kN; however, there were some cracks observed after applying 400 kN. Thus, the ultimate load was determined to be 300 kN. For P163, the critical load was 200 kN, while the ultimate load (H_u) was difficult to read and interpretation was required. The H_{cr} and H_u of P152 were determined to be 240 and 480 kN, respectively.

The interpretation of the horizontal SLTs is provided in Figs. 4.25, 4.26 and 4.27. Through establishing a coordinate system of horizontal loading versus nominated gradient settlement, three lines were discovered in each diagram. Each two intersections between two lines demonstrate the corresponding H_{cr} and H_u. The equations of each line are also illustrated in Figs. 4.25, 4.26 and 4.27. The H_{cr} and H_u of P15, P163 and P152 were 107.5, 154.4 and 175.4 kN (H_{cr}) and 280.9, 447.7 and 470.8 kN (H_u), respectively.

Fig. 4.11 Lateral SLT setup

The results of the H–X curves and interpretation of these SLTs are summarized in Table 4.6. The results were close to each other and all demonstrated that the critical and ultimate loads of these non-placement precast piles increased with pile length. It is worth noting that it is sometimes difficult to determine the critical and ultimate load through H–X curves; thus, comprehensive analysis between SLTs and interpretation is required for geotechnical engineers.

The load–movement curve results of three uplift loaded piles are provided in Fig. 4.28. The maximum uplift movements of these piles' heads were discovered to be 1.73 mm (P138), 1.6 mm (P151) and 1.31 mm (P165) when the maximum loading of 1,900 kN was applied. The corresponding residual uplift displacements of these three piles were 0.47, 0.5 and 0.35 mm, respectively. As shown in Fig. 4.28, all three lines from the loading stages were approximately linear; thus, the maximum loading of 1,900 kN was not the ultimate bearing capacity of these three piles. As such, interpretation of this diagram was required.

Fig. 4.12 Uplift SLT setup

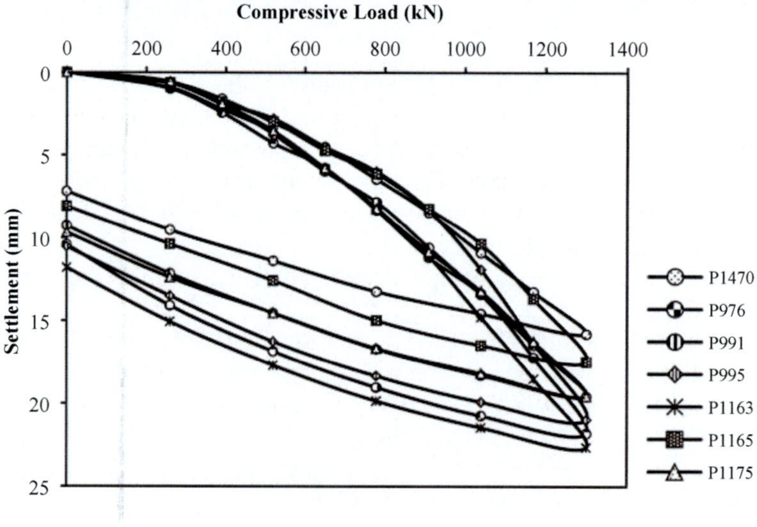

Fig. 4.13 Load–settlement results of 21 m pipe pile

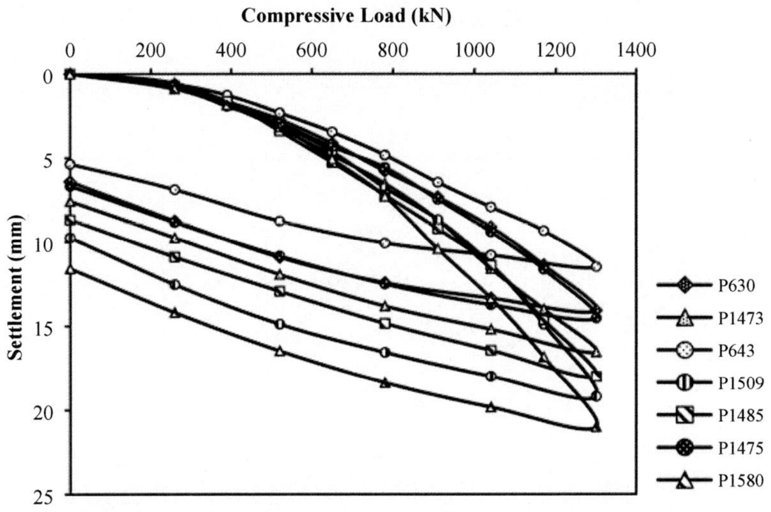

Fig. 4.14 Load–settlement results of 22 m pipe pile

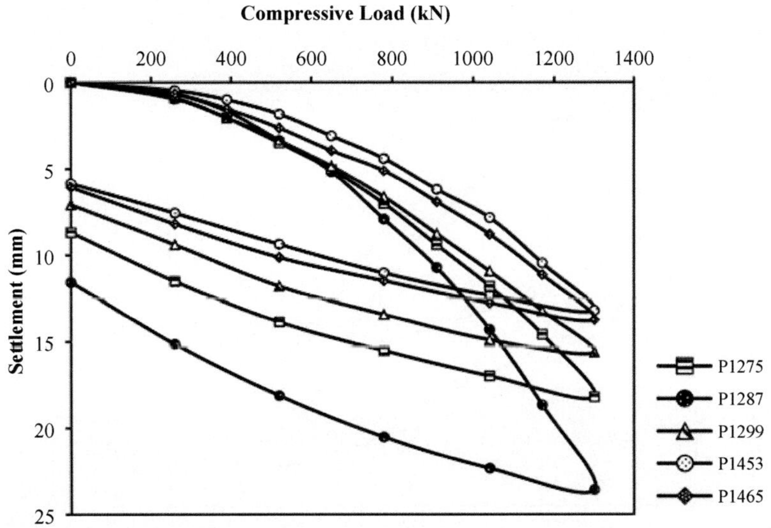

Fig. 4.15 Load–settlement results of 24 m pipe piles

The ultimate uplift bearing capacity of non-failure uplift SLTs can be determined with the modified Mazurkiewicz method. The assumption of this method is that the nominate movement (load/displacement) value is equal to 0 when the loading is small and the displacement is very high, the dash line would be go through by y-axis which illustrates the uplift ultimate bearing capacity of pile.

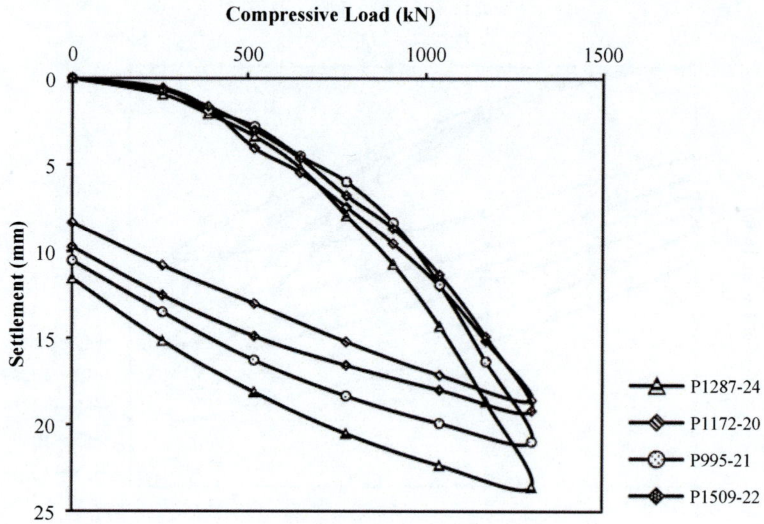

Fig. 4.16 Pipe piles at Area A with different lengths

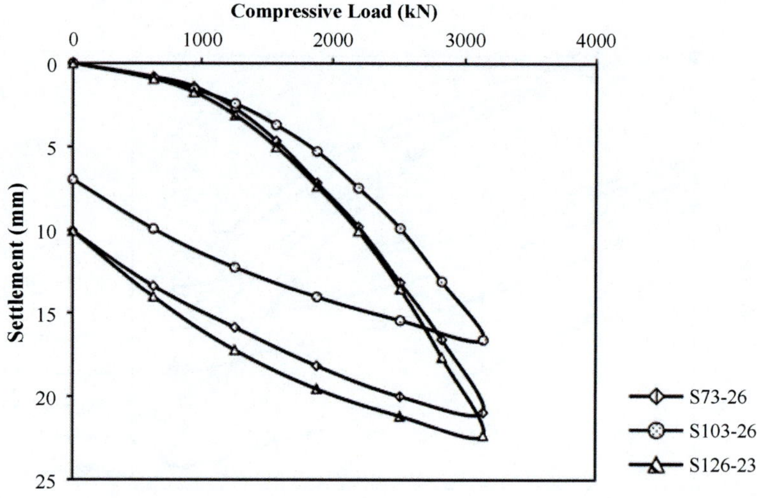

Fig. 4.17 Pipe piles at Area B with different lengths

Based on Fig. 4.29, the equations representing these three lines are provided in Eqs. 4.3, 4.4 and 4.5. When nominate movement is equal to 0, the ultimate uplift bearing capacities of P138, P151 and P165 were determined as 2,903.6, 3,467.0 and 4,724.2 kN, respectively. As summarized in Table 4.7, the uplift bearing capacity of these non-displacement piles increased with the pile length:

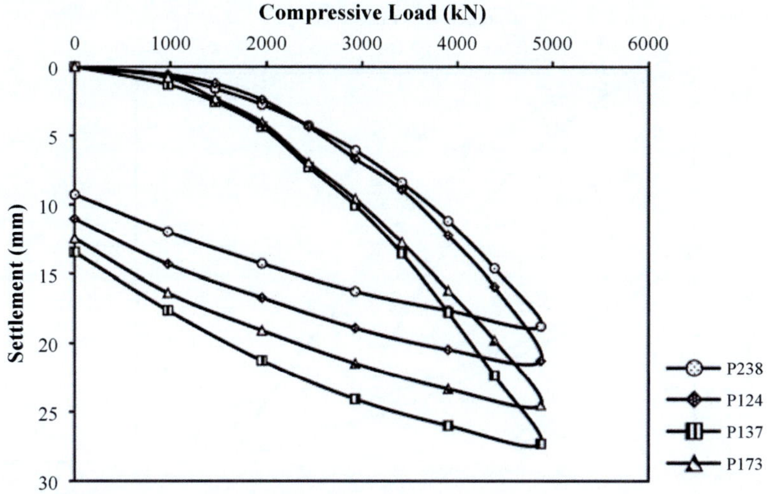

Fig. 4.18 Load–settlement results of 27 m rectangular piles

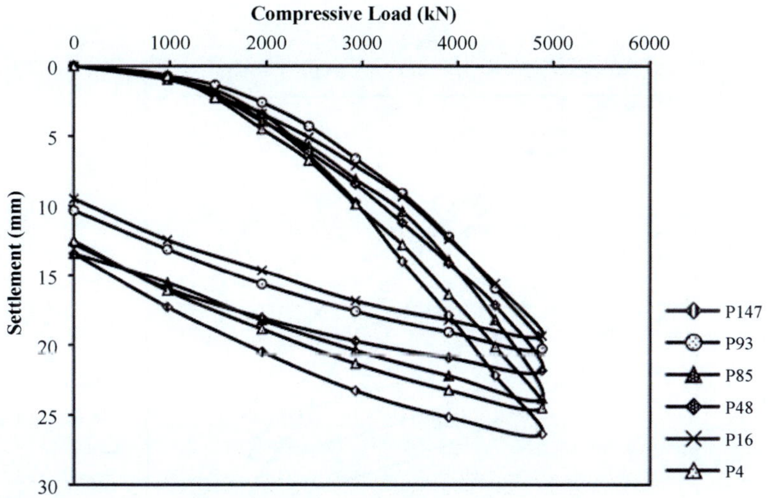

Fig. 4.19 Load–settlement results of 28 m rectangular piles

$$\text{P138: } y = -0.9253x + 2903.6 \tag{4.3}$$

$$\text{P151 : } y = -1.318x + 3467.0 \tag{4.4}$$

$$\text{P165: } y = -1.9371x + 4724.2 \tag{4.5}$$

The vibration frequency obtained from strain gauges in P165 was recorded at each loading stage. Using Eqs. 4.1 and 4.2, the average axial uplift loads were determined

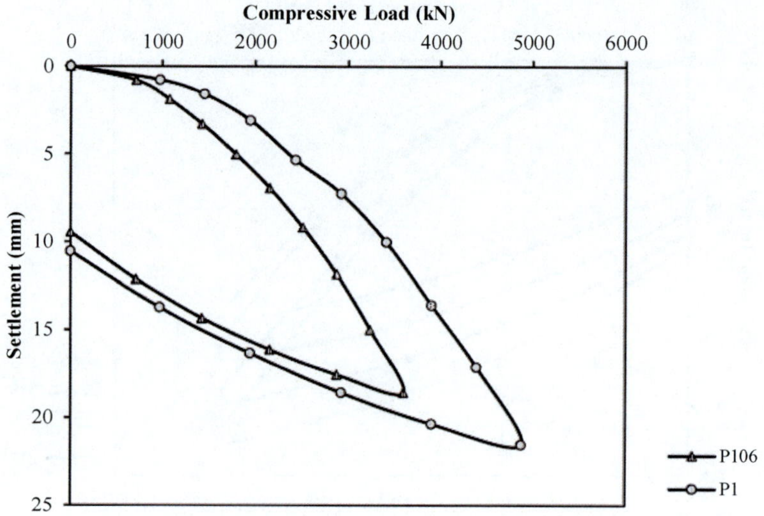

Fig. 4.20 Same pipe with different load increments

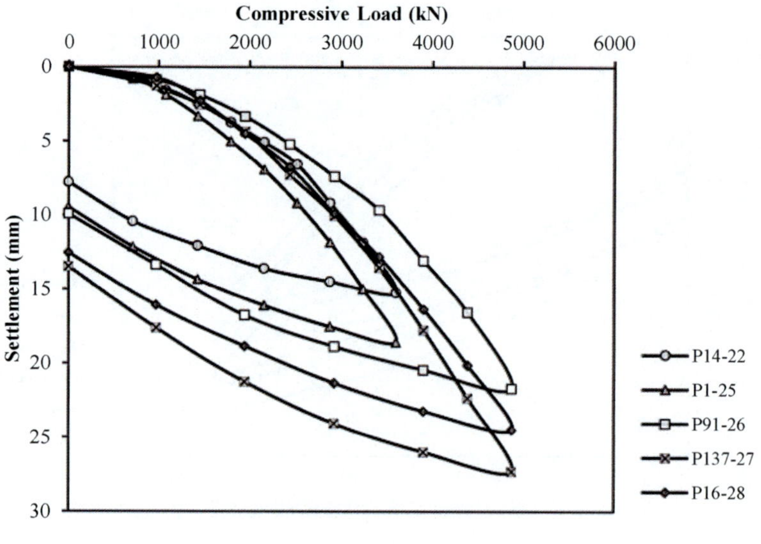

Fig. 4.21 Rectangular piles at Area C with different lengths

along the pile. As shown in Fig. 4.30, the axial force decreased with depth along the pile and increased with applied loads provided by hydraulic jacks. This was caused by friction resistance provided by the soil layers. The accumulated friction resistance along the pile is provided in Fig. 4.31.

The distribution of unit shaft resistance was obtained by calculating the differences between the two adjacent axial loads acquired from the strain gauges. As shown in

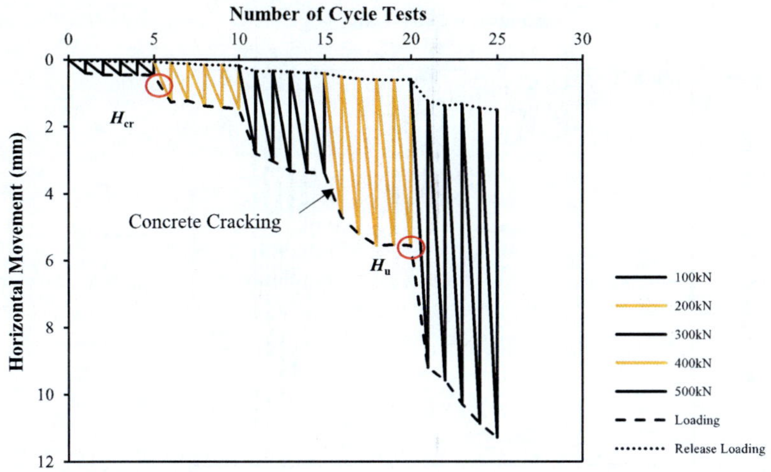

Fig. 4.22 H–X curves of P15

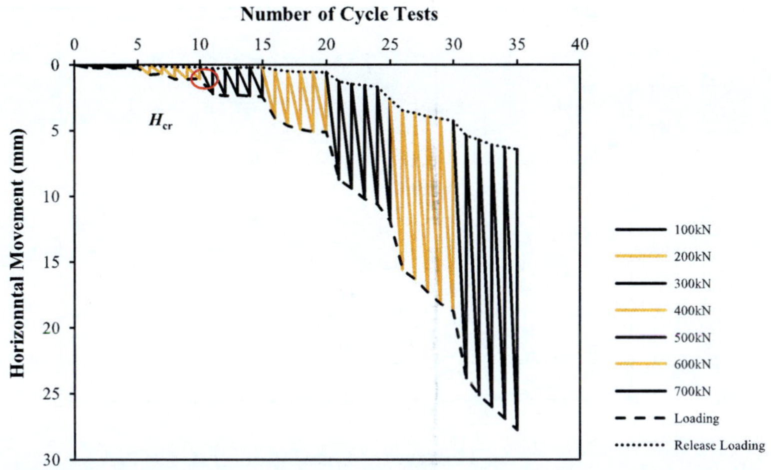

Fig. 4.23 H–X curves of P163

Fig. 4.32, the shaft resistance along the pile shaft is provided. As shown in this figure, when the uplift loads increases, the friction loads of each soil layer increases. It also shows that the soil layer from 7 to 16 m plays the most significant role in resisting the uplift force. Furthermore, it indicates that before the loading stage of 1,330 kN was applied, the friction resistance of soil layer from 7 to 11 m showed a rapid increasing trend, and after this stage, the friction resistance of the soil layer from 21 to 27 m showed a rapid growth.

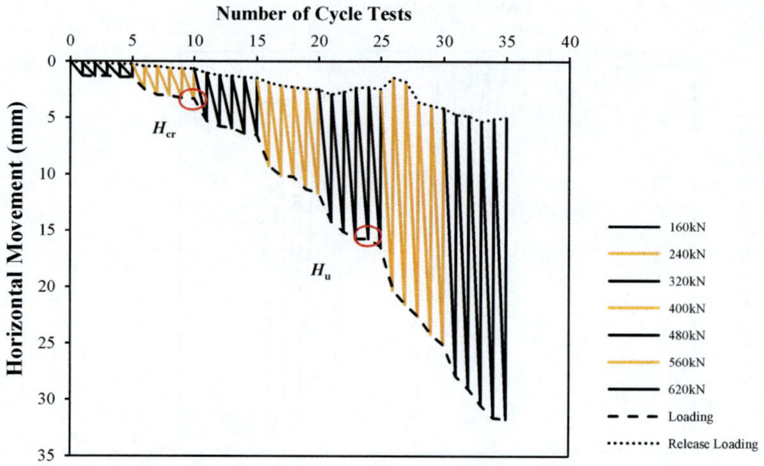

Fig. 4.24 H–X curves of P152

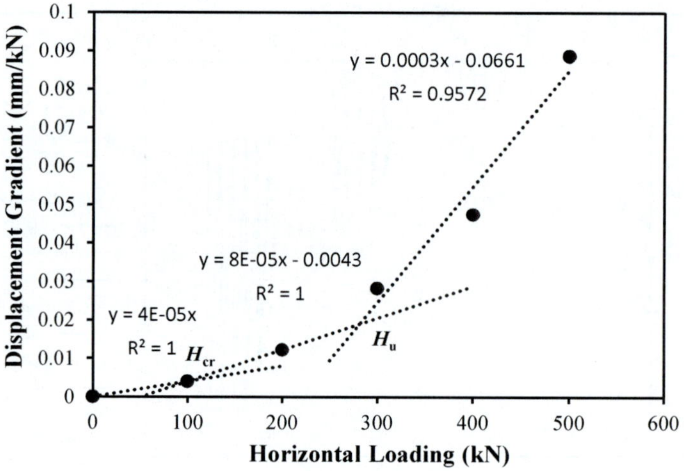

Fig. 4.25 Horizontal SLT interpretation of P15

4.5 Concluding Remarks

This chapter has presented the investigation of precast concrete piles. Three types of piles—small, large and non-displacement piles—were examined through two case studies. For each case study, the background, geotechnical conditions, pile information and test setups were illustrated, and the load settlement results were presented in the final section. For the small and large displacement piles, this chapter focused on the pile behavior under compressive loading, considering various lengths. For

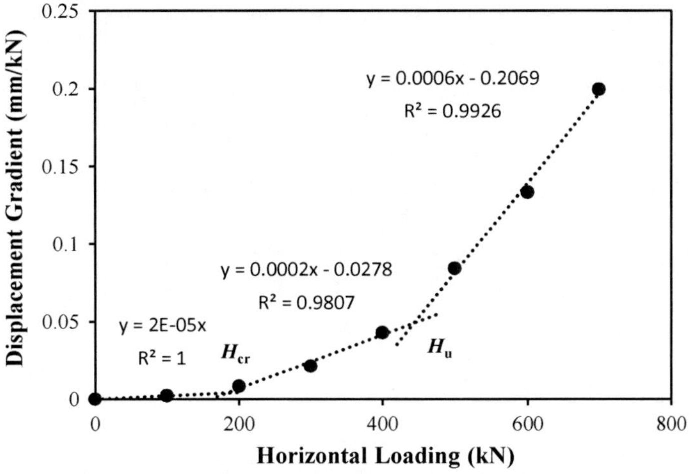

Fig. 4.26 Horizontal SLT interpretation of P163

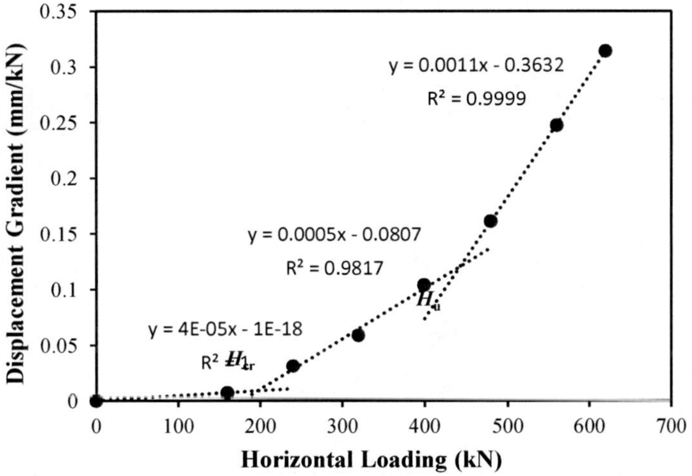

Fig. 4.27 Horizontal SLT interpretation of P152

Table 4.6 Summary of capacity and interpretation of horizontal loaded piles

Pile label	Pile length (m)	Lateral SLT results		Interpretation of SLTs	
		Critical load (kN)	Ultimate load (kN)	Critical load (kN)	Ultimate load (kN)
P15	24.7	100	300	107.5	280.9
P163	26.6	200	N/A	154.4	447.75
P152	28.0	240	480	175.43	470.8

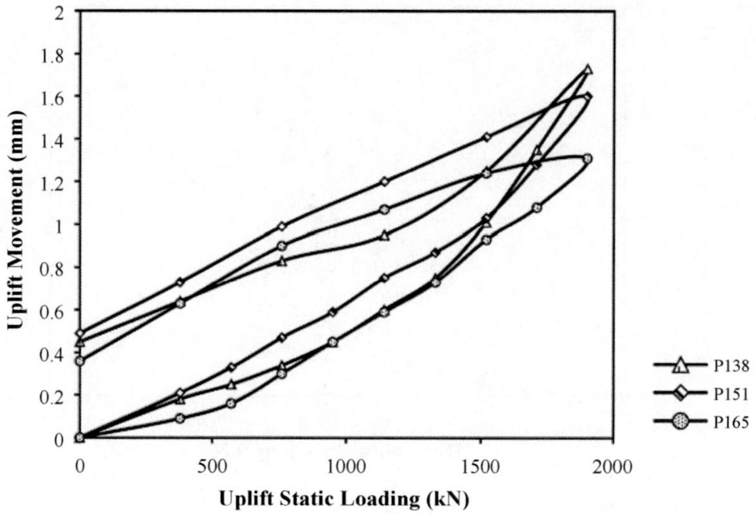

Fig. 4.28 Load–displacement results of uplift load piles

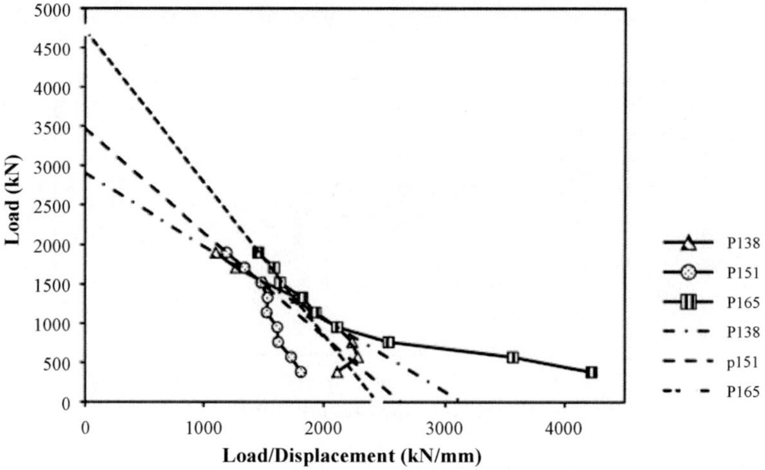

Fig. 4.29 Modified Mazurkiewicz method of uplift loaded piles

Table 4.7 Summary of capacity and interpretation of uplift loaded piles

Pile label	Pile length (m)	Uplift SLT results	Interpretation of SLTs
		Ultimate load (kN)	Ultimate load (kN)
P138	24.7	N/A	2,903.6
P151	26.6	N/A	3,467.0
P165	28.0	N/A	4,724.2

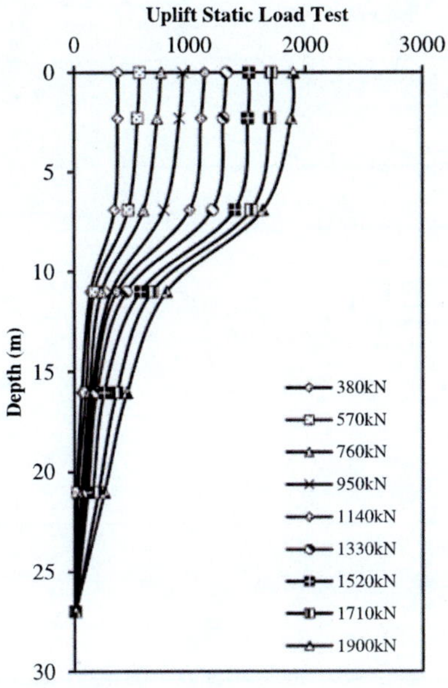

Fig. 4.30 Load transfer mechanism of P156

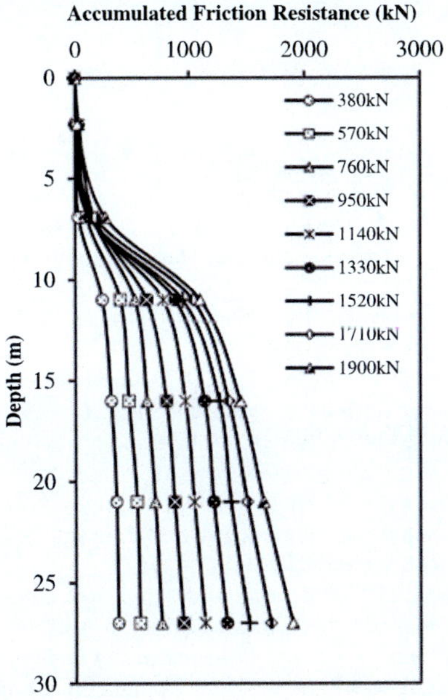

Fig. 4.31 Accumulated shaft resistance

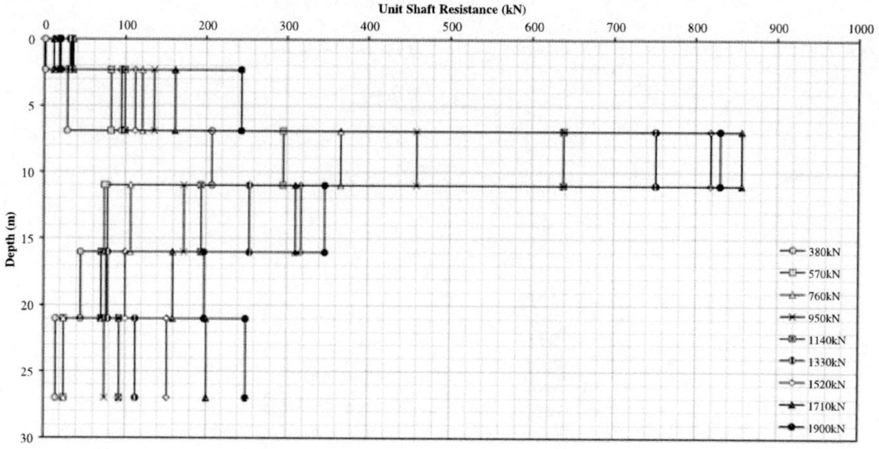

Fig. 4.32 Unit shaft resistance of P156

the non-displacement pile, uplift and lateral loading conditions were considered. Further, the interpretation of uplift and lateral SLTs was provided. Finally, this chapter researched and demonstrated the load transfer mechanism of the non-displacement pile under uplift loading.

References

Abu-Farsakh, M. Y., Haque, M. N., & Tsai, C. (2017). A full-scale field study for performance evaluation of axially loaded large-diameter cylinder piles with pipe piles and PSC piles. *Acta Geotechnica, 12*(4), 753–772.

Ashford, S. A., & Jakrapiyanun, W. (2001). Drivability of glass FRP composite piling. *Journal of Composites for Construction, 5*(1), 58–60.

Balasubramaniam, A., Oh, E., & Phienwej, N. (2009). Bored and driven pile testing in Bangkok sub-soils. *Journal of Lowland Technology International, 11*(1), 29–36.

Chu, E. H. (2004). Base grouted bored pile on weak granite. In *Grouting and ground treatment: ASCE*.

Environtology. (2011a). *China's top ten environmental problems 2011*, viewed 21 February 2017.

Environtology. (2011b). *China's construction waste recycling suffered from serious environmental pollution growing pains 2011*, viewed 21 February 2017.

Fellenius, B. H. and Samson, L. (1976). Testing of drivability of concrete piles and disturbance to sensitive clay. *Canadian Geotechnical Journal, 13*(2), 139–160.

Fellenius, B. H. (2002). *Determining the True Distributions of Load in Instrumented Piles* (p. 116). Paper presented at the ASCE International Deep Foundation Congress: Geotechnical Special Publications.

Gregersen, O. S., Aas, G., & Dibiagio, E. (1975). Load tests on friction piles in loose sand. *International Journal of Rock Mechanics and Mining Sciences and Geomechanics Abstracts, 12*(7), 98–98.

Gwizdala, K., & Wieclawski, P. (2013). Influence of time on the bearing capacity of precast piles. *Studia Geotechnica et Mechanica, 35*(4), 65–74.

Hussein, M. H., Woerner, I., Wayne, A., Sharp, M., & Hwang, C. (2006). Pile driveability and bearing capacity in high-rebound soils. In *GeoCongress 2006: Geotechnical engineering in the information technology age* (pp. 1–4).

Hsu, S.-T. (2014). Behaviors of large-scale driven PC piles. *Journal of Marine Science and Technology, 22*(4), 487–497.

Kumar, S., Alarcon, C., & Hosin, A. (2004). O-cell testing of reinforced concrete driven piles. In*International Conference on Case Histories in Geotechnical Engineering* (pp. 1–7).

Lehane, B. M., Williams, R., & Li, Y. (2013). Shaft capacity of displacement piles in clay using the cone penetration tests. *Journal of Geotechincal and Geoenvironmental Engineering, 139*(2), 253–266.

Liew, S., Ng, H., & Lee, K. (2004). *Comparison of HSDPT and SLT results of driven piles in Malaysian residual soils.* Paper presented at the Malaysian Geotechnical Conference.

Mansur, C. I., & Hunter, A. H. (1970). Pile tests-Arkansas river project. *Journal of Soil Mechanics & Foundations Division.*

Ma, S. C., & Li, G. Z. (2006). The relationship between economic increase of China and environmental contamination based on Kuznets curves analysis. *Statistical Research, 8,* 37–40.

Marcos, M. C. M., Chen, Y.-J., & Kulhawy, F. H. (2013). Evaluation of compression load test interpretation criteria for driven precast concrete pile capacity. *KSCE Journal of Civil Engineering, 17*(5), 1008–1022.

Ministry of Construction of the People's Republic of China. (1999a). *Code for construction and acceptance of metro engineering (GB 50299-1999).* Beijing, China: Ministry of Construction of People's Republic of China.

Ministry of Construction of the People's Republic of China. (1999b). *Standard for soil test method (GB/T 50123-1999).* Beijing, China: Ministry of Construction of People's Republic of China.

Ministry of Construction of the People's Republic of China. (2000). *Technical specifications for construction of highway bridges and culverts (JTJ 041-2000).* Beijing, China: National Standard of the People's Republic of China.

Ministry of Construction of the People's Republic of China. (2001). *Code for investigation of geotechnical engineering (GB 50021-2001).* Beijing, China: Ministry of Construction of People's Republic of China.

Ministry of Construction of the People's Republic of China. (2006). *Code for seismic design of railway engineering (GB 50111-2006).* Beijing, China: Ministry of Construction of People's Republic of China.

Ministry of Construction of the People's Republic of China. (2007). *Code for deformation measurement of building and structure (JGJ 8-2007).* Beijing, China: National Standard of the People's Republic of China.

Ministry of Construction of the People's Republic of China. (2008a). *Code for urban rail transit engineering survey (GB 50308-2008).* Beijing, China: Ministry of Construction of People's Republic of China.

Ministry of Construction of the People's Republic of China. (2008b). *Technical code for building pile foundations (JGJ 94-2008).* Beijing, China: National Standard of the People's Republic of China.

Ministry of Construction of the People's Republic of China. (2009a). *Specifications for global positioning system (GPS) surveys (GB/T 18314-2009).* Beijing, China: Ministry of Construction of People's Republic of China.

Ministry of Construction of the People's Republic of China. (2009b). *Technical code for monitoring of building excavation engineering (GB 50497-2009).* Beijing, China: Ministry of Construction of People's Republic of China.

Ministry of Construction of the People's Republic of China. (2012). *Specification for Dynamic Observation of Groundwater in Urban Area (CJJ 76-2012).* Beijing, China: National Standard of the People's Republic of China.

Ministry of Construction of the People's Republic of China. (2012). *Technical specification for retaining and protection of building foundation excavations (JGJ 120-2012)*. Beijing, China: National Standard of the People's Republic of China.

Ministry of Construction of the People's Republic of China. (2013a). *Code for design of metro (GB 50157-2013)*. Beijing, China: Ministry of Construction of People's Republic of China.

Ministry of Construction of the People's Republic of China. (2013b). *Code for monitoring measurement of urban rail transit engineering (GB 50911-2013)*. Beijing, China: Ministry of Construction of People's Republic of China.

Ministry of Construction of the People's Republic of China. (2013c). *Standard for test methods of engineering rock mass (GB/T 50266-2013)*. Beijing, China: Ministry of Construction of People's Republic of China.

Ministry of Construction of the People's Republic of China. (2014). *Technical code for testing of building foundation piles (JGJ 106-2014)*. Beijing, China: National Standard of the People's Republic of China.

Ministry of construction of the People's Republic of China. (2015). *Code for Tunneling Monitoring Technology Procedures (QCR 9218–200-15)*. Beijing, China: National Standard of the People's Republic of China.

Yu, F., & Yang, J. (2012). Base capacity of open-ended steel pipe piles in sand. *Journal of Geotechnical and Geoenvironmental Engineering, 138*(9), 1116–1128.

Zhusupbekov, A., Alibekova, N., & Morev, I. (2014). Dynamic-penetration method for bearing capacity determination of precast piles. *Soil Mechanics & Foundation Engineering, 51*(2).

Chapter 5
Field Performance of Composite Piles

5.1 General Introduction

After late nineteenth century when the driven piles design was mainly based on the experience, the research of pile behavior which suffered static load started (Paikowsky and Tolosko 1999). Federal Highway Administration provided the static analysis methods for designing of single piles, the capacity of the pile can be primarily calculated through Meyerhof Method which is based on the result obtained from Standard Penetration Tests (SPT), Brown Method, Nordlund Method, α Method (total stress method) and β method (effective stress method) for piles suffered from vertical compression loading (Hannigan et al. 2016).

The ultimate bearing capacity of pile could be primarily estimated based on the static analysis calculation and the empirical method. For example, based on the Chinese code of JGJ 94-2008 (2008), the ultimate capacity of a single pile is related to the end bearing (Q_{pk}) and shaft resistances (Q_{sik}), which are dependent on the types and properties of the soil layers as well as piling technology. Similarly, in Singapore, the ultimate axial load capacity can be determined based on the skin friction coefficient (K_s), end bearing coefficient (K_b) and unit end bearing (q_b) (Xiao and Yang 2011). However, due to the uncertainty such as the properties of soil layers, where the obtained SPT N-value or corrected N value cannot appropriately represent the properties of the soil, these methods can only be used for primarily design.

The SLTs are the mostly accepted method to determine the ultimate bearing capacity of piles. These tests can provide accurate data and are commonly used for investigating the pile behavior. The SLT mostly implements loads from 0 kN to the value equal to 2 times the required anticipated design loads, and later consecutively releases loads back to 0 kN (JGJ 94-2008 2008; ASTM D-1143/D-07 1994). In some local standards, the maximum loading value may be equal to 1.5–4 times the designed loads.

The Load-settlement (Q-s) curve is the typical results obtained after SLTs. However, mostly, even though the maximum loading of SLT applied, there is no

J. Zhou and E. Oh, *Full-Scale Field Tests of Different Types of Piles*,
Advanced Topics in Science and Technology in China 62,
https://doi.org/10.1007/978-981-33-6183-6_5

extraordinary displacement increase observed from in-situ or from the Q-s curve. This is because the SLTs are usually used for checking if the designed pile achieves the project requirement instead of determining the ultimate bearing capacity. In other word, these tests have not been performed to the failure of pile-soil system, which leads to the requirement of interpretation of SLT results to ultimate capacity determination.

In current practice, there are some similarities of pile capacity or pile failure definition among standards. Borel et al. (2004) summarized that there are three types of failures which are mostly used for determining the pile capacity. The first one is $Q_{u,F}$, which is known as plunging failure observed from the SLTs with rapid increase of pile movement without an increase in loading (or increment loading is small). The other two are asymptotic ultimate load ($Q_{u,asym}$) derived mathematically or graphically from the Q-s curve and conventional ultimate load ($Q_{u,conv}$) which is defined as the load causing a gross settlement equal to 0.1 times the diameter of the pile or the load at which the penetration reaches a given value. Some standards may believe that the $Q_{u,conv}$ occurred when different values of gross settlement was discovered, for example, different from the value of 0.1 times the diameter, ASTM D-1143/D-07 (1994) pointed out when the axial settlement achieves to 0.15 times the diameter or width, failure occurs and the increment loading of SLT should stop.

The interpretations of data obtained from SLTs are Davission's offset method, double tangent method, DeBeer's Log–log method, Brinch-Hansen's method and Chin-Kondner method. Methods of interpretation based on maximum allowable gross movements overestimate the allowable capacities of short piles and underestimate the abilities of the long piles. AASHTO (2002) recommended the commonly method for interpreting the pile compression testes results was Davisson's method (1972) for driven piles and double tangent method for drilled shafts (Samtani and Nowatzki 2006). DeBeer's method is the approach that plotting these data using logarithmic scales (DeBeer 1970). There are two methods suggested by Brinch-Hansen (1963) to define the failure conditions. They are 90 and 80% criteria, where the first criteria defines the failure load as the load that is associated with twice the pile head movement as obtained from 90% of the applied load, the 80% criterion defines the failure load as the loading that is corresponding to four times the pile head movement obtained from 80% of the load. Similarly, the Chin's method can also be used to extrapolate the SLTs based on the assumption that the Q-s relationship is hyperbolic and hence the inverse slope of interpreted line is the ultimate bearing capacity (Chin 1970). Through back analysis of the static pile load test in Singapore, Xiao and Yang (2011) pointed out that, theoretically, the Chin's method created two lines which inverse slope of the first line representing the shaft resistance and the inverse slope of the second line representing the ultimate resistance.

Nowadays, the utilization of piles is widespread, however, there are issues when these piles are located in harsh environments, especially in marine or coastal conditions. Piles made with traditional materials can be destroyed due to the corrosion of steel, the deterioration of timber, and the degradation of concrete. The deterioration of the timber, concrete, and steel piling systems costs the United States nearly 2 billion dollars per year for repair and replacement (Hassan and Iskander 1998). The development of concrete piles has continued for more than 200 years, but engineers are now facing pile-related problems because, even though piles are made of concrete and steel, which has good rigidity and high strength, damage can still occur. Under such condition, engineers started to find a new material that has better resistance ability to face with the harsh environments.

Fiber-reinforced Polymer (FRP) is a composite material that contains resin and fiber that can provide high tensile strength. This material is also known as fiber-reinforced plastic or an advanced composite material (Bank and Wiley 2006). FRP is well-known for its high ratio of strength to weight, high ratio of longitudinal/transversal Young's and high ratio of longitudinal/transversal shear modulus (Pecce 2001). This material can be either Carbon-FRP (CFRP), Aramid-FRP (AFRP) Glass-FRP (GFRP), or Basalt-FRP (BFRP), depending on the fiber used. The composition of the FRP product is, therefore, flexible depending on the material properties and the volume ratio of fibers to resin, and the selection of types and orientation of the fiber. Because of this flexibility, FRP composite material products have been widely applied for reinforcement and rehabilitation.

FRP material can be manufactured into FRP laminar (slice/sheet) or FRP bars. FRP bars are usually used for making the GFRP anchors (ribs) and GFRP grilles (Zhang et al. 2001). The FRP bars can not only be used to replace steel bars inside concrete structure, but also can be used as anchors for slope reinforcement and support in deep excavation. The technology of using GFRP reinforcement as an anchor was adopted in the 2008 Chinese Changji Expressway construction built for red sandstone slope reinforcement. The results demonstrated the slope was overall stabilized (Luo 2014).

Moreover, when rehabilitation is required, on either upper structure such as beams, or on columns, and sometimes on walls, the FRP laminar is utilized. The application of the FRP slice using resin onto the surface of the cracked beam or wall on the tensile side recovers the original strength of the components because the FRP material provides good tensile resistance. A large amount of research has focused on analyzing the structural behaviors of beams and walls with consideration of FRP sheet application (Mohammed et al. 2013; Mosallam et al. 2015; Mostofinejad and Tabatabaei Kashani 2013; Mostofinejad and Mohammadi Anaei 2012). A notable method using confinement to concrete columns by FRP has been studied, and the results indicated that this methodology can enhance concrete strength of the column when the structure suffers axial loading (Kwan et al. 2015; Lo et al. 2015; Youssf et al. 2015). The application history of FRP composite materials has surpassed 80 years; however, these previous projects mostly focused on rehabilitating or strengthening the structure members, such as beams and columns.

The history of applying FRP piles is approximately three decades old. Dating back to April 1987, the first prototype recycled pile was driven at The Port of Los Angeles to replace the creosote-treated timber piles which successfully avoided any threat to marine borers (Juran and Komornik 2006). As early as 1998, the Empire State Development Corporation (ESDC) undertook a waterfront rehabilitation project known as the Hudson River Park (Robinson and Iskander 2008). This project involved replacing up to 100,000 bearing piles for lightweight structures. The concrete-filled

FRP composite piles were then employed by the Virginia Department of Transportation (VDOT) in 2000 for the new Route 40 Bridge over the Nottoway River in Sussex County, Virginia (Pando et al. 2004).

Extensive investigation had conducted on the FRP composite piles. 4 types of FRP piles were tested in Elizabeth, New Jersey. These tested FRP piles were concrete filled fiberglass shell piles (Lancaster Pile), polyethylene piles reinforced with steel bars (PPI Pile), polyethylene piles reinforced by fiberglass bars (SEAPILES) and solid polyethylene piles (American Ecoboard Pile). The tests have shown possible applicability of plastic piles to traditional axial loading applications and the further work of long term creep performance and durability of these piles was highlighted (Robinson and Iskander 2008). These piles data are analyzed based on Davisson's method, DeBeer's method, Chin's method and CAPWAP method. In addition, Concrete-filled FRP piles came into notice and were considered the best FRP piles to resist upper loading. From a different perspective, the other types of FRP piles showed low capacity. This is acceptable when using these FRP piles because the steel pipe pile is mostly used under the conditions of being exposed to water, which means the FRP pile is mainly used for protection away from corrosion after replacing the steel pile. Further example like SRP, is mainly used in fendering applications and regarded as potential load-bearing piles (Guades et al. 2012). FHWA proposed the conclusion that the FRP composite piles can be used effectively as vertical load-bearing piles and represent an alternative for deep foundation construction, especially in waterfront environments and aggressive soils (Juran and Komornik 2006).

For composite piles suffering from lateral loading, a test pile program using a statnamic device was conducted and comparisons made between pre-stressed concrete piles and composite piles. It was concluded that the composite pile exhibited a much lower stiffness than the pre-stressed concrete pile. This also illustrated that the composite piles had the ability to sustain lateral load (Pando et al. 2004). To determine the lateral behavior of large diameter composite piles, tests were conducted by Thomann and Zoli using two different tests, the inclinometer measurement and survey measurement instead of a statnamic method. It was concluded that the measured deflections were much higher than the pre-load test calculation results. It was also concluded that when using post-load analysis, 85% reduction in soil strength was required (Thomann et al. 2004). Apart from large diameter composite piles, two other types of piles, namely the concrete-filled GFRP pipe pile and the standard pre-stressed concrete pile, were analyzed to determine the behavior lateral loading and the results compared. These indicated that the concrete-filled GFRP pile was more flexible than the standard pre-stressed concrete pile. Also, the ultimate lateral load capacity of the GFRP pipe pile was greater; however, it exhibited brittle behavior at failure. Additionally, the results concluded that GFRP piles can be modelled using P-y curves and classical beam theory (Weaver et al. 2008). The limitation of all these previous investigations into passive composite piles is that there is no consideration of soil remove during excavation.

As reviewed above, there has been numerous geotechnical applications of FRP materials being performed, however, the research which aims to study the deflection behavior of the FRP bar reinforced concrete piles is limited. In addition, the limitation of the recent research associated with FRP composite material is that there are no tests on precast concrete piles that are partially confined by FRP laminar. This Chapter provides two case studies on the purpose of FRP composite pile's research. The first case study illustrates the capacity improvement of the composite piles under FRP sheet confinement, and the second case study investigates the FRP bar reinforced passive piles used in a deep excavation pit (tunneling project).

5.2 CFRP Laminar Concrete Composite Piles

5.2.1 Background Introduction

To investigate the composite piles' capacity with FRP confinement effect, compressive SLTs were conducted based on the research method illustrated in Chap. 2. The tested piles were rectangular and pipe concrete piles, and each type of tested piles was partially confined by CFRP. For the rectangular piles (P116 and P115), P116 was applied with CFRP laminar with a space of 120 mm (width of CFRP of 100 mm). The tensile strength of this fiber material was 760 kPa. The dimensions of these two piles were the same, with a $400 \times 400\ \text{mm}^2$ cross-section area and total length of 25 m.

For the concrete pipe piles (P2108 and P2054), P2108 was applied with a CFRP with material strength of 760 kPa, and the diameters and length of these two piles were 400 mm and 25 m, respectively. The concrete grade of the pipe pile was C30 ($150 \times 150 \times 150\ \text{mm}^3$ cubic compression strength of 30 N/mm^2) and the concrete grade of the rectangular pile was C40 ($150 \times 150 \times 150\ \text{mm}^3$ cubic compression strength of 40 N/mm^2).

5.2.2 Subsurface Conditions

The subsurface exploration was discovered through laboratory and in-situ tests. The In-situ tests such as standard penetration tests (SPT) and cone penetration tests (CPT) and laboratory tests of consolidation tests, direct shear tests and triaxial tests were conducted. The local standards of Chinese Code of Standard for Soil Tests Method (GB/T 50123-1999 1999), Standard for Test Methods of Engineering Rock Mass (GB/T 50266-2013 2013) and Code for Investigation of Geotechnical Engineering (GB 50021-2001 2001) were used in this project for the subsurface exploration. Based on the borehole logs, the soil layers were discovered as follows.

1. Clay: yellowish brown, plastic, average thickness of 4.55 m; 2. fine sand: loose, grey-dark, contains ferric oxide, average thickness of 1.98 m; 3. silty clay: yellowish brown, plastic, partially contains fine sand, average thickness of 4.57 m; 4. sand: yellowish brown, fine to medium density, average thickness of 3.08 m; 5.

sand: brown, saturated, dense, 8 to 12% loess doll, diameter of 1 to 1.5 cm, average thickness of 6.15 m; 6. strongly weathered glimmerite: average thickness of 2.52 m; 7. medium weathered glimmerite: very dense, average strength of 14.12 MPa, average thickness of 12.6 m.

5.2.3 *Pile Preparation*

As shown in Fig. 5.1, the concrete rectangular and pipe piles were transferred to the required location. The CFRP (Fig. 5.2a) was cut into pieces, and then the glue was prepared by mixing Type A and B resin together. Later, the CFRP pieces were attached to the concrete surface using the admixed resin, ensuring the resin was totally permeated through the FRP laminar. The prepared concrete rectangular and pipe piles are provided in Fig. 5.3. After seven curing days for the composite material,

(a) Concrete pipe pile (b) Concrete rectangular pile

Fig. 5.1 Preparation of the concrete piles

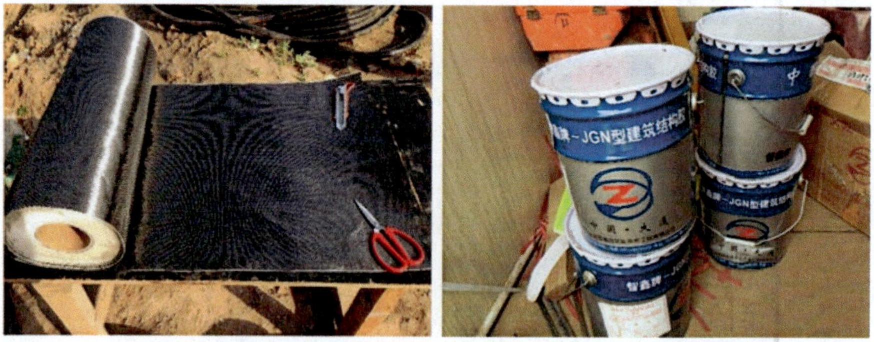

(a) CFRP preparation (b) Resin preparation

Fig. 5.2 Preparation of the composite material

(a) CFRP-confined pipe piles

(b) CFRP-confined rectangular piles

Fig. 5.3 CFRP-confined concrete piles

these piles were ready to install into the ground, as shown in Figs. 5.4 and 5.5. To enable comparisons between the CFRP-confined piles and the piles without FRP application, the driven piles were installed close to each other so that the geotechnical conditions were the same. Mind the overlapping length should be over than the

Fig. 5.4 Installation of composite rectangular pile

Fig. 5.5 Installation of composite pipe pile

recommended value from the product specification. For this case, the overlapping length was 200 mm. Also, do not mixing Type A and B resin in a large amount because chemical reaction is very fast. There will be inadequate time for operation (the glue has already hardened).

5.2.4 Test Setup

Similar to the setup depicted in Chap. 4, these SLTs were conducted based on the Chinese Technical Code for Testing of Building Foundation Piles (JGJ 106-2014 2014). The setup of the SLT is provided in Fig. 5.6. As shown in Fig. 5.7, two hydraulic jacks were used to provide loads to the pile head. The low-speed maintenance methods were used in this case study, and the applied load was maintained until the rate of axial movement did not exceed 0.1 mm. After each load was applied, four dial gauges recorded the vertical movement of piles with intervals of five, 10, 15, 15, 15, 30 and 30 min, and one hour duration was required if accumulated time exceeded two hours.

Fig. 5.6 Compressive SLT

Fig. 5.7 Equipment for SLT

For rectangular piles P116 and P115, the loading started at 976 kN and then progressed in 488 kN increments until the maximum loading of 4,880 kN, and then the load was consecutively released back to 0 kN. For the pipe piles, the loading started at 260 kN, with increments of 130 kN until the maximum loading of 1,300 kN, and then the loading was consecutively released back to 0 kN.

5.2.5 Results of CFRP Laminar–Confined Concrete Piles

5.2.5.1 Load Settlement Curves

The pile head settlements corresponding with each load were recorded, and the load–settlement curves (Q-s) of these four concrete piles are provided in Figs. 5.8 and 5.9.

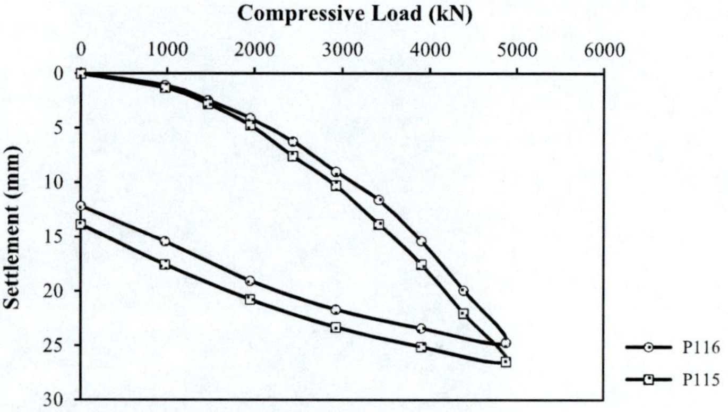

Fig. 5.8 Q-s curves of tested rectangular piles

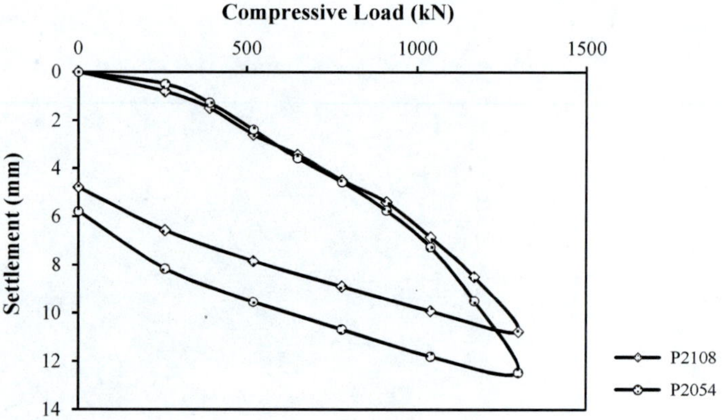

Fig. 5.9 Q-s curves of tested pipe piles

As illustrated in Fig. 5.8, after the maximum loads of 4,880 kN were applied, the maximum settlements of the traditional concrete rectangular pile and the pile with CFRP application were 26.435 and 24.69 mm, respectively. As shown in Fig. 5.9, after the maximum loads of 1,040 kN were applied, the maximum settlements of the traditional concrete pipe pile and the pipe pile with CFRP application were 12.46 and 10.77 mm, respectively.

5.2.5.2 Settlement-lg (Load) Curves

The settlement-lg load (s-lgQ) curves of these four piles are provided in Fig. 5.10. Similar to DeBeer's method, which uses lg settlement and lg load data to plot the

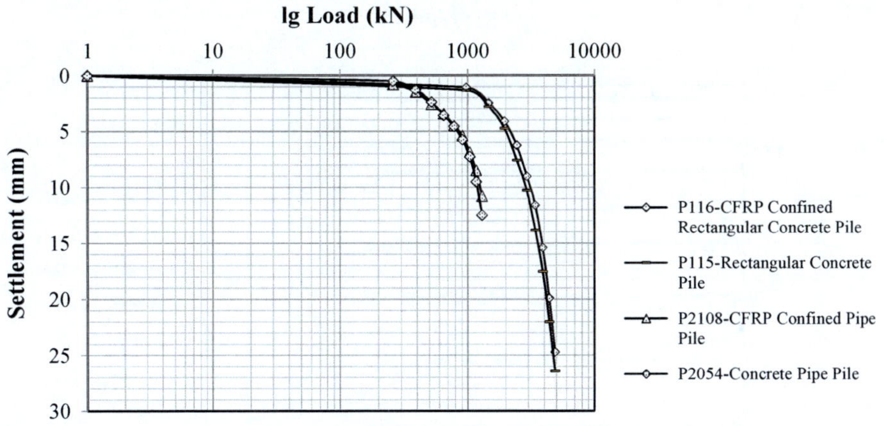

Fig. 5.10 s-lgQ curves of four tested piles

diagram, it was difficult to determine the point that represented the extreme settlement decreasing trend. Thus, the ultimate bearing capacity was difficult to determine and some interpretations were required.

5.2.5.3 Settlement-lg (Time) Curves

The settlement-lg time (s-lgt) curves of these four tested piles are provided in Figs. 5.11, 5.12, 5.13 and 5.14, respectively. These four figures indicate that the CFRP-confined piles presented less settlement during each loading stage than did the piles without CFRP application. If there was an extraordinary settlement distance between two adjacent loading stages, the failure condition occurred, and the ultimate capacity of the pile could be determined. However, as shown in these figures, it was difficult to determine such loading stages. Another method is to determine a load line that illustrates a decreasing trend, which indicates that the soil foundation system is unstable under one loading, and the previous load is then believed to be the failure load; however, there was no such line acquired. This also illustrated that these four piles did not suffer from plunging failure, and interpretation was required to determine the ultimate bearing capacity.

5.2.5.4 Interpretations of the Static Load Tests

Double Tangent Method

The rectangular pile results based on the double tangent method are provided in Fig. 5.15. By using two tangent lines and determining the intersection, it was found

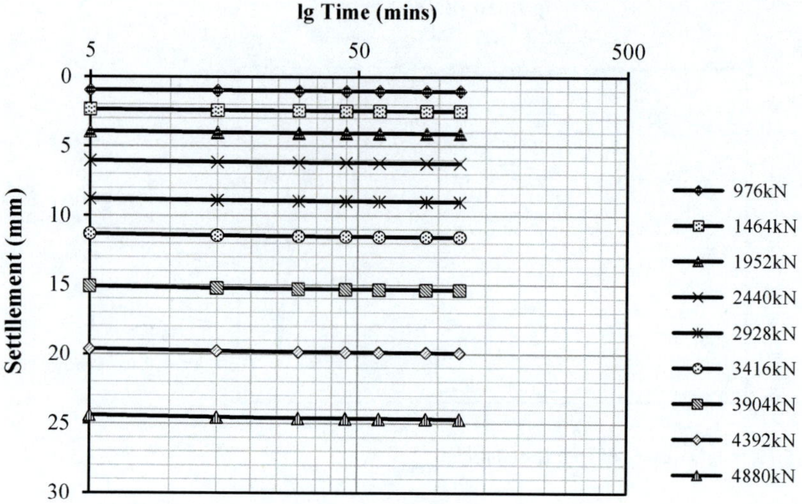

Fig. 5.11 P116 CFRP-confined concrete rectangular pile

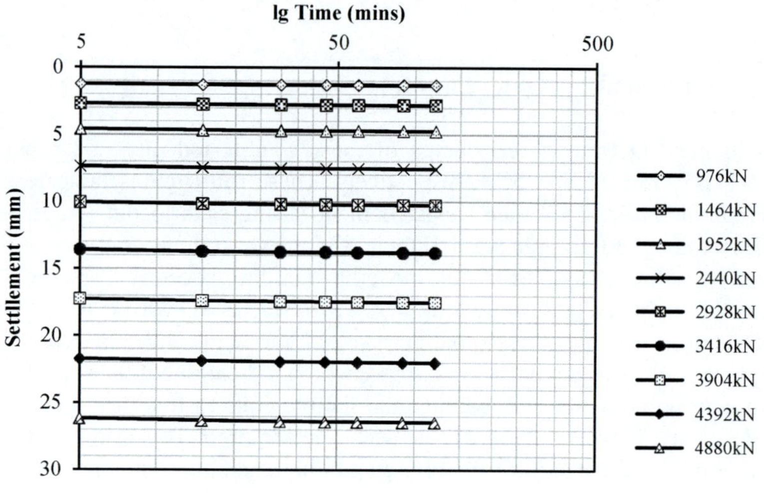

Fig. 5.12 P115 concrete rectangular pile

that the ultimate bearing capacities of the CFRP-confined pile (P116) and normal pile were 2,600 and 2,400 kN, respectively. As illustrated in Fig. 5.16, based on the double tangent method, the ultimate bearing capacities of the CFRP-confined concrete pipe pile (P2108) and normal concrete pipe pile were 850 and 760 kN, respectively. It should be noted that this method uses empirical understanding to find the tangent line; thus, to some extent, the results may be inaccurate.

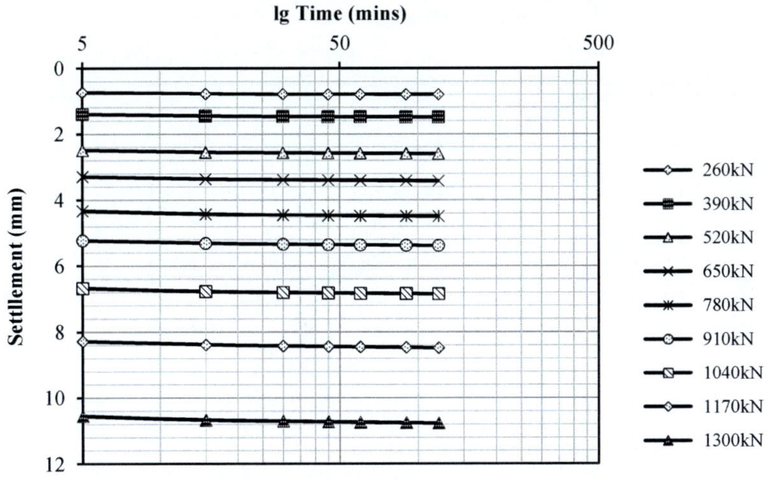

Fig. 5.13 P2108 CFRP-confined concrete pipe pile

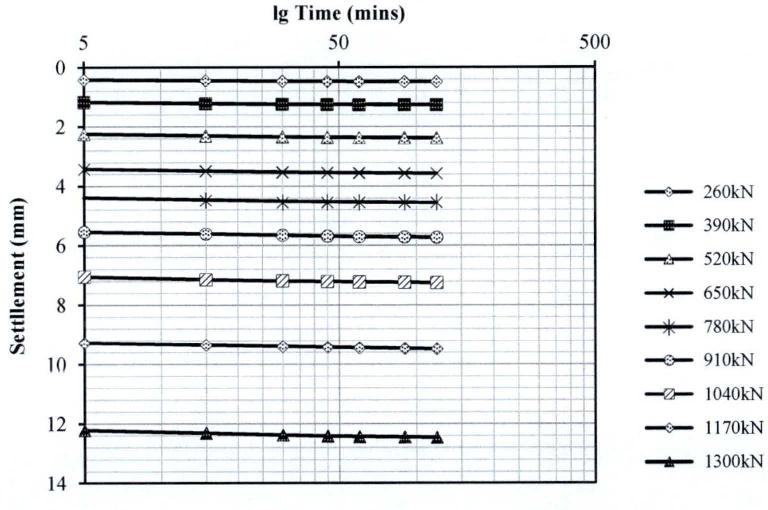

Fig. 5.14 P2054 concrete pipe pile

DeBeer's Method

Based on DeBeer's method (as illustrated in the review in Chap. 2) of plotting the settlement and loading into the lg–lg values, as shown in Figs. 5.17 and 5.18, it was found that the ultimate bearing capacities of these four piles via interpretation were very difficult to determine. This was because DeBeer's method is mostly used in the plunging failure tests.

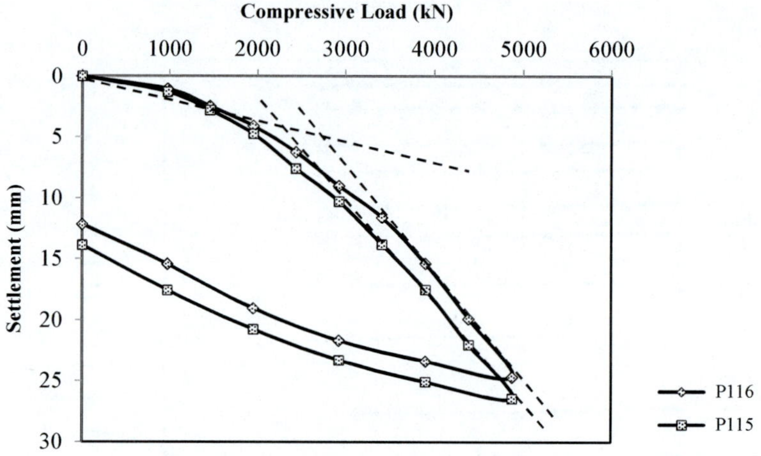

Fig. 5.15 Double tangent method of rectangular piles

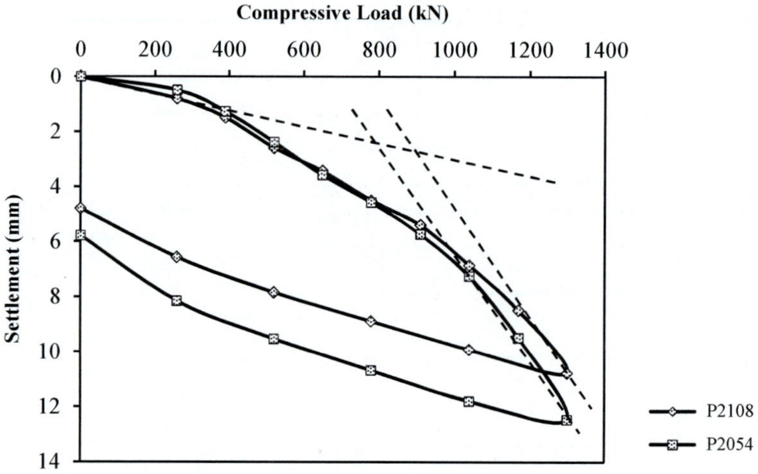

Fig. 5.16 Double tangent method of pipe piles

Davisson's Offset Method

Based on Eq. (2.37), the elastic lines could be determined. After offsetting the elastic lines, the results are provided in Figs. 5.19 and 5.20. It was found that there was no intersection between the offset line and Q-s curve (loading stages), which illustrated that the ultimate bearing capacity was over the maximum applied loading, and other interpretations are required.

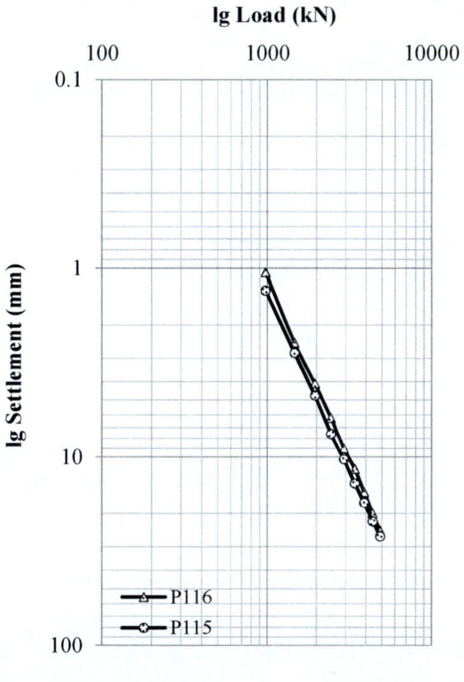

Fig. 5.17 DeBeer's method of rectangular piles

Chin's Method

Chin's (1970) method operates under the assumption that the load–settlement relationship is hyperbolic. By plotting the displacement/load versus the displacement, the trend line provides a slope, and the reverse of this slope gives the ultimate bearing capacity. Theoretically, the curve of settlement versus the ratio (settlement/load) should comprise two straight lines (Chin 1978). In the plotted diagrams for P116 (Fig. 5.21), P115 (Fig. 5.22), P2054 (Fig. 5.23) and P2108 (Fig. 5.24), the reverse of the first line represents the shaft resistance, and the reverse of the second line gives the ultimate resistance. The end bearing could then be determined through calculations with the obtained values of shaft and ultimate bearing resistance.

Eight functions were found and are provided in these figures. Based on these functions, for the traditional precast piles, the slopes of P115 and P2054 are determined as 0.0001 (Qu) and 0.0005 (Qu), respectively, as depicted in Figs. 5.22 and 5.23. For the CFRP confined piles, the slopes of P116 and P2108 are determined as 0.0001 (Qu) and 0.0004 (Qu), respectively, as illustrated in Figs. 5.21 and 5.24. It can be found that, for the concrete pipe pile, the ultimate bearing capacity of the pile with CFRP application is determined as 2500 kN with shaft capacity of 909.1 kN. For the concrete rectangular pile, the ultimate bearing capacity of pile with CFRP application is determined as 10,000 kN with shaft capacity of 1428.6 kN. Table 5.1 summarises the results of these four piles based on different methods.

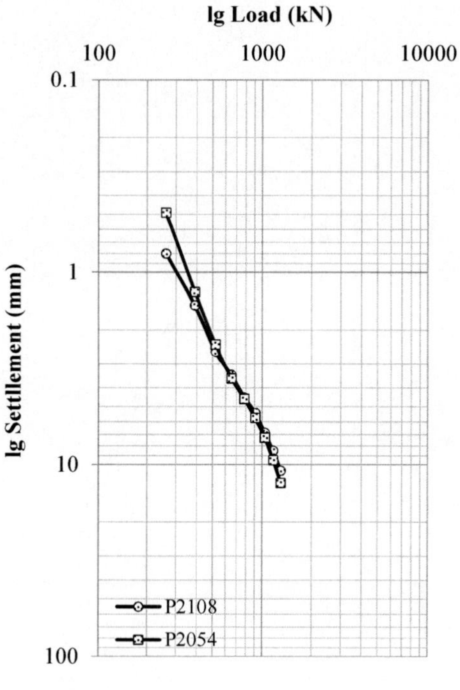

Fig. 5.18 DeBeer's method of pipe piles

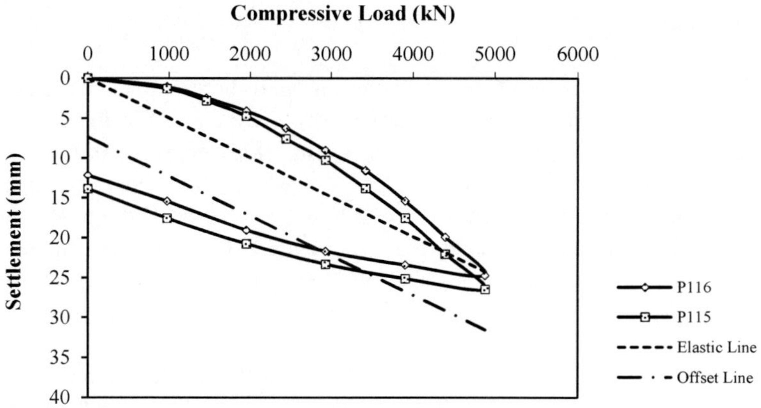

Fig. 5.19 Davisson's offset method of rectangular piles

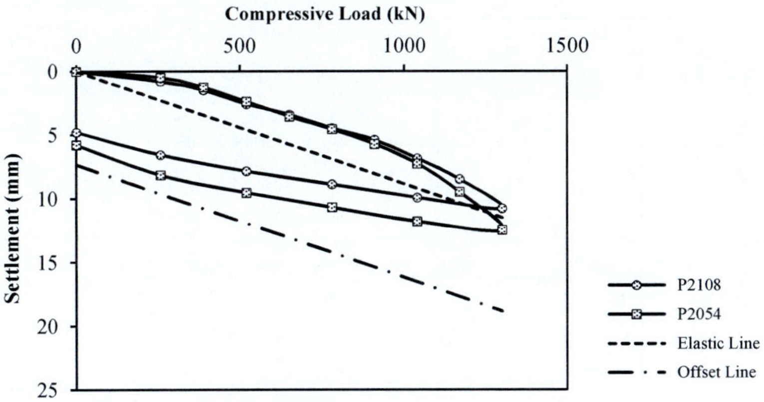

Fig. 5.20 Davisson's offset method of pipe piles

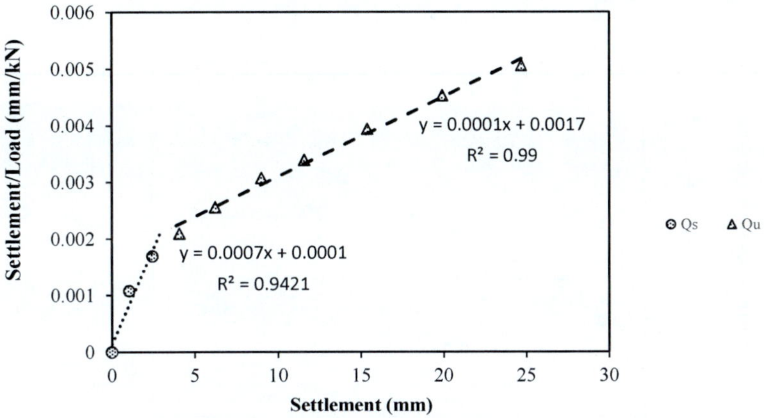

Fig. 5.21 Chin's method of P116

5.3 GFRP Bar–Reinforced Concrete Composite Piles

5.3.1 Project Introduction

As stated in Chap. 2, intensive research has focused on the field test on pile foundation yet has seldom considered the passive pile with FRP bar reinforcement, and has very limited analysis of the deflection behaviour of this pile under changed earth pressure. Further, lateral SLTs are short-term loading, and the research associated with long-term loads and deflection behaviour is limited. As discussed in Chap. 4, the research on non-displacement piles commenced in a tunnelling project, that project

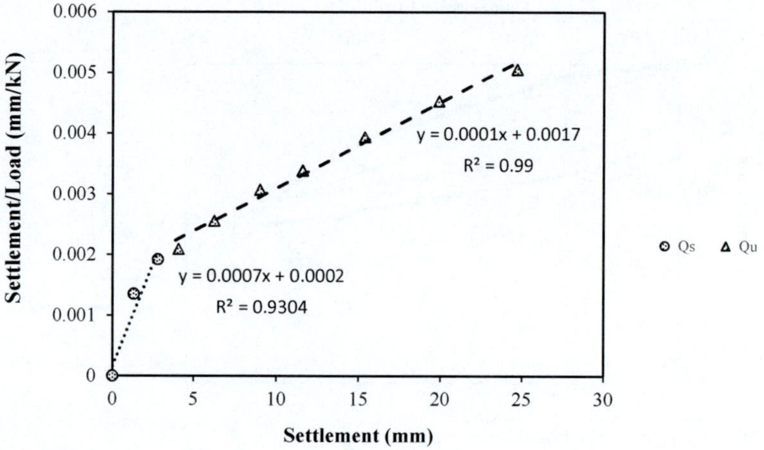

Fig. 5.22 Chin's method of P115

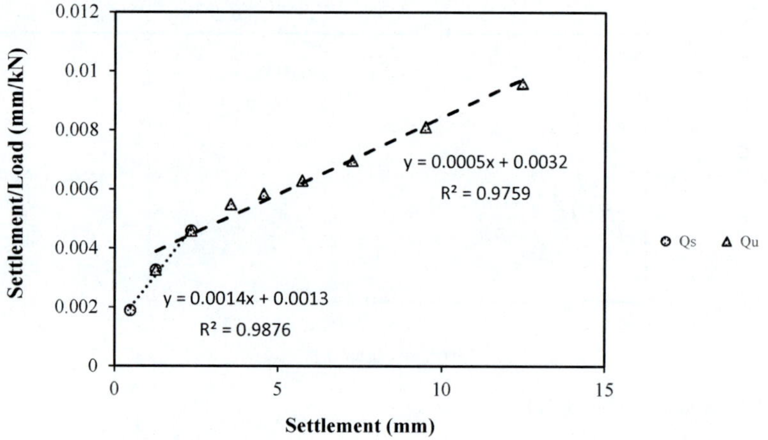

Fig. 5.23 Chin's method of P2054

contained two parts—the first part used non-displacement piles, and the second part used bored piles. The study in this section was based on the project illustrated in Chap. 4, Sect. 4.3.

As shown in Fig. 5.25, the excavation construction required passive piles to resist the lateral loading. As a result of the location of P1 obstructing the Tunnelling Boring Machine (TBM) for tunnelling, where the TBM would have to break through, use of the traditional steel-reinforced piles was not appropriate and even dangerous to do so. The reason is that the TMB can easily be wound with steel reinforcement. The TBM will stop boring underground, and the labour required to untangle the wound steel. In such a case, the potential for loss of life is high because the broken

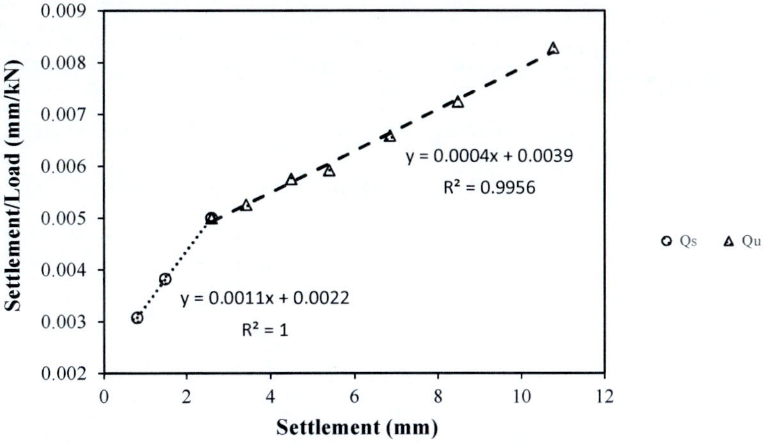

Fig. 5.24 Chin's method of P2018

Table 5.1 Results of piles under different interpretation methods

Methods	Double tangent	Davisson's criteria	Chin's method		
Pile Label	Qu (kN)	Qu (kN)	Qu (kN)	Qs (kN)	Qt (kN)
P116	2600	>4880	10,000	1428.6	8571.4
P115	2400	>4880	10,000	1428.6	8571.4
P2108	850	>1300	2500	909.1	1591.0
P2054	760	>1300	2000	714.3	1285.7

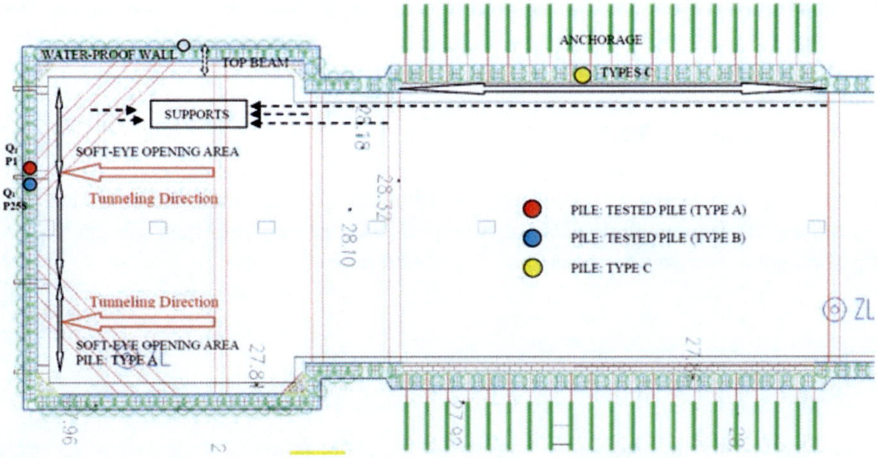

Fig. 5.25 Construction plan map (not to scale)

Table 5.2 Support installation information and excavation duration

Type of support	Support installation date	Excavation date	Excavation duration
Concrete support	18 March 2016	22 April 2016	34 days
Steel support	22 April 2016	27 May 2016	35 days
Steel support	27 May 2016	3 July 2016	36 days
N/A	3 July 2016	29 July 2016	26 days

concrete and the adjacent soil is prone to collapse. By taking advantage of the GFRP property, the project decided to use FRP bar reinforced concrete pile instead of the steel reinforced pile from the soft-eye opening area (Type A). As shown in Fig. 5.25, P1 was a GFRP-reinforced concrete pile, while the adjacent pile P258 was a steel-reinforced pile (Type B). These two piles' deflection were monitored every day by considering the support strut application and excavation process for the deflection behaviour analysis.

To investigate the pile deflection, it is vital to consider the construction process. In this particular construction, three supports were used to help the piles resist the changed earth pressure caused by excavation. The construction process is illustrated as follows: (1) The concrete supports were first installed (no earth pressure) and then the excavation began. These passive piles started to resist the loading caused by soil movement. Note that near pile Type C (Fig. 5.25), anchorage was installed. (2) After 6.5 m of soil was removed, the second support (steel) was installed. Following this, a further 4.8 m of soil was removed, and the third support strut (steel) was installed. (3) After the third steel support was installed, an additional 3 m of soil was removed, and the total excavation depth of 16 m was achieved.

As summarised in Table 5.2, three supports were installed on 18 March, 22 April and 27 May 2016. Once the support installation was finished, the excavation durations were 34, 35 and 36 days, respectively. To monitor the behaviour of the passive pile, with the aim of checking if these passive piles were stable after the final excavation finished, a further 26 days of monitoring was required.

5.3.2 *Subsurface Conditions*

After conducting the laboratory and in-situ tests and interpreting the borehole logs near these tested two piles, the subsurface condition was determined. The simplified soil profile contained seven layers, and the parameters of each layer are summarised in Table 5.3. As shown in Fig. 5.26, these piles reached the seventh layer with embedment of 5.69 m. It should be noted that Table 5.3 was determined from boreholes near these monitored two piles. For other pile investigations, the soil profile may be different.

Table 5.3 Properties of simplified soil layers

Soil layer	Thickness (m)	ω	G_s	ρ	ρ_d	Es	e	w_L	w_P	I_P	I_L	c	φ
		%	–	g/cm^3	g/cm^3	MPa	–	%	%	–	–	kPa	o
Fill	2.5	N/A	N/A	1.8	N/A	N/A	N/A	N/A	N/A	N/A	N/A	N/A	N/A
Loess	2.9	24.5	2.7	1.93	1.55	9.3	0.75	30.7	18.9	11.8	0.55	29	24
Silt	1.2	23.2	2.69	1.96	1.59	15.8	0.69	27.5	18.8	8.7	0.51	19	20
Silty clay	4.5	26.4	2.72	1.95	1.54	9.1	0.77	32.5	19.7	12.8	0.53	36	22
Fine sand	2.6	N/A	N/A	1.96	N/A	N/A	N/A	N/A	N/A	N/A	N/A	0	22
Silty clay	6	27.3	2.72	1.93	1.52	7.2	0.8	34.3	20.7	13.8	0.47	42	16
Clay	4.0	28.6	2.75	1.93	1.51	6.9	0.80	43.5	23.5	20	0.3	38	15

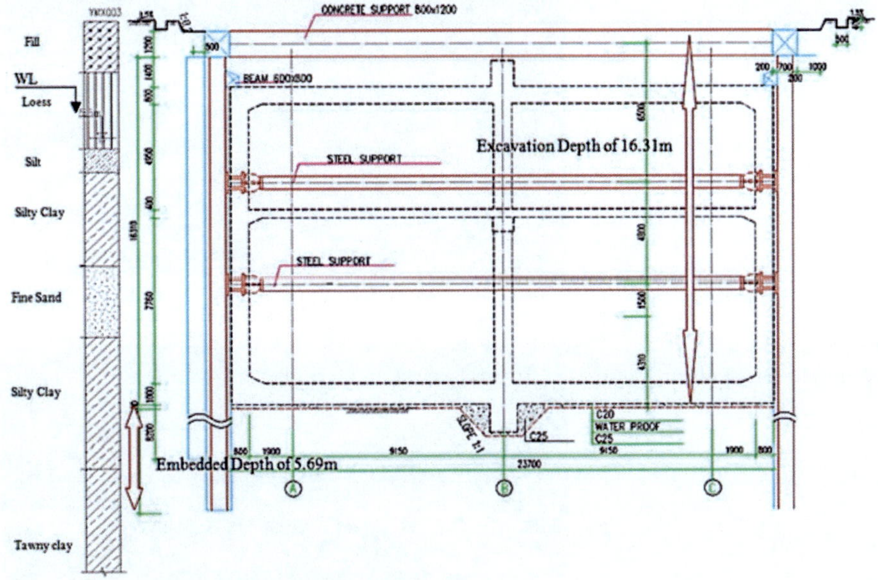

Fig. 5.26 Soil layers adjacent to pile (not to scale)

5.3.3 *Pile Preparation*

These two monitored piles, P1 (GFRP bar reinforcement) and P258 (steel bar reinforcement), were designed to be 22 m long, with pile diameters of 700 mm. The manufacture process of GFRP bar–reinforced concrete piles and steel bar–reinforced concrete piles involves driving holes, assembling reinforcement, and transferring into the holes and casting.

The GFRP stirrups were 10 mm in diameter with spacing of 150 mm (Fig. 5.27). The longitudinal reinforcement was 26 GFRP bars with diameters of 28 mm (Fig. 5.28). The ultimate tensile strength of the GFRP bar was over 500 MPa and the designed grade of the concrete was C30. Based on the local standard of Code for Design of Concrete Structures GB 50010-2002 (2002), the cubic compression strength of C30 was 30 MPa (the dimensions of the cubic concrete sample were $150 \times 150 \times 150\ \text{mm}^3$). The GFRP cage was assembled as shown in Fig. 5.29. The

Fig. 5.27 GFRP stirrups of piles

Fig. 5.28 GFRP reinforcements of piles

Fig. 5.29 Assembled GFRP cage of pile

traditional steel-reinforced concrete pile was also made of C30 concrete. The stirrups were 16 mm diameter HPB300 (hot-rolled plan) bars with spacing of 150 mm. The horizontal reinforcements were 26 HRB400 (hot-rolled ribbed) bars with diameters of 28 mm. The ultimate compression strength of the concrete was 30 MPa, and the yield strengths of HPB300 and HRB400 were 300 and 400 MPa, respectively. The steel cages were welded. In Fig. 5.30, the left side of the figure presents the completed steel cages. Table 6.4 summarises the parameters and prices of the GFRP and steel reinforcements.

As illustrated in Table 5.4, the cost of GFRP material was approximately four times that of the steel reinforcement. If substituting all steel-reinforced concrete piles with GFRP-reinforced concrete piles, this project (516 piles, all with a length of over 20 m) would cost extra ¥3,096,000 (AU$595,384.70) or more. Nevertheless, this GFRP design is underestimated its capacity through applying high safety factor due to its rare usage.

Fig. 5.30 Assembled steel cage of pile

Table 5.4 Parameters and prices of reinforcements

Type of pile	GFRP pile		Steel pile	
Reinforcement	Horizontal reinforcement	Stirrup	Horizontal reinforcement	Stirrup
Brand	GFRP500	GFRP bar	HRB400	HPB300
Diameter	28 mm	26 mm	26 mm	26 mm
Yield strength	500 MPa	300 MPa	400 MPa	300 MPa
Price	¥18.5/m	¥18.5/m	¥4.8/m	¥4.8/m

5.3.4 Test Setup and Procedures

The measurement system to determine the lateral deflection was a plastic PVC tube and a readout probe. The internal diameter of the PVC tube was 75 mm with four grooves cut at 90° intervals. Figures 5.31 and 5.32 illustrate the installation of PVC

Fig. 5.31 PVC tube on GFRP cage

Fig. 5.32 PVC tube on steel cage

Fig. 5.33 Readout probe for deflection measurement of pile

tubes on a GFRP cage and steel cage, respectively.

The readout probe that fit into the grooves is demonstrated in Fig. 5.33. The amount and location of horizontal movement in a deep foundation could be determined through lowering the readout probe into the bottom of the PVC tube and pulling upwards. During system installation, the top of the PVC was covered by a cap to avoid the concrete being poured inside and was then covered again with a weaving bag in case the plastic cap being damaged by coarse aggregate.

After the reinforcement cages were assembled and the PVC tubes were installed, the cages were transferred to the acquired location. The cages of each pile were lifted by two cranes, and were then filled into the guide slots. To ensure the concrete cover met the requirements, concrete cushion blocks were used with spacing of 2 m. Figure 5.34 shown the installation of the steel-reinforced concrete bored passive pile. After the installation of the GFRP and steel reinforcements into the bored holes, fresh concrete was poured into the holes.

Because of the installation of concrete support which ultimate compression strain is very small (0.003), assuming that pile head should not move and the accumulated deflection value changes from pile head to the pile toe. The lateral displacement measurement was recorded every day during the support installation and excavation process.

5.3.5 Results of GFRP Bar–Reinforced Concrete Piles

The concrete supports were cast before excavation. As shown in Fig. 5.35, the excavator dug out soil after the concrete support was cured (cured date: 18 March 2016). It was believed that there was no lateral displacement of the pile at this stage, since

Fig. 5.34 Installation of steel-reinforced concrete pile

Fig. 5.35 Excavation after first concrete support being installed

the earth pressure was relatively small during the first excavation. However, there were three excavators working together, which created extra pressure to push the pile head outside the excavation site.

The obtained lateral displacement of the upper part of the GFRP pile was discovered within +0.5 mm (negative value represents load direction pointing to foundation pit), possibly caused by the excavators. This small value could also have been caused by the accuracy of the inclinometer.

Under the assumption of zero movement of the pile head, the accumulated displacement values were overestimated by calculation, especially from the pile bottom. As shown in Fig. 5.36, the data showed that, during the five days of exca-

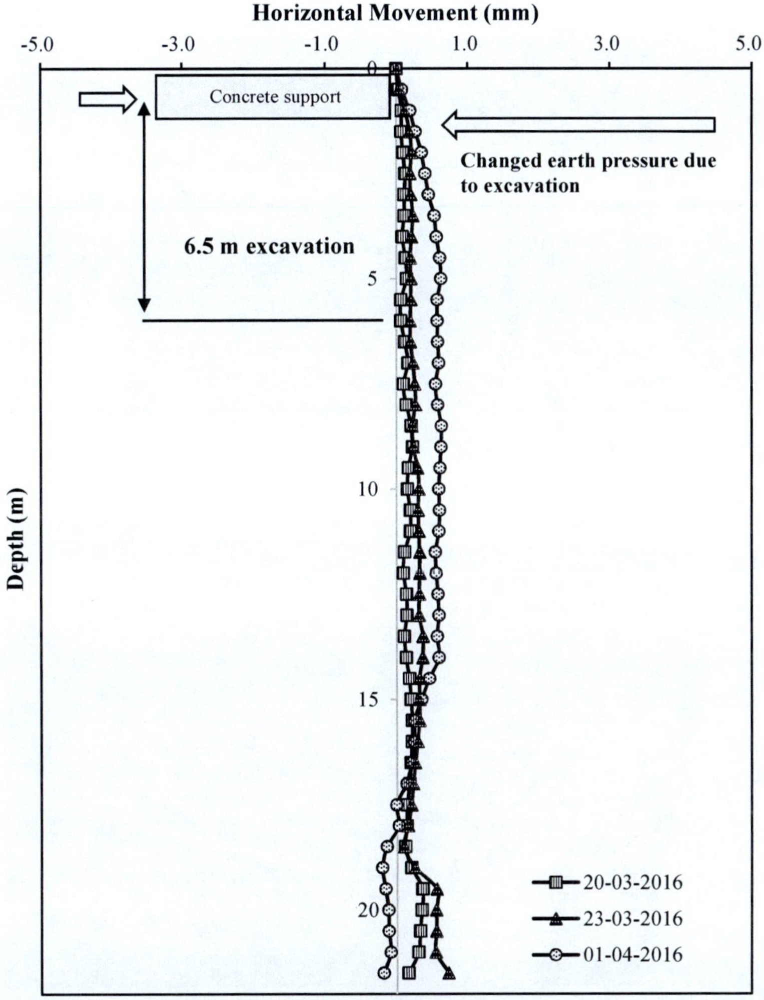

Fig. 5.36 Deflection of CFRP pile at the beginning of excavation

vation from 18 to 23 March 2016, lateral deflection was detected within +0.4 mm from the pile head (error because of equipment or effect of the excavators), and the accumulated lateral deflection was +0.73 mm at the bottom of pile, which was calculated by adding the error values from the pile top.

As the excavation continued, the earth pressure started to increase, which meant that the data could be accurately captured by the equipment. As shown in Fig. 5.36, after 1 April 2016, the recorded data could be trusted because a negative value occurred, which represented that the earth pressure had pushed the pile to the excavation pit.

The first and second steel supports were installed on 22 April and 27 May 2016, respectively, as shown in Figs. 5.37 and 5.38. After these three supports

Fig. 5.37 First steel support applications

Fig. 5.38 Second steel support applications

being installed, the deflection behaviours of the GFRP-reinforced concrete pile were recorded, as shown in Fig. 5.39a, with labels a, b and c representing the collected data after installation of the three supports. As demonstrated in Fig. 5.39b, during 26 days of continuous data collection (no soil excavation), the horizontal movement remained stable.

Through analysis of the data obtained from the inclinometer (Fig. 5.39a), it was detected that the lateral deflection of the FRP-reinforced concrete pile was increasing as the excavation depth was increasing. It was detected that the maximum horizontal deflection was −1.56 mm at the depth of 12 m when the second support was installed (the first steel support). After the first and second steel supports were installed, the maximum deflections were discovered at the depths of 12.5 m, with values of −4.73 and −6.02 mm, respectively.

Additionally, after the last steel support was installed, the data collection continued. The deflection became stable after 3 July 2016 (Fig. 5.39b). It was also detected that the horizontal movement of the pile toe was stable at the value of −1.1 mm, which led to the heave in the excavation pit (also discovered by total station). This phenomenon proved the correct assumption of zero movement from the pile head.

The steel-reinforced concrete pile was also simultaneously monitored. After approximately 15 weeks of monitoring, the lateral deflection data of the traditional steel-reinforced concrete pile obtained from the inclinometer system demonstrated that, after the concrete support and two steel supports being installed, the maximum lateral deflections were −4.70 mm (at the depth of 12.5 m), −7.95 mm (at the depth of 12.5 m) and −10.56 mm (at the depth of 12.5 m), as shown in Fig. 5.40. For comparison, the deflection behaviour of these two piles is summarised in Table 5.5.

5.4 Concluding Remarks

This chapter has investigated composite piles with FRP application. Two types of concrete piles (small and large displacements piles) were examined, with consideration of the CFRP confinement effect. The conventional results such as load-settlement curves and interpretations of the compressive SLTs were presented to determine the ultimate bearing capacity. Furthermore, in this Chapter, another two bored piles were monitored to research the deflection behaviour of the traditional steel reinforced and GFRP-bar reinforced pile foundations. The long-term load deflection behaviour with consideration of support strut installation and excavation process was provided. In addition, the behaviours between the GFRP-bar reinforced concrete pile and traditional steel bar-reinforced concrete pile were compared in the last section.

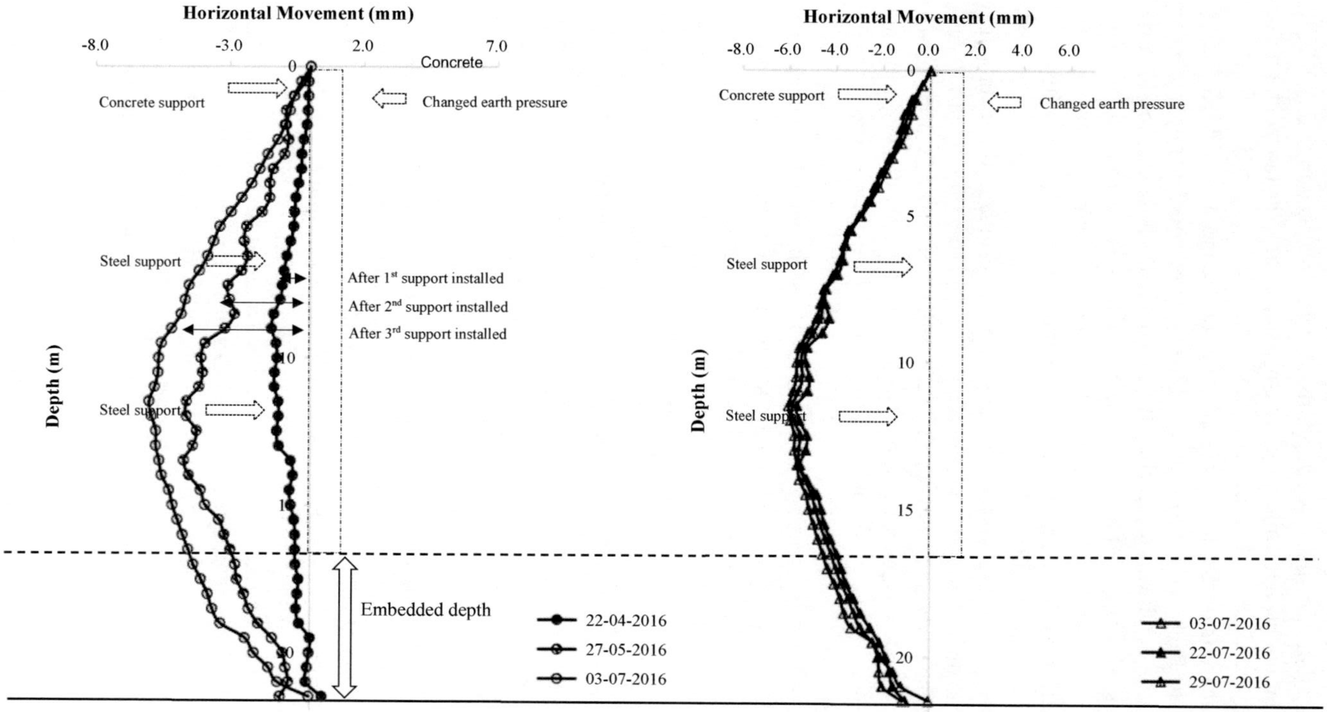

Fig. 5.39 **a** Lateral deflection of GFRP pile during three stages; **b** Total horizontal deflection of FRP pile

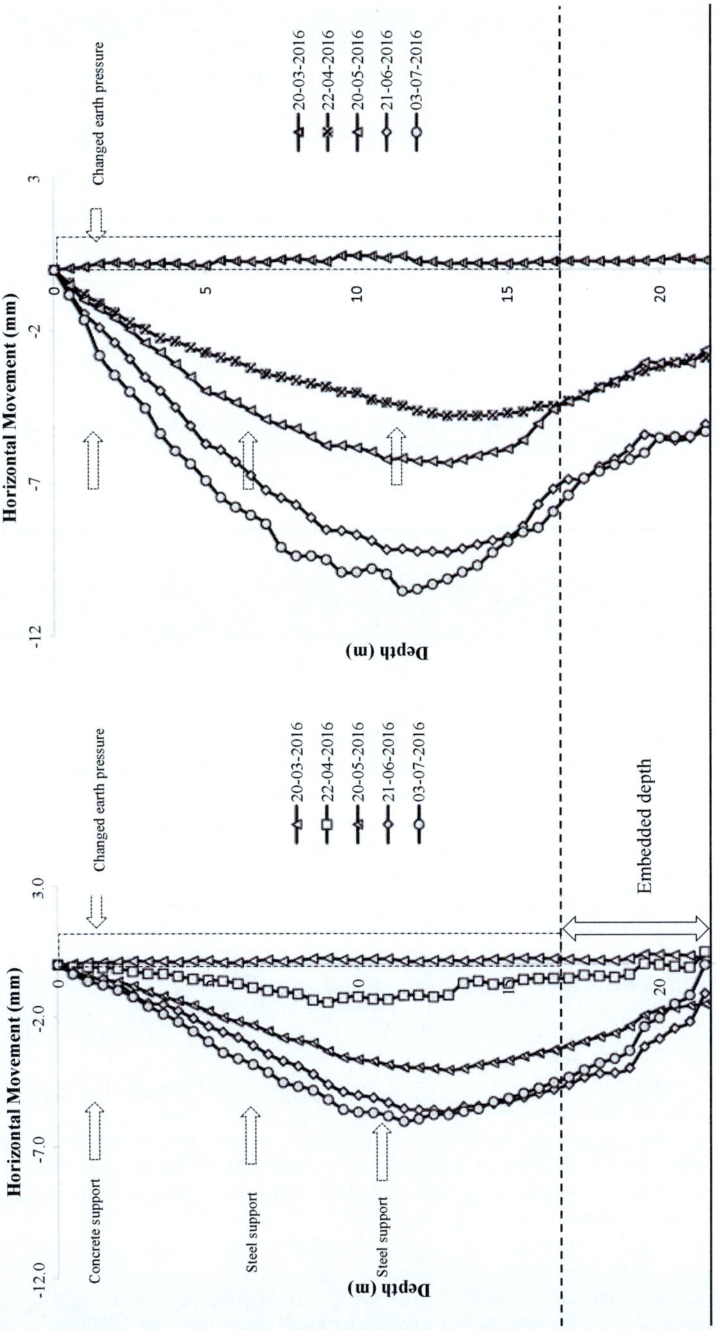

Fig. 5.40 Deflection of **a** GFRP bar–reinforced and **b** steel bar–reinforced concrete bored piles

Table 5.5 Deflection behaviours of CFRP- and steel-reinforced concrete piles

Construction condition (after)	GFRP-reinforced Concrete Pile		Steel-reinforced concrete pile	
	Maximum deflection (mm)	Depth (m)	Maximum deflection (mm)	Depth (m)
First concrete support installed	−1.56	12.0	−4.70	12.5
First steel support installed	−4.73	12.5	−7.95	12.5
Second steel support installed	−6.02	12.5	−10.56	12.5

References

American Association of State Highway and Transportation Officials (AASHTO). (2002). *Standard specifications for highway bridges*, Washington, DC.

ASTM International. (1994). *Standard test methods for deep foundation under static load compressive load (ASTM-D-1143/D-07)* (pp. 1–15).

Bank, L. C., & Wiley, B. (2006). *Composites for construction: Structural design with FRP materials*. Hoboken, N.J: Wiley.

Borel, S., Bustamante, M., & Gianeselli, L. (2004). An appraisal of the Chin method based on 50 instrumented pile tests. *Ground Engineering, 37*(1), 22–26.

Brinch-Hansen, J. (1963). Hyperbolic stress-strain response: Cohesive soils discussion. *American Society of Civil Engineers Journal of Soil Mechanics and Foundation Division, 89*(SM4), 241–242.

Chin, F. K. (1970). Estimation of the ultimate load of piles not carried to failure. In *Proceedings of the 2nd Southeast Asian Conference on Soil Engineering* (pp. 81–91).

Chin, F.K. (1978). Diagnosis of pile condition. In *Guest Lecture on 5th Southeast Asian Conference on Soil Mechanics* (Vol. 9, pp. 85–104). Bangkok: Geotechnical Engineering.

Davisson, M. T. (1972). High capacity piles. In *Proceedings of the soil mechanics lecture series on innovations in foundation construction* (pp. 81–112). IIlinois Section, Chicago: American Society of Civil Engineers.

DeBeer, E. E. (1970). Experimental determination of the shape factors and the bearing capacity factors of sand. *Geotechnique, 20*(4), 387–411.

Guades, E., Aravinthan, T., Islam, M., & Manalo, A. (2012). A review on the driving performance of FRP composite piles. *Composite Structures, 94*(6), 1932–1942.

Hannigan, P. J., Goble, G. G., Likins, G. E., & Becker, M. L. (2016). *Design and construction of driven pile foundations (FHWA-NHI-16-009) (FHWA-NHI-16-009)*. Federal Highway Administration: United States.

Hassan, M., & Iskander, M. G. (1998). State of the practice review in FRP composite piling. *Journal of Composites for Construction, 2*(3), 116–120.

Juran, I., & Komornik, U. (2006). *Behavior of fiber-reinforced polymer composite piles under vertical loads (No. FHWA-HRT-04-107)*. United States Department of Transportation.

Kwan, A. K. H., Dong, C. X., & Ho, J. C. M. (2015). Axial and lateral stress-strain model for FRP confined concrete. *Engineering Structures, 99*, 285–295.

Lo, S. H., Kwan, A. K. H., Ouyang, Y., & Ho, J. C. M. (2015). Finite element analysis of axially loaded FRP-confined rectangular concrete columns. *Engineering Structures, 100*, 253–263.

Luo, L. (2014). Development and application of FRP materials in the structural in China. In *Recent advances in material, analysis, monitoring, and evaluation in foundation and bridge engineering* (pp. 126–132).

Ministry of Construction of the People's Republic of China. (1999). *Standard for soil test method (GB/T 50123-1999)*. Beijing, China: Ministry of Construction of People's Republic of China.

Ministry of Construction of the People's Republic of China. (2001). *Code for investigation of geotechnical engineering (GB 50021-2001)*. Beijing, China: Ministry of Construction of People's Republic of China.

Ministry of Construction of the People's Republic of China. (2002). *Code for design of concrete structures (GB 50010-2002)*. Beijing, China: National Standard of the People's Republic of China.

Ministry of Construction of the People's Republic of China. (2008). *Technical code for building pile foundations (JGJ 94-2008)*. Beijing, China: National Standard of the People's Republic of China.

Ministry of Construction of the People's Republic of China. (2013). *Standard for test methods of engineering rock mass (GB/T 50266-2013)*. Beijing, China: Ministry of Construction of People's Republic of China.

Ministry of Construction of the People's Republic of China. (2014). *Technical code for testing of building foundation piles (JGJ 106-2014)*. Beijing, China: National Standard of the People's Republic of China (pp. 13–51).

Mohammed, B. S., Ean, L. W., & Malek, M. A. (2013). One way RC wall panels with openings strengthened with CFRP. *Construction and Building Materials, 40,* 575–583.

Mosallam, A., Elsanadedy, H. M., Almusallam, T. H., Al-Salloum, Y. A., & Alsayed, S. H. (2015). Structural evaluation of reinforced concrete beams strengthened with innovative bolted/bonded advanced FRP composites sandwich panels. *Composite Structures, 124,* 421–440.

Mostofinejad, D., & Mohammadi Anaei, M. (2012). Effect of confining of boundary elements of slender RC shear wall by FRP composites and stirrups. *Engineering Structures, 41,* 1–13.

Mostofinejad, D., & Tabatabaei Kashani, A. (2013). Experimental study on effect of EBR and EBROG methods on debonding of FRP sheets used for shear strengthening of RC beams. *Composites Part B: Engineering, 45*(1), 1704–1713.

Paikowsky, S. G., & Tolosko, T. A. (1999). *Extrapolation of pile capacity from non-failed load tests* (No. FHWA-RD-99-170). US Department of Transportation.

Pando, M. A., Filz, G. M., & Brown, D. (2004) Performance of a laterally loaded composite pile at the Nottoway River bridge. *Geotechnical Engineering for Transportation Projects* 1317–1326.

Pecce, M. (2001). Structural behaviour of FRP profiles. *Composites in Construction* 241–249.

Robinson, B., & Iskander, M. (2008). Static and dynamic load tests on driven polymeric piles. In *GeoCongress 2008: Geosustainability and Geohazard Mitigation* (pp. 939–946).

Samtani, N. C., & Nowatzki, E. A. (2006). *Soils and foundations—Volume II* (No. FHWA-NHI-16-009). US Department of Transportation.

Thomann, T. G., Zoli, T., & Volk, J. (2004). Lateral load test on large diameter composite piles. *Geotechnical Engineering for Transportation Projects*, 1239–1247.

Weaver, T. J., Ashford, S. A., & Rollins, K. M. (2008). Lateral load behavior of a concrete-filled GFRP pipe pile. *GeoCongress, 2008,* 931–938.

Xiao, D., & Yang, H. (2011). Back analysis of static pile load test for SPT-based pile design: A Singapore experience Advances in Pile Foundations. *GeoHunan International Conference 2011* (pp. 144–152).

Youssf, O., ElGawady, M. A., & Mills, J. E. (2015). Displacement and plastic hinge length of FRP-confined circular reinforced concrete columns. *Engineering Structures, 101,* 465–476.

Zhang, B. D., Sun, J. Z., & Song, F. (2001). Briefing the application of resin material to boring tunnels. *Coal Technology, 20*(2), 61–62 (In Chinese).

Chapter 6
Field Tests of Super-Long and Large Diameter Piles

6.1 General Introduction

A compressive loaded pile is a slender element that transfers the loading from the upper structure to the soil layers. Nowadays, because pile foundations with large capacities are required, increasing numbers of super-long and large-diameter piles are being designed for high-rise buildings and bridge projects. Usually, in China, a pile longer than 50 m with a diameter greater than 0.8 m is recognised as a super-long and large-diameter pile foundation (Yang and Zhong 1998). Lin et al. (1999) pointed out that the pile with length to diameter ratio of 100 or more, though the length may be less than 50 m, can also be deemed to super long pile in practice.

Numerous studies have been conducted to examine the compressive loaded piles including encompassing H-steel piles, pretensioned spun high-strength concrete (PHC) piles and cast-in-situ piles with post grouting (Zhou et al. 2012a, b; Wang et al. 2015; Feng et al. 2016, Li et al. 2016; Michael and Jae 2017). Most of the research has focused on the behaviour of these piles with respect to different geotechnical conditions, driving process, analytical method, theoretical load transfer method and finite element method (Zhou et al. 2014; Zhang et al. 2014; Zou and Zhao 2016, Sun et al. 2016; Oliaei and Ghotbi Siabil 2017; Lu et al. 2019). Previous investigations also include self-balanced testing method, determination of the effective length, effect of soil stiffness and negative skin friction of the large diameter and supper long pile foundation (Zhang et al. 2012, 2014; Zhou et al. 2015; Xing et al. 2017).

Static field tests are one of the most accurate methods to provide valuable information, such as pile capacity and the load transfer mechanism. However, it is preferable to perform these tests on piles less than 40 m in length. For a super-long pile, the capacity always achieves up to 10,000 kN, which means that more than 1000 t of reaction system are required. Thus, when using a weighted platform, the platform is unsafe and prone to collapse. Moreover, if using reaction beams, the beams are prone to fail because the bending movement is too great. Under such conditions, other test methods are preferred, such as dynamic load tests and O-cell tests. However, these

J. Zhou and E. Oh, *Full-Scale Field Tests of Different Types of Piles*,
Advanced Topics in Science and Technology in China 62,
https://doi.org/10.1007/978-981-33-6183-6_6

tests have some limitations. Some types of pipe piles or H-steel piles, which do not contact the rock, may behave different under dynamic and static loading conditions (Hannigan et al. 2016). For O-cell testing, many engineers believe that the method is too costly because the O-cell is cast inside the pile (and is not recyclable) and the pile will be damaged after load is applied.

This chapter presents a case study referring to super-long and large-diameter piles, and introduces an improved static load test (SLT) method. In this research, three drilled shaft pile foundations with various lengths and diameters were tested in a bridge project. Different to the previous chapters which focused on the compressive, uplift or horizontal ultimate bearing capacity, this chapter however, aims to study the behaviour relating to load transfer mechanism and shaft resistant development and distribution of the super-long and large diameter pile foundation. It is expected that the results of this field tests' research will provide practical information for geotechnical engineers.

6.2 Compressive Static Load Tests

6.2.1 Geotechnical Conditions

A bridge project was constructed in Henan Province, China. The subsurface exploration was conducted based on the Specifications for Highway Engineering Geological Remote Sensing (JTG/T C21-01-2005 2005) and Standard for Soil Test Method (GB/T 50123-1999 1999). Based on the interpretation of borehole logs, the soil layers near these three tested piles were determined as illustrated in Fig. 6.1. It was found that the main layers were dense to very dense sand and silty sand. Further, it was found that the thickness of the soil layers varied, and the soil layers were loose in the very upper region and dense to very dense in the rest of the region. As shown in Fig. 6.1, all three piles were embedded in the bearing stratum with depth of 1 m. The bearing stratum are dense sand, very dense grait and dense grait for piles with labels of P12, P66 and P105, respectively. Further, in order to determine the shaft resistance, strain gauges were installed in each soil layer. An example in relation to the strain gauge location is also demonstrated in this figure.

6.2.2 Description of Tested Drilled Shaft Piles

Figure 6.2 presents information of the tested piles. As shown in the figure, the diameter of P12 and P66 are the same of 1500 mm, and the diameter of P105 is 1800 mm. The pile lengths varied, the shortest pile is 52 m, the medium length pile is 73 m, and the longest pile is 83 m. All pile foundations were made of concrete with the same concrete strength of 30 MPa. The designed capacity for the P12, P66 and P105

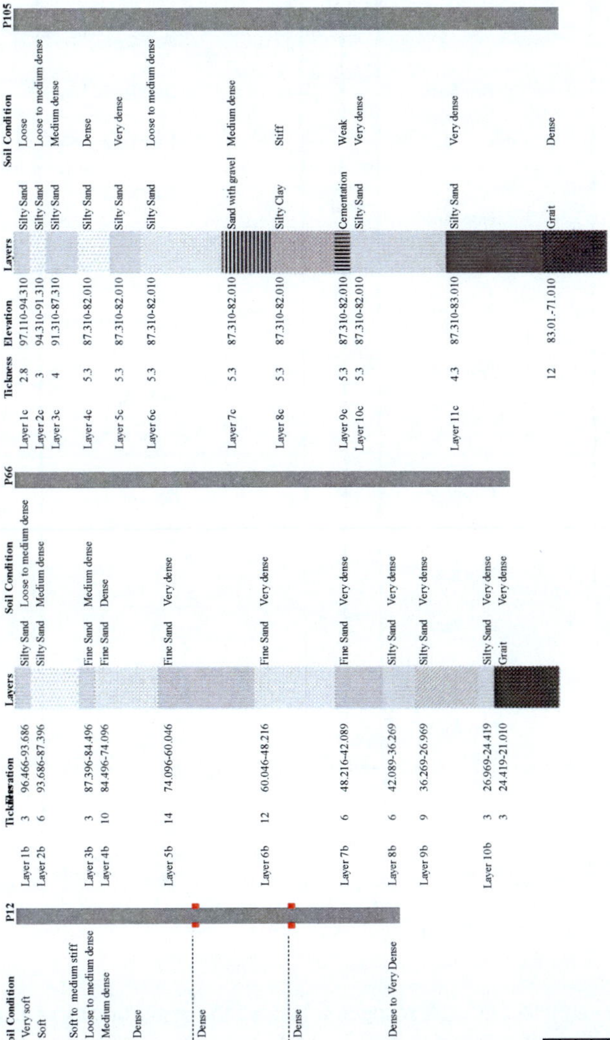

Pile Label	Dia.	Length	Concrete Strength
	(mm)	(m)	(MPa)
P12	1,500	52	30
P66	1,500	73	30
P105	1,800	83	30

Fig. 6.1 Subsurface conditions and tested piles (not to scale)

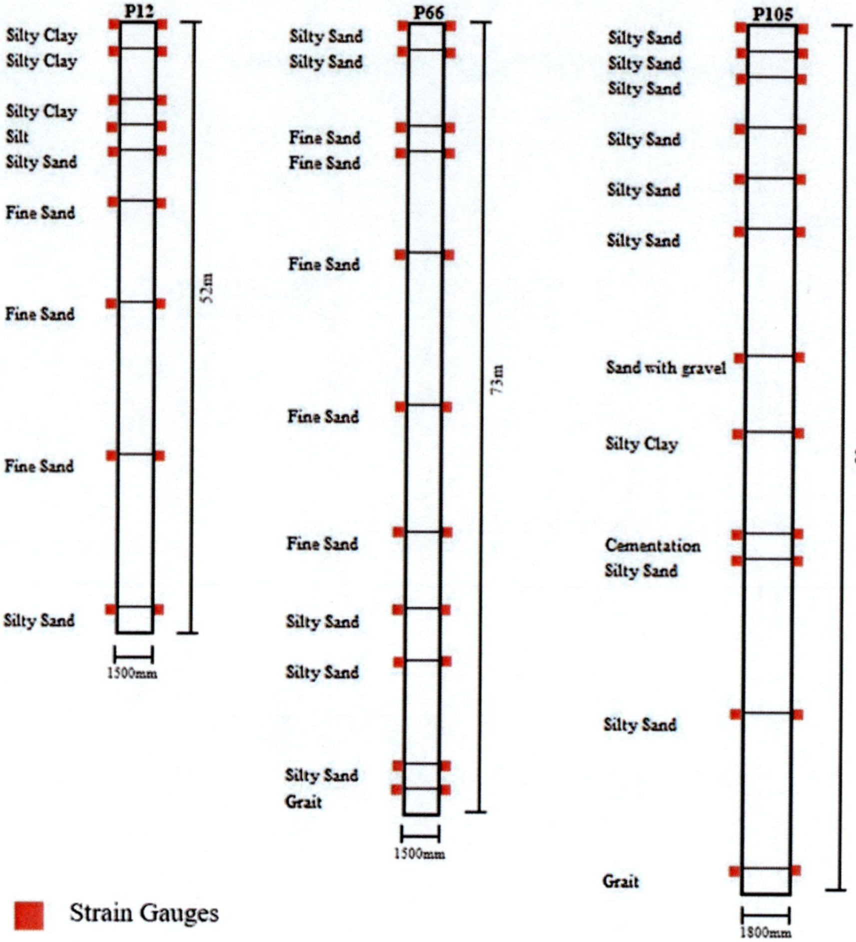

Fig. 6.2 Simplified diagram of the tested piles (not to scale)

are 28,000, 40,000 and 64,000 kN, respectively, and the reaction frame capacity is designed to be 1.2 times the designed capacity which are 33,600, 48,000 and 76,800 kN, respectively.

The locations of the strain gauges are also illustrated in Fig. 6.2. These locations are selected according to the soil layers' locations. All the strain gauges are installed symmetrically in case one of the strain gauges being failed (if a strain gauge failed, the other symmetrically installed strain gauge with the same depth can be used for data collection; if there is no failure of strain gauges, the average values can be used). The telltale is also installed in each pile so that the pile end displacement can be determined.

In each pile, the reinforcement ratio from the upper part is twice the lower part. The diameter of the auxiliary steel bar and the main longitudinal steel are all 28 mm in diameter, and the diameter of the stirrup is 25 mm (the upper part spacing is 100 mm, other part is 150 mm). Though total reinforcement ratio of these three tested piles are different, the reinforcement effect is ignored in this research. This is because this case study is researching the capacity under compressive loads, all these applied loads are resisted by concrete material, the compressive contribution by the steels are very small.

6.2.3 *Test Setup*

6.2.3.1 Reaction System

The traditional reaction system (SLT) is mostly used to test piles with a capacity lower than 10,000 kN (or 1000 t) to avoid the reaction beams suffering too much bending moment. In this case study, the maximum reaction frame capacity required 76,800 kN (1.2 times the maximum test load); thus, the traditional reaction beam could not be used. Under this condition, a reaction system was designed. As shown in Fig. 6.3, this reaction system contained eight anchoring piles that were connected with the concrete beams and steel strand. The load on the reaction device (provided by the hydraulic jacks) could be resisted by the steel strand, and the strand force could be resisted by the vertical friction force (by an anchoring pile) and horizontal force (by a concrete beam). There was a small bending moment acting on the reaction device; hence, the reaction frame capacity was improved. As shown in Fig. 6.4, hydraulic jacks were used to provide loads, and a reaction device was employed. Two reference steel beams were also used, so that dial gauges could be installed.

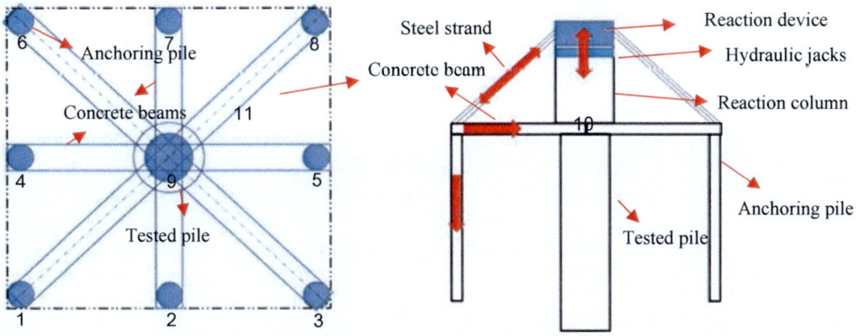

(a) Plain view of test set-up (not to scale) (b) Mechanical mechanism

Fig. 6.3 Reaction system design

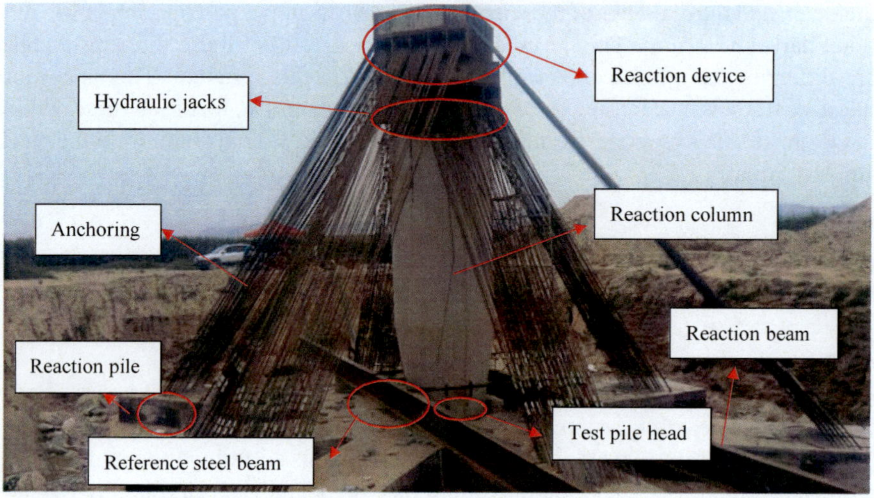

Fig. 6.4 Anchoring reaction system

6.2.3.2 Experimental Principle

The strain of the rebar can be determined through the ratio between the stress and modulus of the rebar, as illustrated in Eq. (6.1). The stress of the rebar at a time of i can be calculated by using the force values transferred through the rebar over the rebar cross-sectional area, as depicted in Eq. (6.2). Before the pile foundation is loaded, the initial coefficients (f_0 and T_0) should be determined. As shown in Eq. (6.3), the force transferred from the pile can be determined after recording the coefficients at a time of i. It should be noted that these two initial factors should be recorded after the concrete is cured and the test set-up is finished.

$$\varepsilon_{ri} = \sigma_{ri}/E_r \tag{6.1}$$

$$\sigma_{ri} = P_{ri}/A_{ri} \tag{6.2}$$

$$P_{ri} = K\left(f_i^2 - f_0^2\right) + K_T(T_i - T_0) \tag{6.3}$$

where:

ε_{ri} = the strain at a time of i developed inside the rebar;

σ_{ri} = the rebar stress at a time of i from the cross area of A_{ri} (kN/m^2);

E_r = individual rebar modulus provided from the manufacturer (kN/m^2);

P_{ri} = vertical force transferred to the rebar (kN);

A_{ri} = area of the rebar at a time of I (m^2);

K = calibration coefficient of rebar (kN/Hz^2);

K_T = temperature compensation factor (kN/°C);

f_0 = initial frequency recorded from rebar (Hz);

f_i = frequency recorded at a time of i (Hz);

T_0 = initial temperature (°C);

T_i = temperature at time of i (°C).

Theoretically, as shown in Eq. (6.4), the strain of the rebar is equal to the strain of the steel reinforcement and the concrete strain after the loading is applied because these three materials act as one element. The concrete cross-sectional axial force can be determined based on Eqs. (6.1)–(6.4), as illustrated in Eq. (6.5). The force difference (ΔP_i) between different concrete cross-sections represents the soil friction forces, and the friction stress between the concrete pile and soil can be determined using force difference divided by friction area (ΔA_{sh}), as depicted in Eq. (6.6).

$$\varepsilon_{ci} = \varepsilon_{ri} = \varepsilon_{si} \tag{6.4}$$

$$\begin{aligned} P_i &= A_{ci}E_c\varepsilon_{ci} + A_{si}E_s\varepsilon_{si} = \frac{E_cA_{ci} + E_sA_{si}}{E_sA_{ri}}P_{ri} \\ &= \frac{E_cA_{ci} + E_sA_{si}}{E_sA_{ri}}\left[K\left(f_i^2 - f_0^2\right) + K_T(T_i - T_0)\right] \end{aligned} \tag{6.5}$$

$$q_s = \Delta P_i / \Delta A_{sh} \tag{6.6}$$

where:

P_i = concrete cross-sectional axial force.

$A_{ci}, E_c, \varepsilon_{ci}$ = concrete section area, modulus and strain at a time of i, respectively.

$A_{si}, E_s, \varepsilon_{si}$ = steel section area, modulus and strain at a time of i, respectively.

q_s = friction stress between concrete pile and soil (kN/m^2).

ΔP_i = force difference between two cross-section axial forces (kN).

ΔA_{sh} = column friction area (m^2).

6.2.3.3 Construction Process

Figure 6.5 displays the preparation of the pile test, eight anchor piles were initially cast (Fig. 6.3a), and later the test pile was cast. After 28 days of concrete curing,

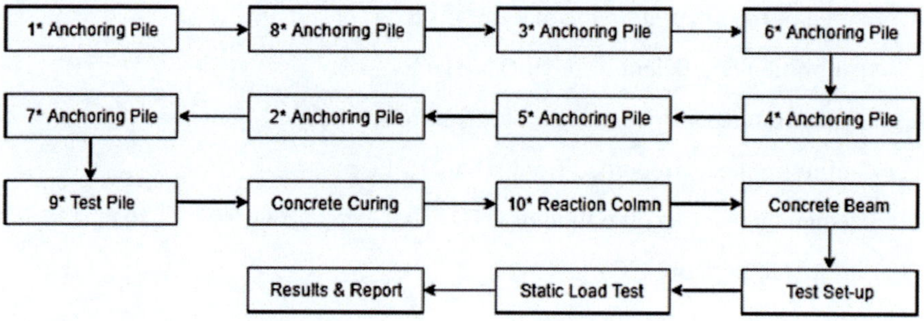

Fig. 6.5 Construction process

the reaction column was cast. Later, the reaction column, concrete beam, anchoring strand and test pile were connected, as demonstrated in Fig. 6.3b. After the installation of hydraulic jacks and dial gauges, the compressive SLT could be performed and, finally, the results could be obtained.

6.2.4 *Test Process*

The maintained compressive load tests were selected to perform. As shown in Table 6.1, there were 14, 10 and 16 loading stages for P12, P66 and P105, with the first loadings of 4000 kN, respectively. The number of unloading stages was half of the loading stages, and the decrement load was twice the increment load. During each loading stage, the vertical settlement was recorded at time intervals of 15 min, and when the accumulated time exceeded to one hour, the time interval changed to 30 min. For the unloading stages, the time interval was 30 min. Among these recorded values during loading stages, if the difference settlement value was <0.1 mm, it could be primarily recognised as a stable condition. If the difference value of 0.1 mm occurred again, it could be viewed as stable and the next load could be applied.

Table 6.1 Information of test process

Pile label	Loading stage	Unloading stage	First load (kN)	Increment (kN)	Decrement (kN)
P12	14	7	4000	2000	4000
P66	10	5	4000	4000	8000
P105	16	8	4000	4000	8000

6.3 Test Results and Discussion

6.3.1 Settlement–lg Time and Load–Settlement Curves

Figures 6.6, 6.7, and 6.8 display the settlement–lg time of the three tested piles. As indicated in these figures, there were no points that illustrated a dramatic turning trend, and all settlement values during each applied load were close, which demonstrated that these pile foundations were stable during each applied load.

As shown in Fig. 6.6, after the maximum loading of 28,000 kN was applied, the maximum vertical pile head settlement of P12 was determined as 87.42 mm. Comparing to P66, as demonstrated in Fig. 6.7, the stable settlement of the P66 was found as 22.92 mm when loads of 28,000 kN was applied and the total head settlement was small (55.49 mm) when maximum loads of 40,000 kN was applied. This represents that the capacity of P66 is greater than P12. Also, this phenomenon also illustrated that increasing the pile length, especially when reaching a harder bearing stratum, can effectively increase the ultimate bearing capacity of the single pile.

As shown in Fig. 6.8, when loading of 28,000 kN was applied, the pile head displacement of P105 was determined as 17.10 mm, which was less than P66 (22.92 mm) and P12 (87.42 mm). Furthermore, when loads of 40,000 kN was applied, the vertical settlement of P105 was determined as 27.33 mm, which was also less

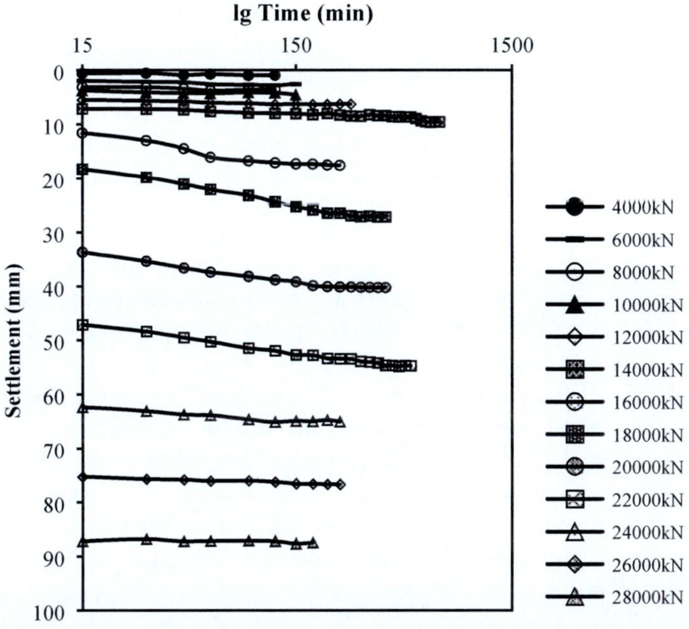

Fig. 6.6 s–lgt results of P12

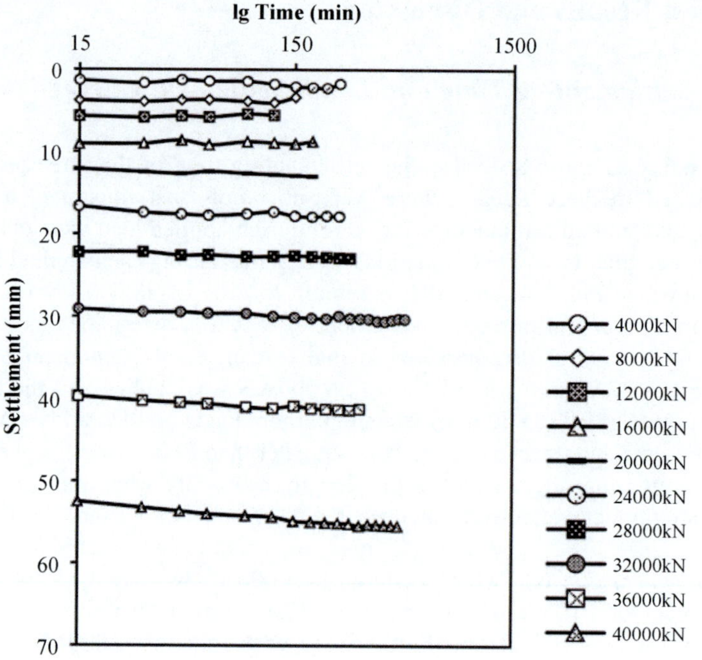

Fig. 6.7 s-lgt results of P66

than the settlement of P66 (55.49 mm). This is primarily due to the increment of the pile diameter and pile length that improves the total capacity of a single pile. In addition, when the pile was loaded 64,000 kN, the total settlement of P105 was determined as 45.07 mm as shown in Fig. 6.8, this represents that this pile possesses greatest load capacity. From another point of view, the design of this pile foundation is too conservative (45.07 mm < 0.1 * D = 0.1 * 1800 mm = 180 mm).

The load–settlement curves of these tested piles in relation to the pile toe, pile head and pile deformation are presented in Figs. 6.9, 6.10 and 6.11. As shown in Fig. 6.9, after the maximum loads were applied, the pile toe displacement of P12 was determined as 74.2 mm, which represents that the compression of concrete is very small. This is because the bearing stratum being the silty sand (compressible) as depicted in Fig. 6.1. This also illustrates that the loads have transferred to the pile toe and the transferred loads are resisted by the bearing stratum.

As shown in Fig. 6.10, the pile toe movement of P66 was determined as 21.73 mm (ratio between toe and head settlement: 21.73/55.49 = 39%). Two reasons may have led to this phenomenon, the first one is that the bearing stratum is the very dense grait; the second reason is that the shaft friction decreases the transferred load (P66 is longer than P12).

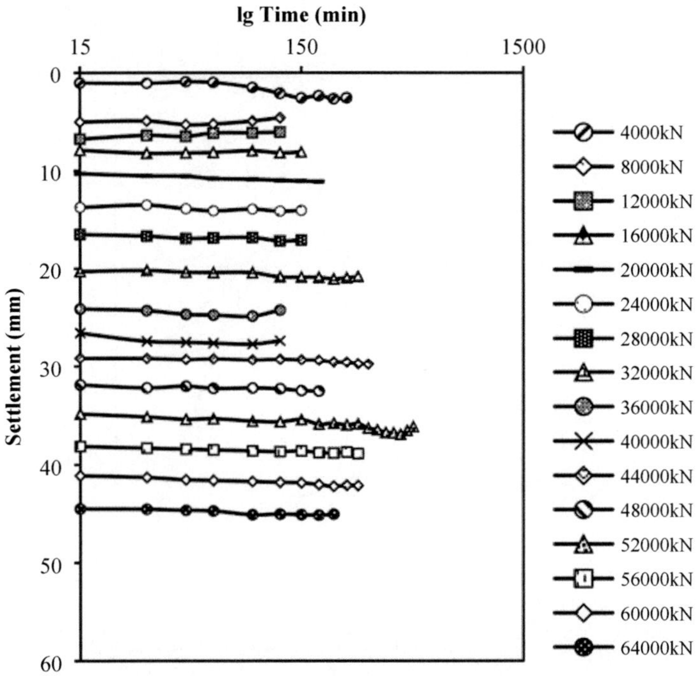

Fig. 6.8 s-lgt results of P105

Fig. 6.9 Load–settlement curves of P12

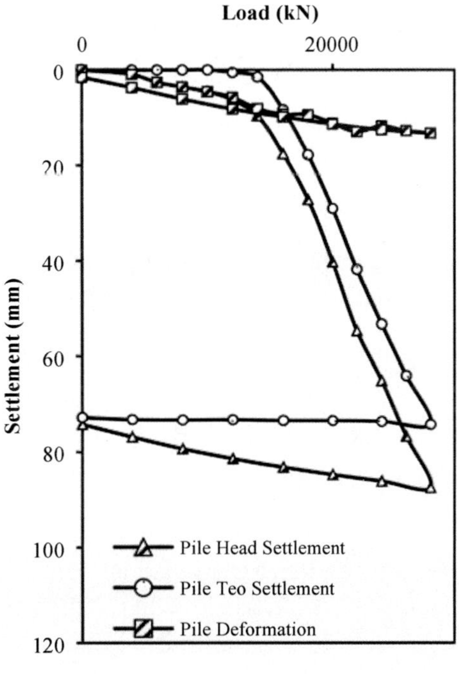

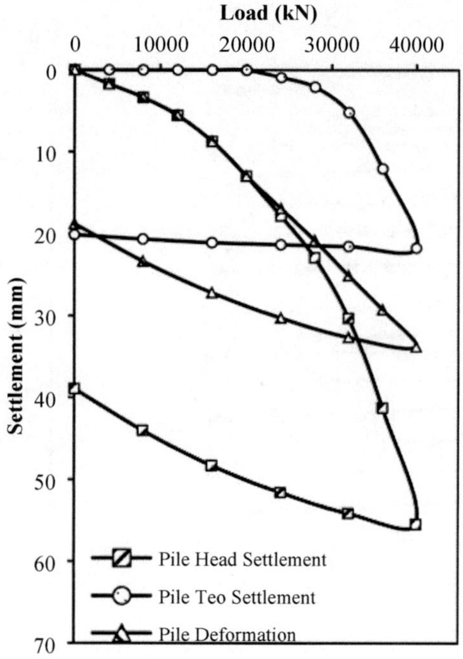

Fig. 6.10 Load–settlement curves of P66

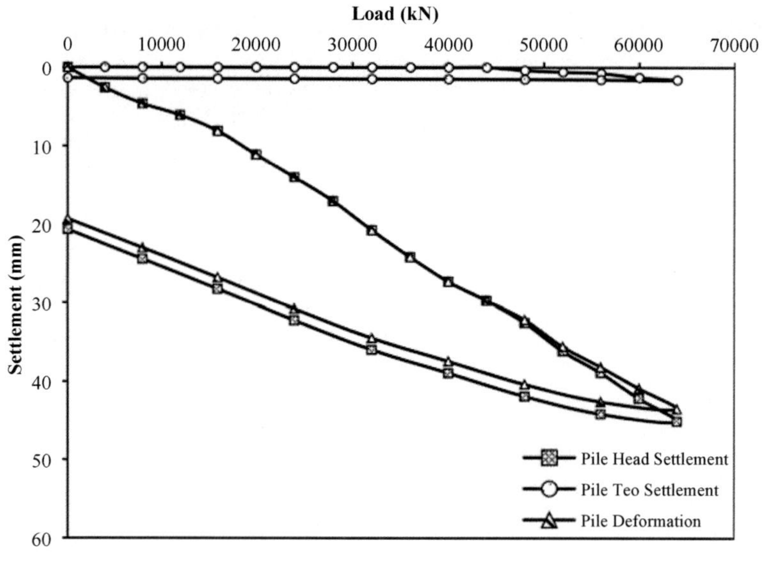

Fig. 6.11 Load–settlement curves of P105

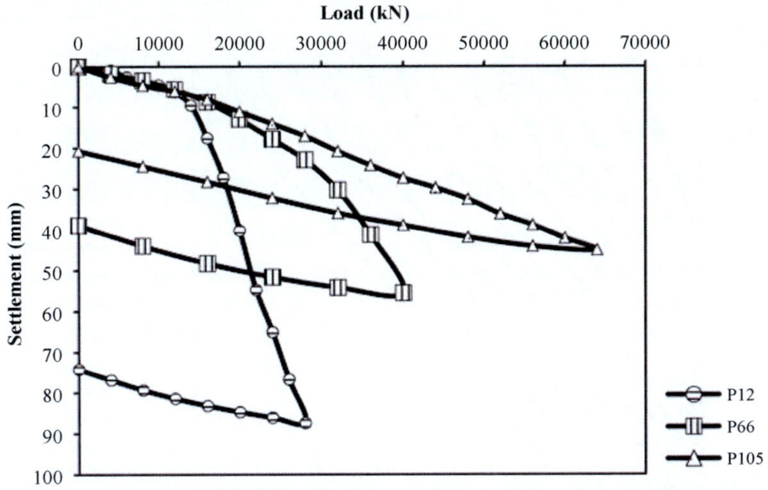

Fig. 6.12 Load–settlement curves of tested piles

As demonstrated in Fig. 6.11, the pile toe displacement of P105 was very small (only 1.59 mm). This illustrates that the load did not transfer to the pile toe because this pile is too long, in other word, this pile is designed very conservative.

The load–settlement curves of these tested piles in relation to the pile head are shown in Fig. 6.12. It can be found that, after the maximum loadings of 28,000, 40,000 and 64,000 kN were applied, the maximum settlements of P12, P66 and P105 were determined to be 87.42, 55.49 and 45.07 mm, respectively. This demonstrates that the pile capacity of P105 is greater than that of the other two piles, and the smallest capacity is for P12. Further, it was found that, for P105, a linear line was observed (close to elastic deformation of PL/EA), which represented that the settlement of P105 was mainly due to compression deformation. Given that the deformation of the concrete element is the main factor indicating the ultimate capacity of a pile (P105), the ultimate bearing capacity would not effectively increase if applying base grouting technology.

6.3.2 Load Transfer Mechanism

The load transferred to the pile cross-section along the pile length can be determined using Eq. (6.5). Figures 6.13, 6.14, and 6.15 present the load transfer mechanism of these three tested piles. It can be seen that the transferred load decreased along the pile length during each applied load, but the gradients were different. This was because the pile shaft area and soil layers were different.

As shown in Fig. 6.13, the change of the axial force of P12 from 0 to 15 m was smaller than rest of the part. This was because the layers from 0 to 15 m were mostly silty clay with soil condition of 'very soft and soft'. Also, it can be found that below

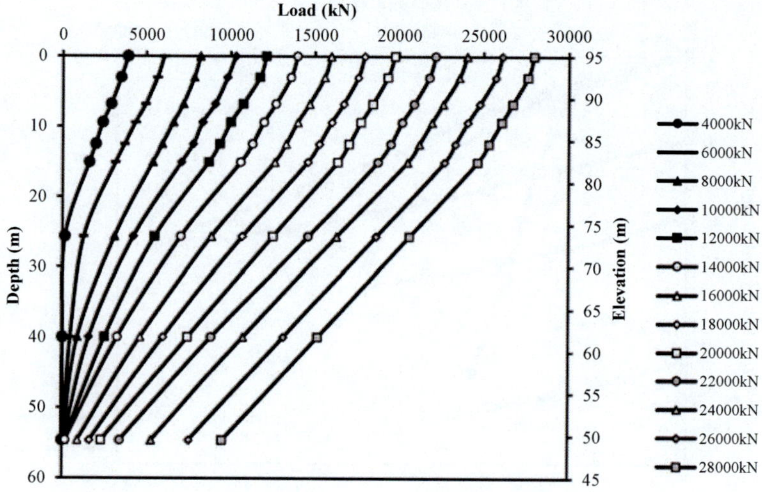

Fig. 6.13 Load transfer mechanism of P12

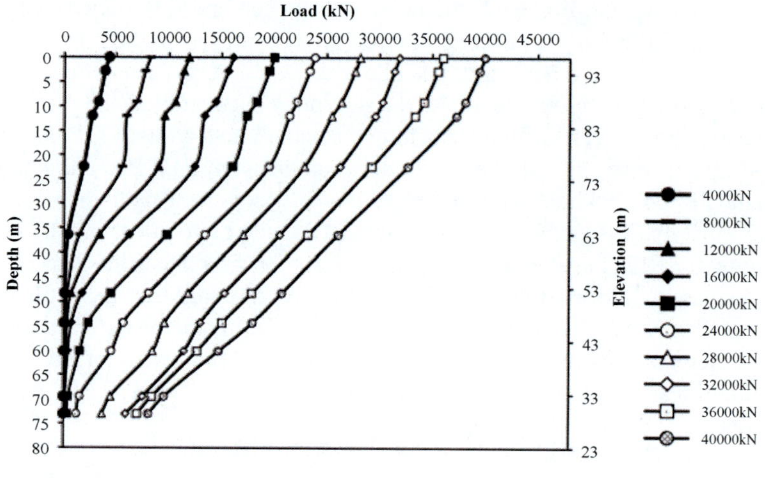

Fig. 6.14 Load transfer mechanism of P66

15 m, the gradients from each layer were similar, this was due to similar soil layers and conditions being existed (Fig. 6.1, dense fine sand). As shown in Fig. 6.14, the gradients of P66 from 0 to 12 m was greater than the rest part, this was because the existence of loose sand and soft clay. As shown in Fig. 6.15, the gradient from 0 to 10 m was large, which illustrates small change of axial force. This was because the loose to medium dense sand existed in this range. Also, the smallest gradient was found from 30 to 38 m, this was due to the stiff soil layer existing in this range and more shaft resistance was provided by this layer.

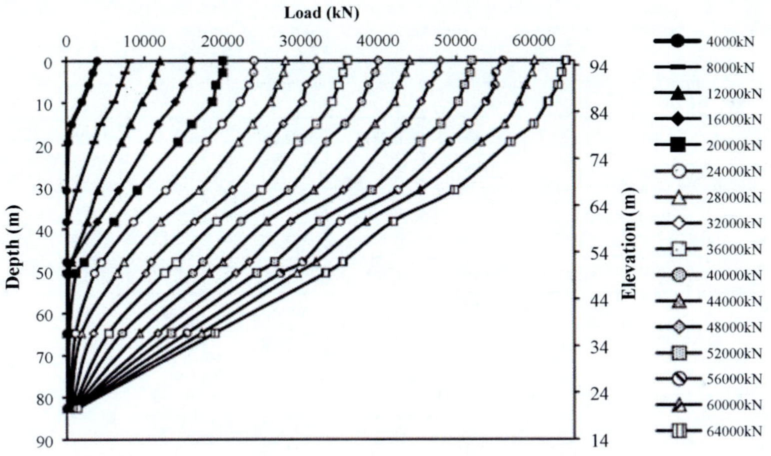

Fig. 6.15 Load transfer mechanism of P105

For P12 and P66, the bearing capacity included end bearing as demonstrated in Figs. 6.13 and 6.14, respectively. Further, it can be found that the end bearing percentage of P12 (9523/28,000 = 34%) was greater than P66 (8117/40,000 = 20.2%), which represented that the increase of pile length can make contribution to the increase of the shaft resistance and hence reduce the transferred loads to the bearing stratum. However, for P105, the end bearing was very small. This indicated that P105 was a shaft-resistant pile (Fig. 6.15). Under this condition, seeking to improve the ultimate bearing capacity by base grouting would be pointless for P105. As illustrated in Fig. 6.1, the soil layers around P105 were mostly sandy material; thus, shaft grouting technology was preferred to increase the total ultimate bearing capacity. In addition, since no loads were transferred to the pile end, increasing the pile length would not increase the pile capacity, and the shaft resistance near the pile toe would not be fully developed.

Table 6.2 provides the proportion of shaft and toe resistance under each working load. It can be seen that, for P105, a very small end bearing was observed. For P12 and P66, before R_W was <80%, the shaft and toe resistance proportions were similar. Further, when R_W was 100%, the shaft resistance proportion of P12 was less than P66, which was mainly because the pile length of P66 was greater than that of P12.

6.3.3 Shaft Resistance Development and Distribution

Figures 6.16, 6.17, and 6.18 present the load–shaft resistance curves of the three tested piles in relation to each soil layer. As shown in Fig. 6.16, for P12, it was observed that, for most layers, when increasing the load, the shaft resistance developed from zero to a maximum value and was then maintained. Moreover, the shaft resistance

Table 6.2 Proportion of shaft and toe resistance under working load

P105			P66			P12		
R_W (100%)	R_E (100%)	R_S (100%)	R_W (100%)	R_E (100%)	R_S (100%)	R_W (100%)	R_E (100%)	R_S (100%)
6.25	0.00	100.00	10.00	0.00	100.00	14.29	0.00	100.00
12.50	0.00	100.00	20.00	0.19	99.81	21.43	0.00	100.00
18.75	0.00	100.00	30.00	0.39	99.61	28.57	0.63	99.38
25.00	0.13	99.87	40.00	0.66	99.34	35.71	0.83	99.17
31.25	0.21	99.79	50.00	1.80	98.20	42.86	1.15	98.85
37.50	0.27	99.73	60.00	5.18	94.83	50.00	1.46	98.54
43.75	0.30	99.70	70.00	13.32	86.68	57.14	5.84	94.16
50.00	0.33	99.67	80.00	18.70	81.30	64.29	9.25	90.75
56.25	0.53	99.47	90.00	19.56	80.44	71.43	11.84	88.17
62.50	0.75	99.25	100.00	20.29	79.71	78.57	15.76	84.24
68.75	0.92	99.08				85.71	22.18	77.83
75.00	1.02	98.98				92.86	29.07	70.93
81.25	1.19	98.81				100.00	34.01	65.99
87.50	1.26	98.74						
93.75	1.47	98.53						
100.00	1.80	98.20						

Note R_W = working load proportion; R_E = end bearing percentage; R_S = shaft resistance percentage

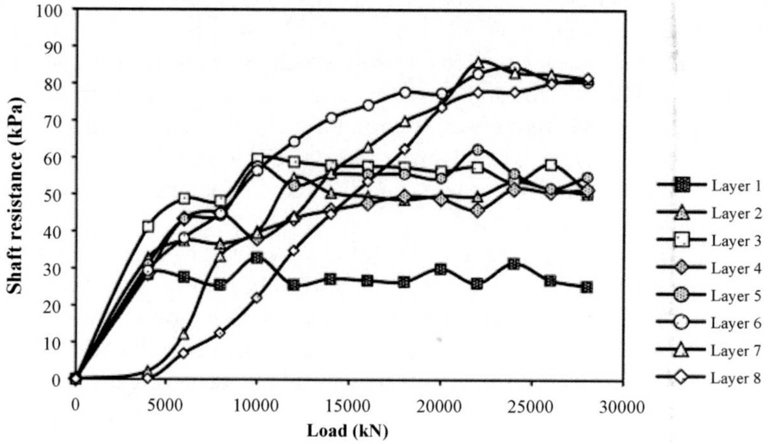

Fig. 6.16 Shaft resistance development of P12

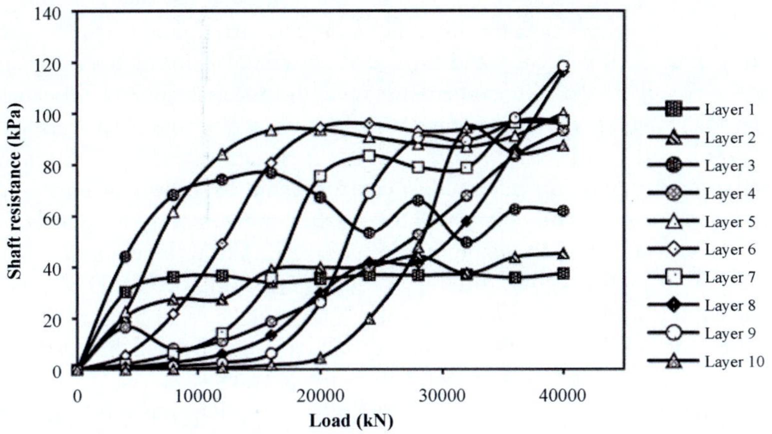

Fig. 6.17 Shaft resistance development of P66

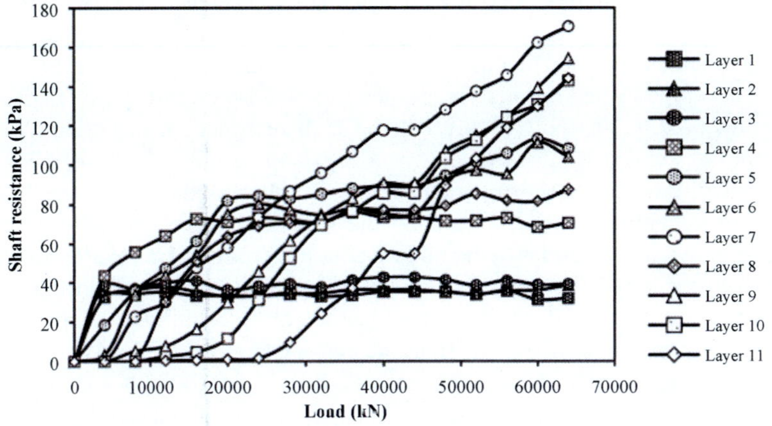

Fig. 6.18 Shaft resistance development of P105

from the upper layers developed first (the shaft resistance from layer 7 and layer 8 did not well developed before loading of 8000 kN was applied). Also, it can be found that, after loads of 8000 kN were applied, the shaft resistance from layer 6, 7 and 8 increased dramatically, and the shaft resistance of layer 6 and 7 maintained after loads 22,000 kN were applied, but the resistance was keep going for layer 8, which represented that the total shaft resistances were almost fully developed.

As illustrated in Fig. 6.17, similar to P12, the shaft resistance of P66 from the upper layers developed first. The shaft resistance from layer 1, 2 and 3 was fully developed after 8000 kN, and the resistance from layers 8, 9 and 10 started to increase after 16,000 kN was applied. In addition, it was found that some curves fluctuated. As for layer 3, the shaft resisting softening was found when loading of 16,000 kN was applied, and after 24,000 kN, the stress hardening occurred. The reason was that the pile was compressed vertically and expanded horizontally during load was applied, and the sandy soil of layer 3 (medium dense) near the pile rearranged the location. The slippage of the sand particles decreased the development of the shaft resistance, but later, the soil particles became stable, the shaft resistance developed. Interestingly, the soil softening and soil hardening occurred again as shown in the figure, the reason may be related to the slurry wall thickness. Unfortunately, that data is limited, further research referring to the amount of slurry admixture is required. It is worth to notice that, from the second softening of layer 3 (loading of 32,000 kN), the shaft resistance of layer 10 increased dramatically, and from the second hardening of layer 3 (loading of 36,000 kN), the shaft resistance of layer 10 decreased dramatically. This phenomenon can be explained as 'Mutual Compensation'.

As demonstrated in Fig. 6.18, it is evidently to see that the shaft resistance of layer 1, 2 and 3 were fully developed after 4000 kN, and the shaft resistance of 4, 5, 6, 8 were fully developed after 20,000 kN. Later, with the increasing of pile head loading, the layers of 8, 9, 10 and 11 played a significant role in resisting the transferred load. Also, it can be seen that the shaft resistance of layer 7 was keep going to increase (though tress softening occurred which was due to slippage of sand and gravel), this illustrated that the shaft resistance from the upper layer did not have to fully developed when the shaft resistance from the lower part started to develop.

Figures 6.19, 6.20, and 6.21 present the shaft resistance distribution along the pile depth of the tested piles. As shown in Fig. 6.19, before loading of 10,000 kN was applied, the shaft resistance of P12 was mainly provided by layers 3, 5 and 6. Moreover, with the loads increasing, the shaft resistance generally developed along the pile length. Similar to the finding obtained from Fig. 6.16, after loading of 22,000 kN was applied, except for the bottom layer (the shaft resistance have kept increasing), the shaft resistance from the other layers almost fully developed (in each layer, the distances among vertical lines are very close to each other).

As shown in Fig. 6.20, for the P66, before loading of 16,000 kN was applied, the shaft resistance was mainly provided from upper layers, the bottom layer provided small shaft resistance, which was less than 5 kPa. Moreover, if connecting the shaft resistances with a smooth line, the resistance along the pile illustrated an '***R***' shape. Figure 6.20 also presented that the fluctuation curves of layer 3, the shaft resistance softening and hardening observed.

The shaft resistance distribution of P105 is represented in Fig. 6.21. It can be found that there was almost no shaft resistance from the bottom layer before 28,000 kN was applied. Also, it can be seen that the layers 1, 2 and 3 were fully developed firstly and 4, 5, 6 and 8 were developed later because the vertical lines' distances are very close. Similar to the findings obtained from Fig. 6.18, the shaft resistance of layer 7 kept going to develop during the whole loading tests.

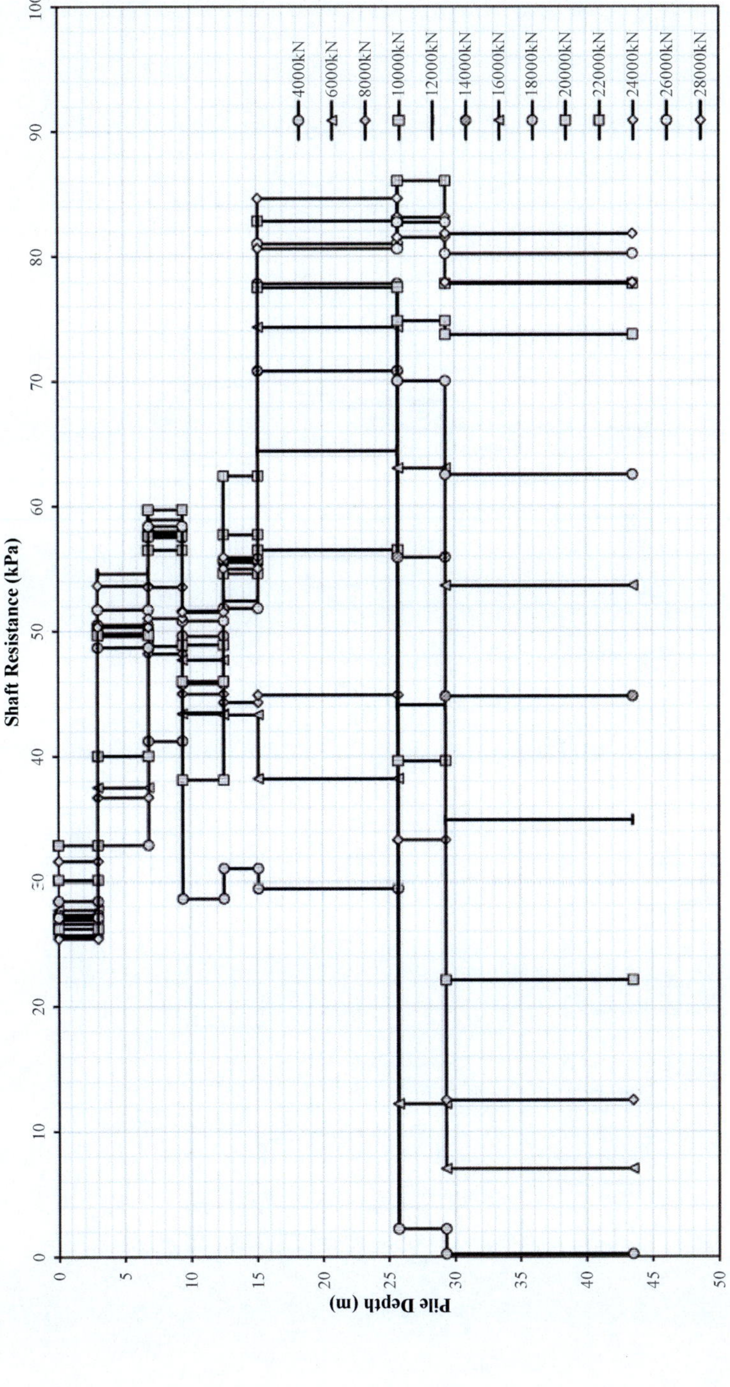

Fig. 6.19 Shaft resistance distribution along the pile for P12

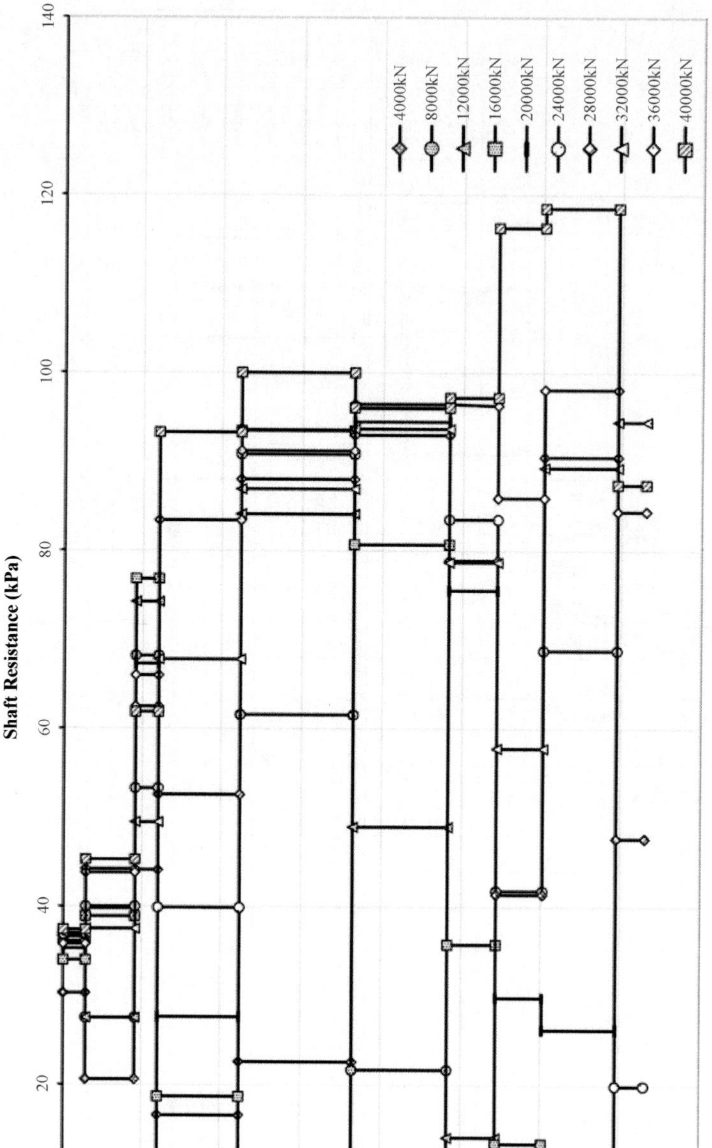

Fig. 6.20 Shaft resistance distribution along the pile for P66

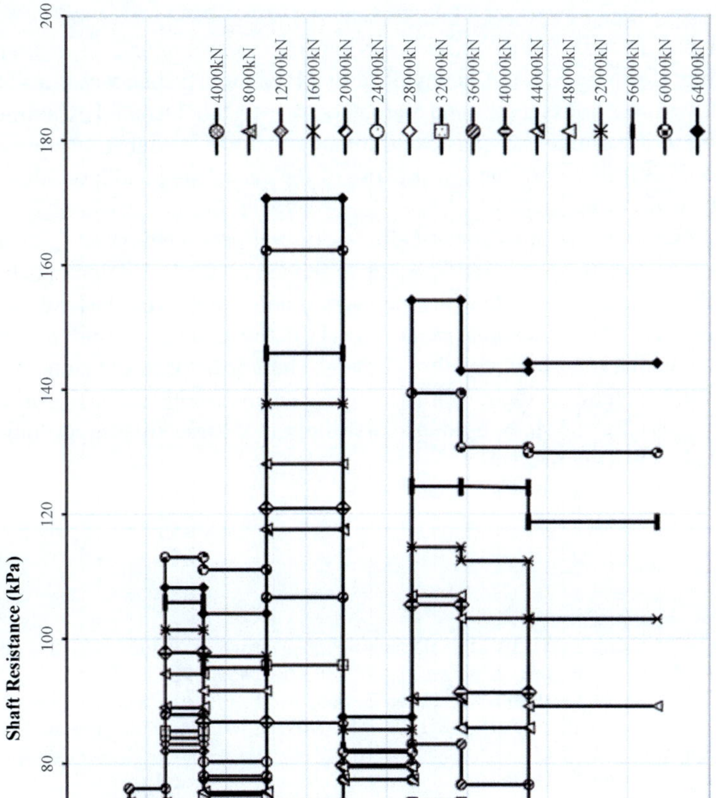

Fig. 6.21 Shaft resistance distribution along the pile for P105

6.4 Concluding Remarks

This chapter has investigated the axial behaviour of super-long and large-diameter piles through performing SLTs. A newly developed reaction system was introduced, and three piles with various lengths and diameters were tested. Detailed experimental principle used to determine the transferred load by using sister rebar was provided with formula derivation. Also, the test process of the super-long and large-diameter was demonstrated in this chapter.

For investigation of the pile capacity, the traditional presentations of Q-s curves including the load versus pile head and pile toe settlement as well as settlement-lg time curves were illustrated. By analysing the load versus depth diagrams, the load transfer mechanism of these tested piles was researched. Also, the load versus shaft resistance was provided, which was used for the study of shaft resistance development. In detail, with load increment, the development of shaft resistance in each soil layer was researched. Lastly, the shaft resistance distribution of these three super-long and large-diameter piles were presented.

References

Feng, S. J., Lu, S. F., & Shi, Z. M. (2016). Field investigations of two super-long steel pipe piles in offshore areas. *Marine Georesources & Geotechnology, 34*(6), 559–570.

Hannigan, P. J., Goble, G. G., Likins, G. E., & Becker, M. L. (2016). *Design and construction of driven pile foundations (FHWA-NHI-16-009) (FHWA-NHI-16-009).* Federal Highway Administration: USA.

Li, S. C., Zhang, Q., Zhang, Q. Q., & Li, L. P. (2016). Field and theoretical study of the response of super-long bored pile subjected to compressive load. *Marine Georesources & Geotechnology, 43*(1), 71–78.

Lin, T. J., Xiong, H. J., & Wang L. Q. (1999). *Pile foundation design manual.* Beijing: China Architecture & Building Press (In Chinese).

Lu, Y., Tan, Y., & Lan, H. (2019). Full-scale load testing of 75–90-m-long post-grouted drilled shafts in Suzhou stiff clay. *Journal of Testing and Evaluation, 47*(1), 284–309.

Michael, C. M., & Jae, C. (2017). *Evaluation of static resistance of deep foundations* (BDV31-977–05).

Ministry of Construction of the People's Republic of China. (1999). *Standard for soil test method (GB/T 50123-1999).* Beijing, China: Ministry of Construction of People's Republic of China.

Ministry of Construction of the People's Republic of China. (2005). *Specifications for Highway Engineering Geological Remote Sensing (JTG/T C21–01-2005).* Beijing, China: Ministry of Construction of People's Republic of China.

Oliaei, M., & Ghotbi Siabil, S. M. A. (2017). Dynamic behavior of large-diameter piles considering liquefaction under clay layer. *Scientia Iranica a, 24*(6), 2665–2683.

Sun, L. Q., Jia, T. Q., Yan, S. W., Guo, W., Ren, Y. X., & Lei, Z. M. (2016). Prediction of pile running during the driving process of large diameter pipe piles. *Ocean Engineering, 128,* 48–57.

Wang, D. D., Wang, L. F., & Zhang, L. P. (2015). Experimental study on post grouting bearing capacity of large diameter bored piles. *MATEC Web of Conferences, 22,* 1–7.

Xing, H. F., Wu, J., & Luo, Y. (2017). Field tests of large-diameter rock socketed bored piles based on the self-balanced method and their resulting load bearing characteristics. *European Journal of Environmental and Civil Engineering,* 1–15.

Yang, J. B., & Zhong, Z. X. (1998). Research on load transfer mechanism of super-long pile. *Chinese Journal of Geotechnical Engineering, 20*(6), 76–79. (In Chinese).

Zhang, J. W., Wang, Z. L., & Qiao, D. Q. (2014). Model test on negative skin friction for super-long pile under surcharge load considering time effect. *Applied Mechanics and Materials, 470,* 1105–1108.

Zhang, M. Y., Gao, H. Y., Li, Y., Song, H. Z., & Peng, W. T. (2012). Effect of soil stiffness on bearing characteristics of large-diameter and super-long steel pipe pile based on FLAC3D. *Applied Mechanics and Materials, 170–173,* 743–746.

Zhou, L. J., Lu, Z. A. I., & Wang, W. J. (2012a). Vertical load mechanism analysis and experimental research on super-long PHC pipe pile in soft clay. *Applied Mechanics and Materials*, 204–208, 471–478.

Zhou, D., Lam, K. H., Chen, Y., Zhang, Q., Chiu, Y. C., Luo, H., & Chan, H. L. W. (2012b). Lead-free piezoelectric single crystal based 1–3 composites for ultrasonic transducer applications. *Sensors and Actuators A: Physical, 182,* 95–100.

Zhou, X. L., Fan, X., & Ji, Y. B. (2014). Optimization design and analysis of super-long pile in deep soft soil foundation for Huzhou commercial and residential building. *Electronic Journal of Geotechnical Engineering, 19,* 3089–3098.

Zhou, Z. J., Wang, D. D., Zhang, L. P., & Ma, W. S. (2015). Determination of large diameter bored pile's effective length on Mindlin's solution. *Journal of Traffic and Transportation Engineering,* 2(6), 422–428.

Zou, X. J., & Zhao, M. H. (2016). Axial bearing behaviour of super-long piles in deep soft clay over stiff layers. *Journal of Central South University of Technology, 20*(7), 1–16.

Chapter 7
Piles Under Ultimate Loads

Check for updates

7.1 General Introduction

During the construction of a bored pile, the remove of soils by machines may cause the soil residue to remain in the drilled hole. This leads to a decrease in the end resistance capacity of the pile and an increase of pile settlement. Because of these remaining soil deposits from the base of hole, it has been found that the capacity of some piles from a mansion was much lower than the designed requirement. It has also been found that the settlement of these piles was much greater than the treated piles through performing the SLTs in Taiyuan City, China, (Shi 2005).

For this particular problem, it can be solved by applying the slurry support technology. However, it is reported that, sometimes, this slurry support admixture layer (bentonite or polymeric material, soils and water) deceased the friction resistance between soil layers and concrete pile, which consequently decreased the ultimate bearing capacity of the piles. It is reported that this admixture layer or composite layer decreases 30–40% of the pile bearing capacity (Li et al. 2000).

Under condition of the existed admixture layer, the problem can be solved by applying the post grouting technology. During this process, cement admixture is pressured down, and the slurry support layer is then forced out (Zhou et al. 2017). There are intensive researches that focused on the ultimate bearing capacity of these piles which suffered from compressive, uplift and horizontal forces during SLTs (Castelli and Wilkins 2004; Nguyen et al. 2012; Sinnreich and Simpson 2013; Patel et al. 2015). The ultimate bearing capacity of post pile is mostly determined by interpretation of non-plunging curve result. Other investigations paid attention to the methods of grouting to find out the best way to increase the bearing capacity under various subsurface condition (Ho 2003). Some of the researchers take the load transfer mechanism into account. However, the SLTs of these piles under failure load are rarely conducted, and the research of these piles' behaviours is limited.

This chapter provides an understanding of pile failure criteria under SLTs to enable continuous improvement in construction practices and primary foundation

J. Zhou and E. Oh, *Full-Scale Field Tests of Different Types of Piles*,
Advanced Topics in Science and Technology in China 62,
https://doi.org/10.1007/978-981-33-6183-6_7

design. There is very limited information referring to the behaviour of a pile under ultimate loading in AS 2159-2009 (Standards Australia 2009). Thus, the objectives of this chapter are to illustrate the behaviours of piles suffering from ultimate failure condition.

Three dominant factors can result in pile failure. Commonly in practice, pile failure occurs because of inadequate rigidity of the concrete. Inadequate concrete vibration may lead to decrease of the concrete strength. If the vibration is insufficient, the deficient concrete will crack. If the vibration is too great, the lower part of concrete will achieve the required strength, but the pile head (which is mainly made of cement, sand and water without coarse aggregate) will fail to achieve the required strength. The second common reason for pile failure is that the rigidity of the concrete pile is overdesigned, or the engineer overestimates the soil rigidity. The pile foundation element will not break, but the soil fails when the ultimate load is applied. The final reason for pile failure is that eccentric loading is applied, and the pile is broken by bending moment.

To obtain the pile behaviours under ultimate loading, three piles were cast and SLTs were performed. Piles P60, P40 and P12 were loaded until failure occurred for the aforementioned conditions. Specifically, P60 failed because of inadequate concrete rigidity, P40 failed because of bending movement and P12 failed because of the insufficient rigidity of the soil. The other failure conditions depicted in Figs. 7.1 and 7.2—failure of the tension bar system in the reaction pile and platform bearing failure—are not included because there was no pile soil destruction.

Fig. 7.1 Failure of tension bar system in reaction pile (Handley et al. 2006)

Fig. 7.2 Platform bearing failures under Kentledge test (Handley et al. 2006)

7.2 Subsurface Explorations

The construction site was located in western Jinan city, Shandong Province, China. The subsurface exploration was discovered through laboratory and in-situ tests. The in-situ tests of SPTs and CPTs and laboratory tests of consolidation, direct shear and triaxial tests were conducted based on the Chinese Standard for Soil Test Method (GB/T 50123-1999 1999), Standard for Test Methods of Engineering Rock Mass (GB/T 50266-2013 2013) and Code for Investigation of Geotechnical Engineering (GB 50021-2001 2001), respectively. Based on the borehole logs and soil samples, ten layers were discovered as follows:

1. fill: yellowish-brown: plastic clay with ash concrete and brick crumbs, thickness ranges from 0.6 to 3.4 m; 2. silty clay: yellow, loess-shaped, saturated, thickness ranges from 1.2 to 3.2 m; 3. silty clay: yellowish-brown, plastic, thickness ranges from 0.4 to 4.4 m; 4. clay: grey-brown to beige, plastic, thickness ranges from 1.2 to 2.5 m; 5. clayey silt: it contains ferric oxide, thickness ranges from 0.6 to 1.6 m; 6. clay, brown: partially contains loess-doll, particle size between 1 and 2 cm, thickness ranges from 1.2 to 4.1 m; 7. silty clay: yellowish-brown, plastic, contains iron and manganese oxides, thickness ranges from 3.7 to 6.8 m; 8. clay: yellow to pale brown, plastic, thickness ranges from 3.4 to 6.2 m; 9. silty clay to clay: pale brownish red, plastic stiff, thickness ranges from 12.3 to 34.2 m; 10. clay: pale brownish red, very stiff, contains gravel with particle size between 2 and 3 cm.

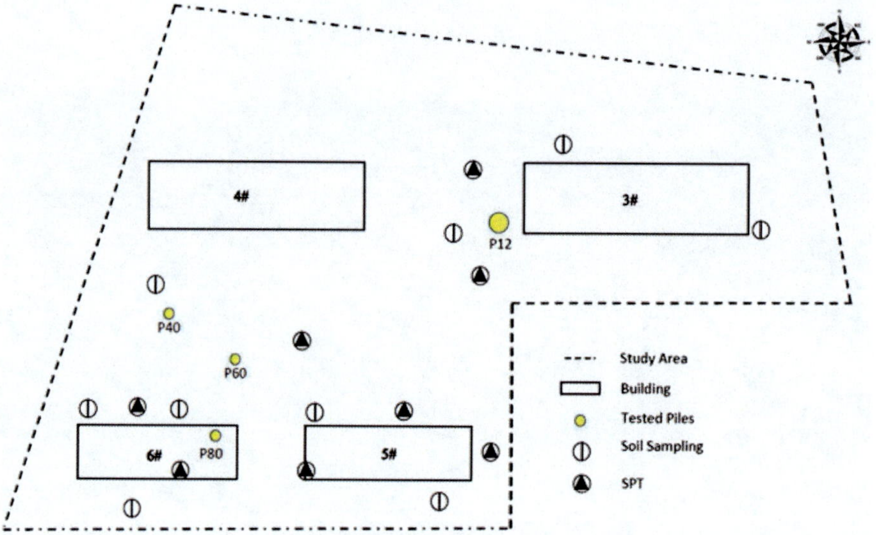

Fig. 7.3 Plan view of the tested piles (not to scale)

As shown in Fig. 7.3, three piles with labels of P80, P60 and P40 are next to building $6^{\#}$. These piles are in the same diameter and length. Pile 12 which is near building $3^{\#}$ has a larger diameter and length. Based on the borehole logs from the study area that shown in Fig. 7.3, it is discovered that the subsurface condition near building $6^{\#}$ is similar. The soil parameters of the test piles are summarized in Figs. 7.4 and 7.5.

7.3 Designed Ultimate Bearing Capacity

JGJ 94-2008 (Technical Code for Building Pile Foundations) has depicts some of the methods for determining the ultimate bearing capacity of pile foundation when designing a pile foundation. The equations used in this project are illustrated in Eqs. 7.1 and 7.2 which are calculations for the traditional pile foundation and post grouted pile foundation, respectively. Equation 7.1 represent the ultimate bearing capacity of single pile without any grouting application hence the total capacity (Q_{uk}) is the sum up of shaft capacity (Q_{sk}) and end capacity (Q_{pk}). The values of shaft stress (q_{sik}) and end stress (q_{pk}) are relating to types of soils, state of soils (such as void ratio, liquid index, SPT N-value), types of pile foundation and construction method. Different to Eq. 7.1, the total capacity of pile as demonstrated in Eq. 7.2 includes the parts where resistances with grouting strengthening effect from shaft (Q_{gsk}) and base (Q_{gpk}). The strengthening coefficients of shaft (β_{si}) and base (β_p) are relating to the types of soils.

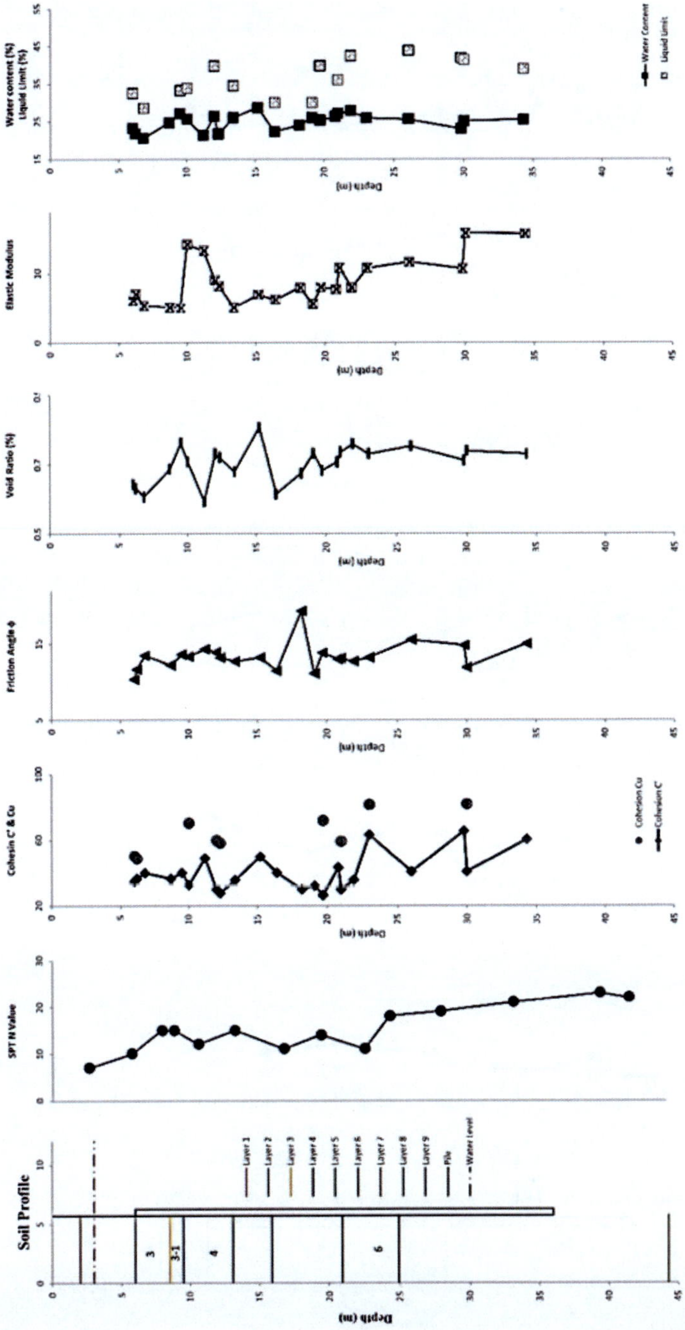

Fig. 7.4 Subsurface conditions of P80/P60/P40

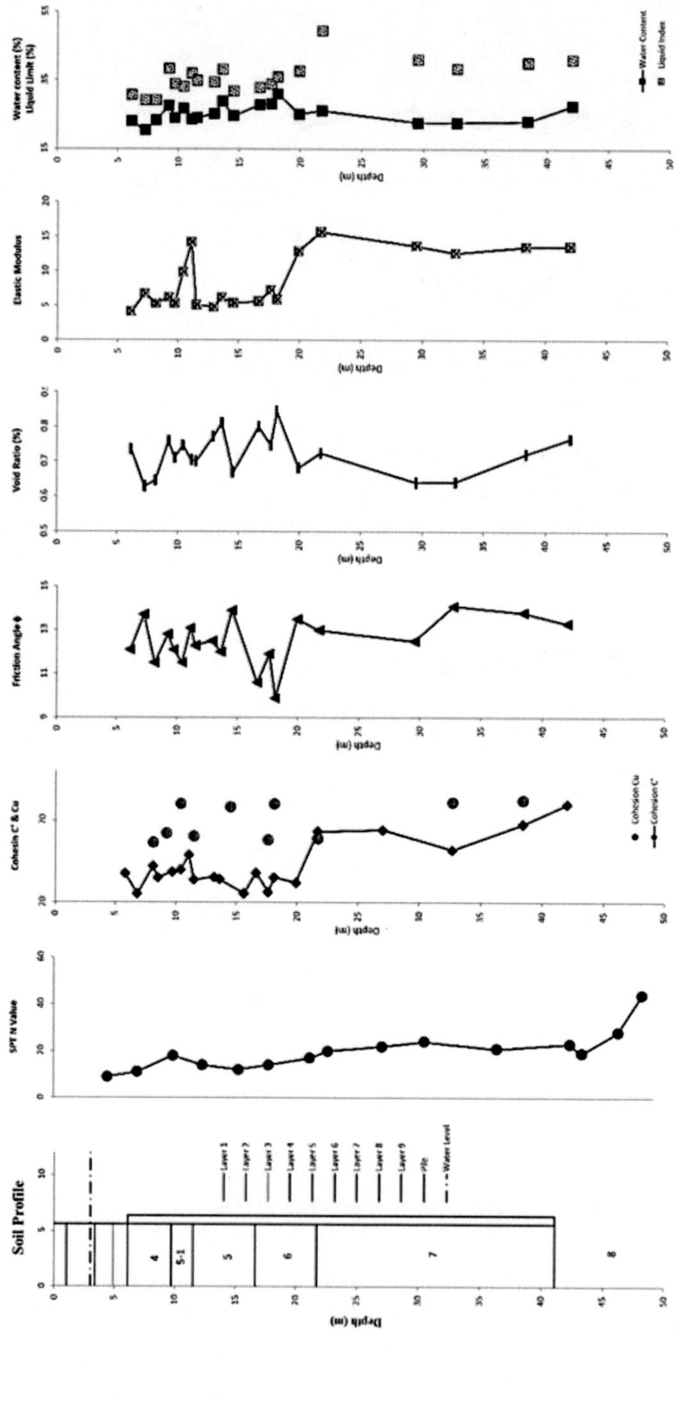

Fig. 7.5 Subsurface condition of P12

$$Q_{uk} = Q_{sk} + Q_{pk} = u \sum q_{sik} l_i + q_{pk} A_p \quad (7.1)$$

$$Q_{uk} = Q_{sk} + Q_{gsk} + Q_{gpk} = u \sum q_{sjk} l_j + u \sum \beta_{si} q_{sik} l_{gi} + \beta_p q_{pk} A_p \quad (7.2)$$

where:

Q_{uk}—Ultimate bearing capacity of single pile.
Q_{sk}—Shaft capacity without grouting strengthening.
Q_{pk}—Base capacity without grouting strengthening.
q_{sik}—Shaft resistance of each soil layers.
q_{pk}—End resistance from the pile toe.
A_p—End area from the pile toe.
Q_{gsk}—Shaft capacity with grouting strengthening.
Q_{gpk}—Base capacity with grouting strengthening.
q_{sjk}—Shaft resistance without grouting strengthening.
u—Perimeter of pile.
l_j—Thickness of layers within soil strengthening part.
β_{si}—The strengthening coefficients of shaft.
β_p—The strengthening coefficients of base.

The capacity of single piles based on the In-situ method and API method are also considered in this case study. Different to the JGJ 94-2008, In-situ method considers the SPT N-value and the API method considers α coefficient which is relating to the overburden pressure. The ultimate bearing capacity of these four piles are summarized in Table 7.1. It can be seen that, the total capacity obtained from

Table 7.1 Ultimate bearing capacity of piles

Standards	Shaft resistance (kN)	End resistance	Total capacity	Designed load	Reaction weight	Pile label
JGJ 94 (Non-grouting)[a]	4973	804	5777	8400	10,080	P12
In-situ method	3610	1055	4665			
API method	4285	904	5190			
JGJ 94 (Grouting) [b]	5696	1768	7465			
JGJ 94 (Non-grouting) [a]	3156	382	3538	5200	6240	P80/P60/P40
In-situ method	2669	466	3135			
API method	3335	203	3538			
JGJ 94 (Grouting)[b]	3699	877	4577			

Note [a]JGJ 94 (non-grouting)
[b]JGJ 94 (Grouting) represent the ultimate bearing capacity based on Eqs. 7.1 and 7.2, respectively

Table 7.2 Information of tested piles

Pile label	Pile condition	Before cutting (m)	Pile length (m)	Pile diameter (mm)	Concrete strength (MPa)
P40	Eccentricity	32	$32 - 2 = 30$	600	35
P60	Inadequate concrete strength	30	$30 - 0 = 30$	600	Upper 2 m: 25 – 30 Lower part: 35
P80	Standard pile	32	$32 - 2 = 30$	600	35
P12	Inadequate soil strength	37	$37 - 2 = 35$	800	50

In-situ method is conservative, and the capacity gained from JGJ 94-2008 and API method are similar. Finally, the designed maximum loads for SLTs are determined as 8400 and 5200 kN for P12 and P80, P60, P40 respectively. For security consideration, the reaction weight (1.2 times the designed testing loads) are defined to be 10,080 kN (P12) and 6240 kN (P80/P60/P40) for the testing piles.

7.4 Pile Construction

There are 4 piles designed to be tested. As depicted in Table 7.2, after the concrete were cured, the pile length of P40, P60, P80 and P12 were 32, 30, 32 and 37 m, respectively (before pile cut). After cutting of 2 m from the pile head as shown in Fig. 7.6, all three piles (P40, P60, P80) were in the same dimensions of length of 30 m. Except for P12, the diameter of the other three piles was 600 mm. Because there was no cutting for P60, it was believed that the top 2 m' concrete did not achieve the required strength of C35. Note that, after the concrete cured, the strength of the concrete from the upper pile was discovered from 25 to 35 MPa via conducting concrete rebound tests by resiliometer.

7.5 Test Setup

The SLTs were conducted based on the Chinese Technical Code for Testing of Building Foundation Piles (JGJ 106-2014 2014). The setup of the SLTs is provided in Fig. 7.7. As shown in Fig. 7.8, four hydraulic jacks (QF630T-20) were used to provide loads to the pile head. The low-speed maintenance methods were used in this case study, and the applied load was maintained until the rate of axial movement did not exceed to 0.1 mm. After each load was applied, the vertical movement of the piles was recorded with intervals of five, 10 and 15 min, and 30 min if the accumulated time exceeded 1 h.

Fig. 7.6 Percussion of pile head

Fig. 7.7 Compressive SLT

For piles P40, P60 and P80, the loading started at 1,040 kN and progressed with 520 kN increments. The vertical settlements corresponding to each loading were recorded with four automatic dial gauges. For the piles that achieved the maximum loading of 5,200 kN, the applied loads decreased with decrements of 1,040 kN, and the unloading terminated at 0 kN. For pile P12, the loading started at 1,680 kN and progressed with 840 kN increments.

Fig. 7.8 Equipment of SLT

7.6 Observation of Pile Under Failure Condition

7.6.1 Piles with Inadequate Rigidity

As shown in Fig. 7.9, a concrete crack in P60 occurred when the loading was at 3,640 kN. The crack was discovered at the pile head, and the maximum width of the crack was up to 30 mm. Through a low-strain integrity test, this pile was found to be broken at the depth of 1.2 m underground. After digging out the pile and breaking the concrete from the pile head, it was found that the reinforcement was bent outside at a depth of 0.9–1.2 m, as shown in Fig. 7.10. This occurred because, after the concrete was damaged, the reinforcement tried to resist the vertical loading, and, for a second, the reinforcement was bent. It could not be bent inside the pile because the concrete core resisted this deflection.

7.6.2 Piles Suffering from Eccentric Loads

As shown in Fig. 7.11b, pile P40 was tested with eccentricity of 200 mm. After a load of 4,680 kN was applied, the pile seemed to successfully resist the loading; however, the concrete actually crushed at the compression zone, and the reinforcement at this zone (Fig. 7.11a) tried to resist this compressive load, yet failed, and the orientation was outside. At the same time, the concrete ruptured in the tensile side, but the reinforcement in this side provided tensile resistance for a couple of minutes and failed later. There was no reinforcement hump discovered in this zone (Fig. 7.10c).

Fig. 7.9 Concrete crack from pile head

Fig. 7.10 Deformed reinforcement

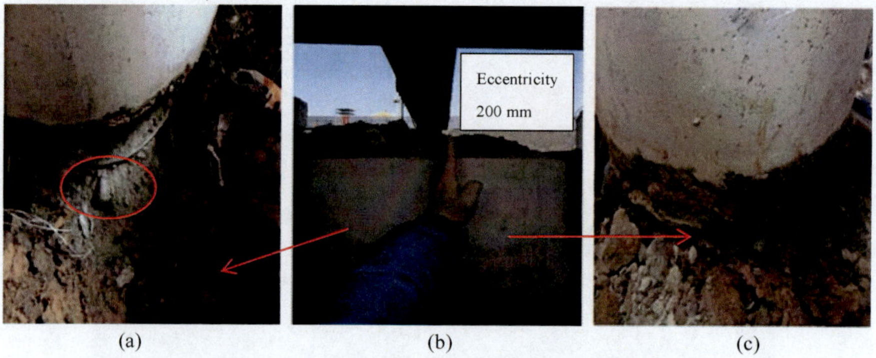

Fig. 7.11 Pile suffering from eccentric loads. **a** Bent reinforcement of left part **b** Pile suffering from eccentricity **c** Fractured concrete of right part

7.6.3 Failure from Inadequate Soil Rigidity

For pile P12 that suffered from plunging failure, in which concrete was not damaged. After a load of 8,400 kN applied, nothing occurred until 115 min, when the pile was suddenly driven into the ground. This phenomenon can be primarily explained by the fact that the shaft resistance was fully developed, and, after 115 min, the loading transferred to the pile tip. As a result of the inadequate stiffness of the clay-bearing stratum, the pile was plunged into the ground. As shown in Fig. 7.12, cracks in the ground were discovered up to 100 mm.

Fig. 7.12 Plunging failure of pile

7.7 Test Results

7.7.1 Pile with Achieved Design Requirements

The results of pile P80 are provided in Figs. 7.13, 7.14 and 7.15. The interpretation results based on the double tangent method, Chin's method and offset method are provided in Fig. 7.16a–c, respectively.

From the Q-s curve (Fig. 7.13), the ultimate loading was conservatively determined as 4,680 kN, with a corresponding maximum settlement of 4.9 mm. For some local standards, the ultimate bearing capacity could be determined as 5,200 kN because the corresponding settlement of 7.88 mm was relatively small. According to the Chinese Standard JGJ 106-2014 (2014), Clause 4.4.2.1, through analysing thes-lgt results at each loading, the capacity of the pile could be determined where there was a deep downwards curve occurring at the end of s-lgt curve. As shown in Fig. 7.14, there were two evident downward trends from the loading stages of 4,680

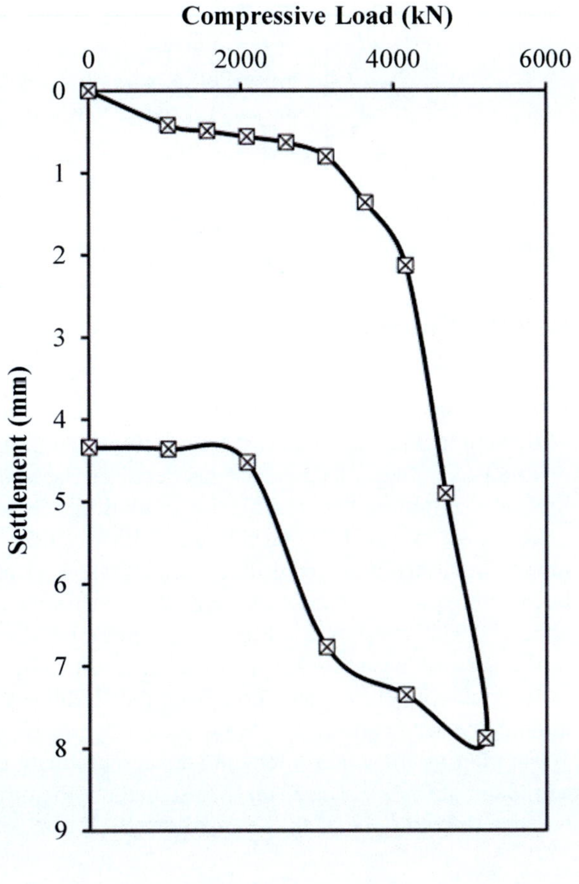

Fig. 7.13 Q-s curve of P80

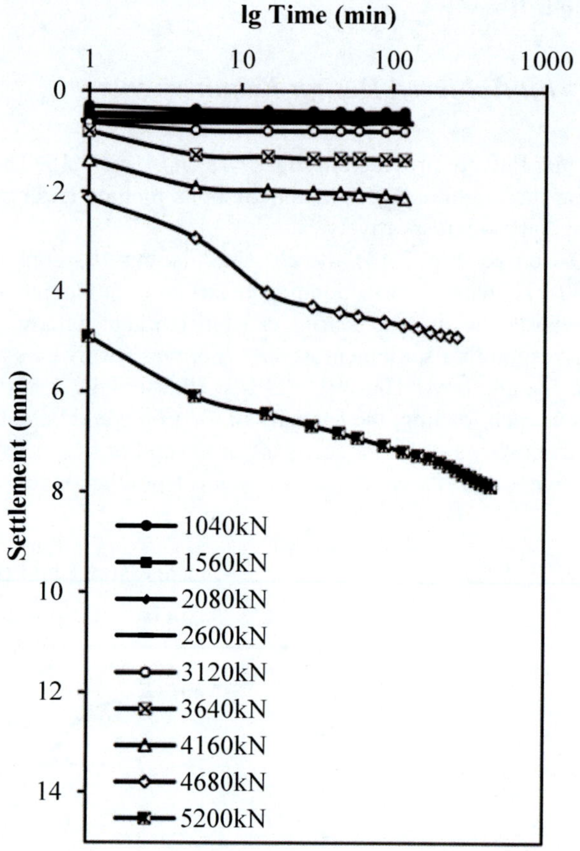

Fig. 7.14 s-lgt curve P80

and 5,200 kN, but the settlement remained stable in the end, without an extreme settlement increasing. Hence, the capacity of the pile was determined as 5,200 kN, or 4,680 kN for the conservative consideration. Based on the s-lgQ curve, the ultimate loading was determined as 4,680 kN, with difficulty (Fig. 7.15).

As stated in AASHTO (2002) and FHWA (1992), the double tangent method is more commonly used for drilled shafts. Based on plotting two tangent lines, the intersecting point that presents the ultimate bearing capacity of P80 was determined to be 3,900 kN, as shown in Fig. 7.16a. Through Q-s curve, plotting elastic line, and offset this line based on Eq. 2.38. As shown in Fig. 7.16b, there was no intersect between the offset line and curve from the loading stages, which indicated that the ultimate bearing capacity of P80 was over the maximum applied load of 5,200 kN. When plotting the load/settlement versus settlement, the result indicated a line with a gradient of 0.0002. Based on Chin's method, the ultimate bearing capacity of P80 was determined to be 5,000 kN, as shown in Fig. 7.16c.

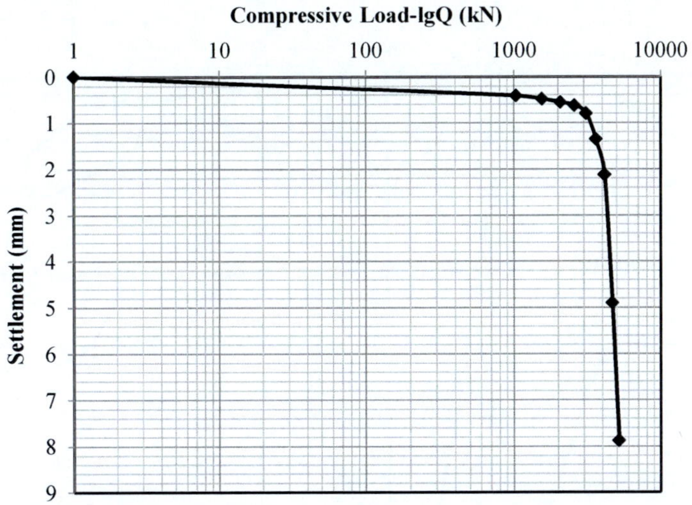

Fig. 7.15 s-lgQ curve P80

7.7.2 Failure from Inadequate Concrete Strength

The Q-s curve, s-lgt curve and s-lgQ curve of P60 are provided in Figs. 7.7, 7.17, 7.18 and 7.19. As shown in Fig. 7.17, a small settlement 14.13 mm was discovered from a loading of 3,640 kN, and there was an extremely increasing (from 14.13 to 83.2 mm) discovered when the next loading applied, the ultimate bearing capacity was determined to be 3,640 kN for P60 based on the Q-s curve.

As shown in Fig. 7.18, there was a dramatic decreasing trend when the loading of 4,160 kN was applied, which showed increments in settlement with increasing time, this represented that the failure criteria of the soil occurred. So, the ultimate bearing capacity of P60 was then determined to be 3,640 kN. This figure also provides valuable information illustrating the pile failure behaviour. As shown in Fig. 7.18, when loads of 3,640 kN were applied, the concrete from the upper part of pile foundation broken and resulted in the settlement increment. Later, under such vertical loads, the pile foundation could still resist the applied vertical load and thus the settlement data remained stable. In practice, when finding a curve similar to the 3,640 kN-curve, the pile is expected to be broken, checking the pile integrity is indispensable, thus low strain detection test is strongly recommended to perform.

As shown in Fig. 7.19, a point with corresponding load of 3,640 kN was found to be the ultimate load because, after this point, the settlement of the pile head decreased dramatically (the settlement changed from 14.13 to 83.2 mm with small load increment of 520 kN).

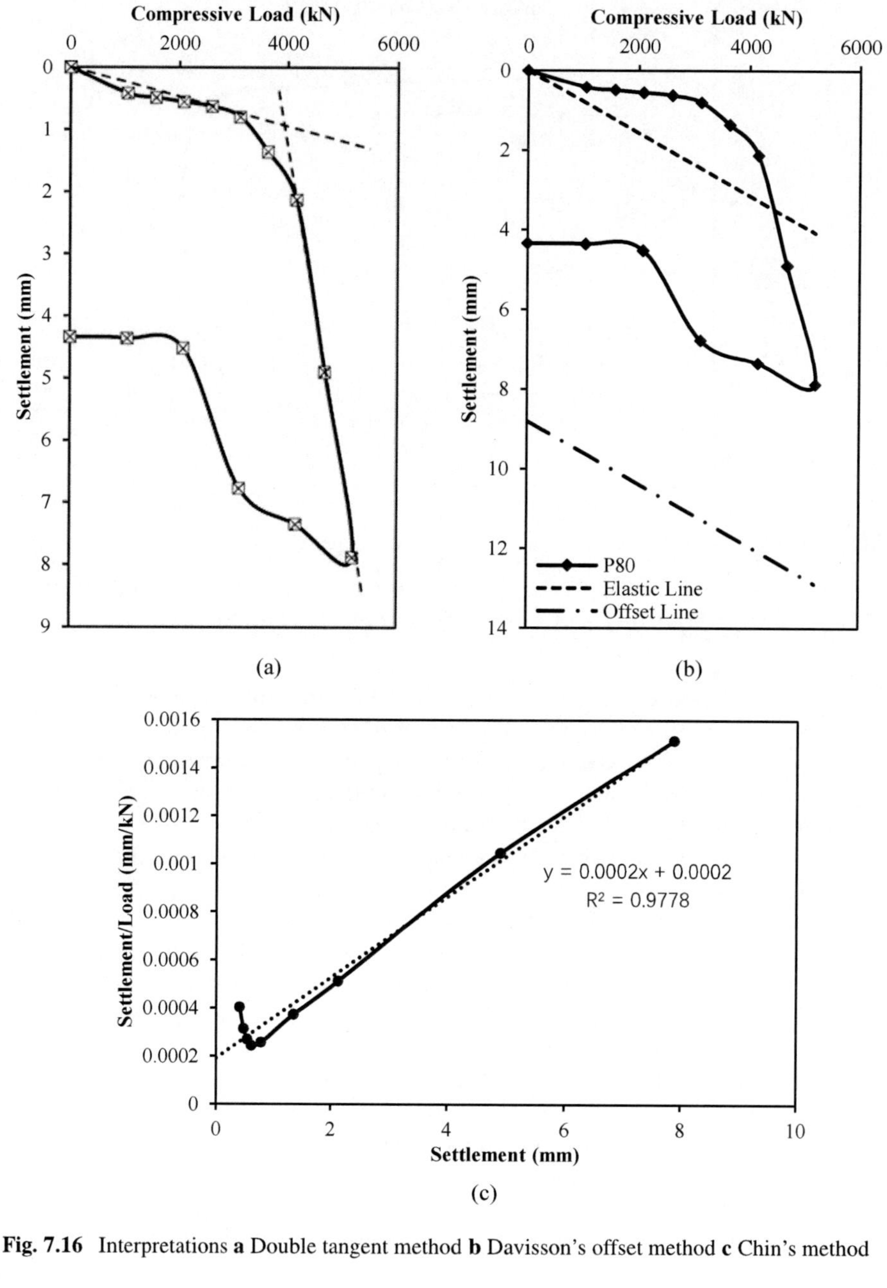

Fig. 7.16 Interpretations **a** Double tangent method **b** Davisson's offset method **c** Chin's method

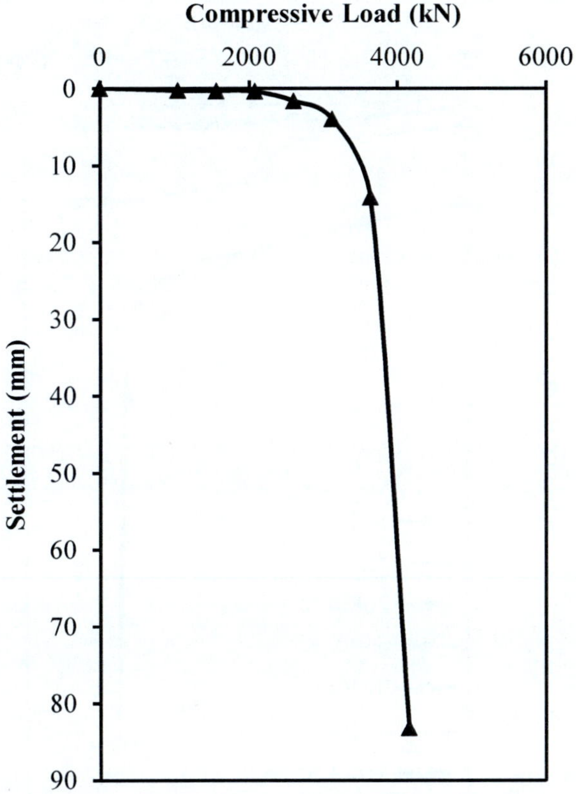

Fig. 7.17 Q-s curve of P60

Interpretations of the Q-s curve based on the double tangent method, Davisson's offset method and Chin's method are provided in Fig. 7.20a–c. As shown in Fig. 7.20a, the intersection of the two tangent lines was determined and the corresponding load was found to be 3,600 kN. For Davisson's offset method, using the concrete modulus of 31,500 N/mm^2 to determine the elastic line (PL/EA). As shown in Fig. 7.20b, the ultimate capacity was determined to be 3,600 kN. However, because the concrete modulus from the upper part was 30,000 N/mm^2, the actual capacity should be slightly over 3,600 kN. By using the data obtained during the SLT before failure occurred, and then plotting the settlement/load versus settlement, a gradient of 0.0003 was discovered, as shown in Fig. 7.20c. Finally, the bearing capacity of P60 was determined to be 3,333 kN.

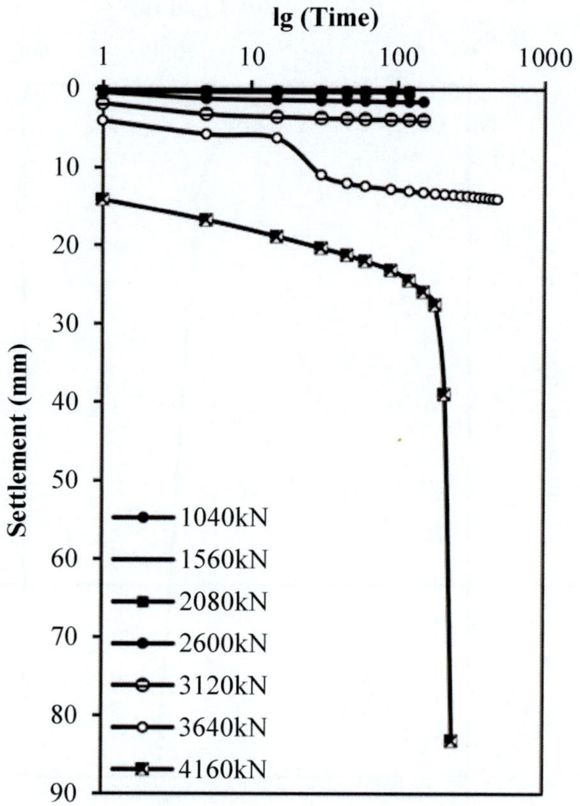

Fig. 7.18 s-lgt curve of P60

7.7.3 Failure from Eccentricity

For the pile that suffered from eccentric loading, the typical results presentation is provided in Figs. 7.21, 7.22 and 7.23. As shown in Fig. 7.21, after the loading of 4,680 kN was applied, a huge increase in settlement was discovered. The ultimate capacity was then determined to be 4,160 kN. As shown in Fig. 7.22, after loading of 4,160 kN was applied, the s-lgt curve demonstrated that the soil settlement was relatively stable with time increasing; however, after the next loading was applied (4,680 kN), there was an extreme increase in settlement, which represented the failure condition occurring. The ultimate bearing capacity was then determined as 4,160 kN. It is worth noting that there are two 'drops' discovered under loads of 4680 kN. The first settlement drop is referring to the crush of concrete from the compression zone (around 14 mm: 6 mm to 20 mm settlement change); the second drop relates to the failure of the pile-soil system. As shown in Fig. 7.23, beyond the point of 4,160 kN loading being applied, a dramatic downwards trend was discovered, which

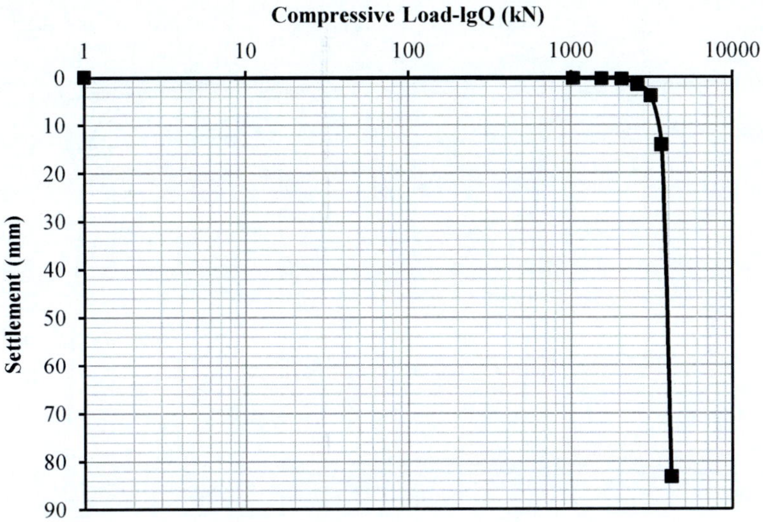

Fig. 7.19 s-lgQ curve of P60

represented the ultimate condition. The ultimate load was then determined as 4,160 kN.

The interpretations based on the double tangent method, Davisson's method and Chin's method are illustrated in Fig. 7.24a–c, respectively. The intersection of two tangent lines was discovered from Fig. 7.24a, and the ultimate bearing capacity of P40 was then discovered to be 4,190 kN. As illustrated in Fig. 7.24b, the intersection between the offset line and Q-s curve was determined, and the ultimate capacity of P40 was obtained as 4,250 kN. As shown in Fig. 7.24c, Chin's method was not applicable to the piles suffering from failure caused by eccentric loading.

7.7.4 Plunging Failure

The typical presentations are provided in Figs. 7.25, 7.26 and 7.27. For pile P12, the capacity was determined to be 7,560 kN, as shown in Fig. 7.25, because, after this 7,560 kN point, the Q-s curve indicated a dramatic downwards trend, which illustrated the failure condition. As shown in Fig. 7.26, under loading of 7,560 kN, the pile was relatively unstable, but failed when 8,400 kN was applied. The ultimate capacity of this pile was determined to be 7,560 kN, based on the s-lgt curve. In addition, from the s-lgQ curve, the ultimate bearing capacity was determined to be 7,560 kN (Fig. 7.27).

The results based on the double tangent method, Davisson's method and Chin's method are presented in Fig. 7.28a–c, respectively. As shown in Fig. 7.28a, the intersection between two tangent lines was found, and the ultimate bearing capacity

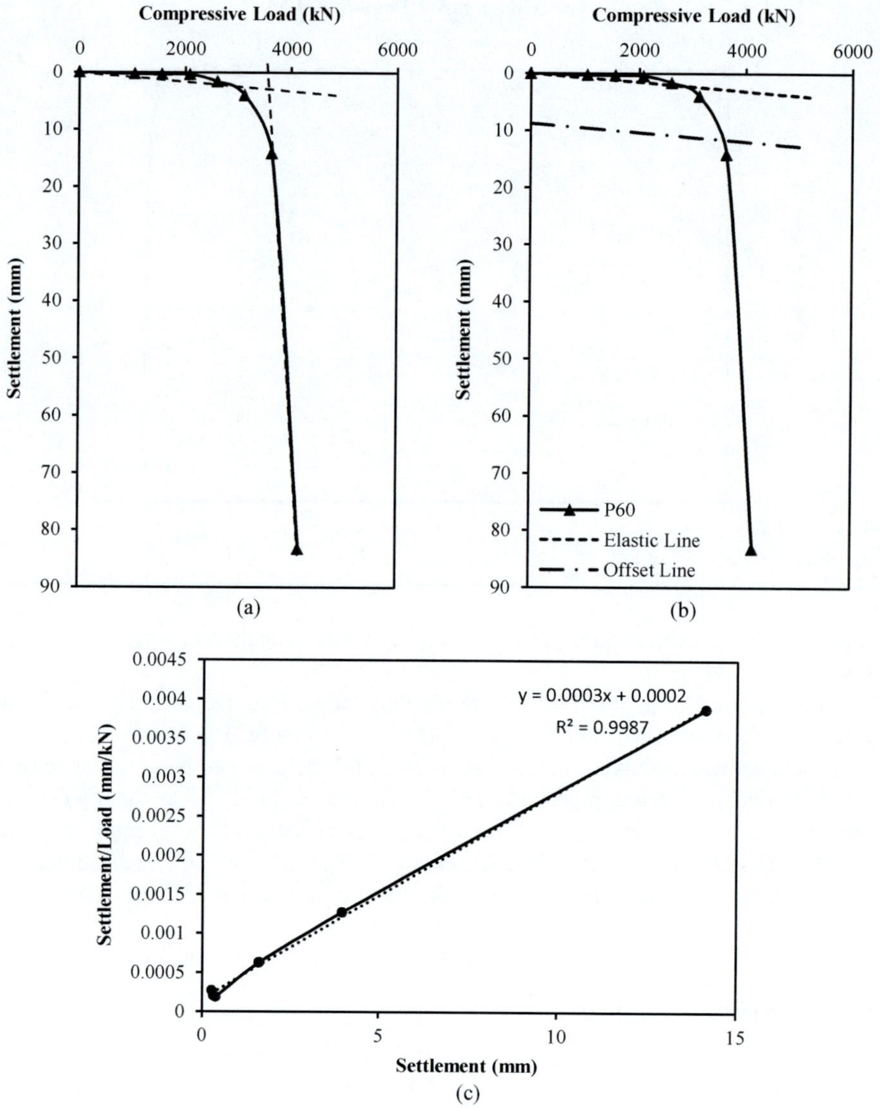

Fig. 7.20 Interpretations **a** Double tangent method **b** Davisson's offset method **c** Chin's method

was determined to be 7,560 kN. For Davisson's method, Eq. 2.39 was required instead of Eq. 2.38 to determine the offset line because this was a large pile with an 800 mm diameter. The intersection between the offset line and Q-s curve was discovered (Fig. 7.28b), and the capacity was determined to be 7,900 kN. By plotting the data obtained from the tested pile before failure occurred, based on Chin's method, a function with a slope of 0.001 was determined (Fig. 7.28c) and the ultimate bearing

Fig. 7.21 Q-s curve of P40

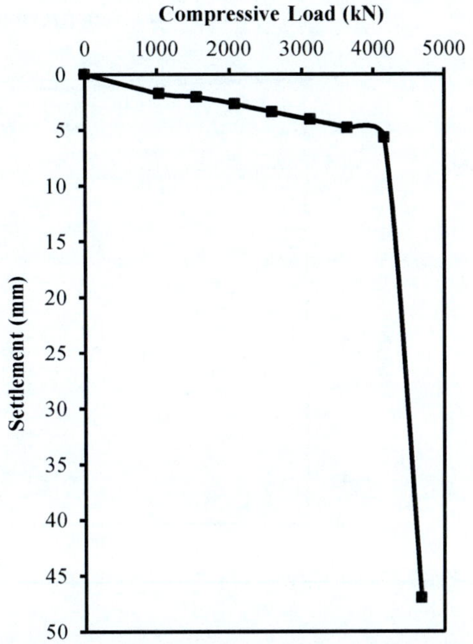

Fig. 7.22 s-lgt curve of P40

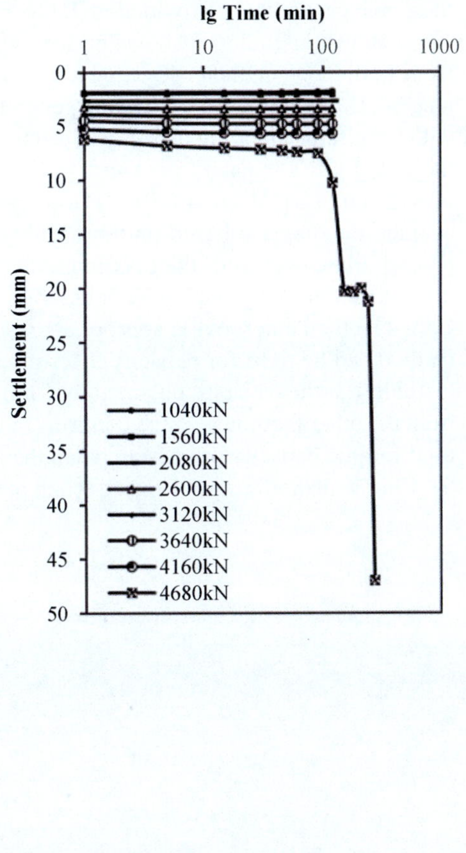

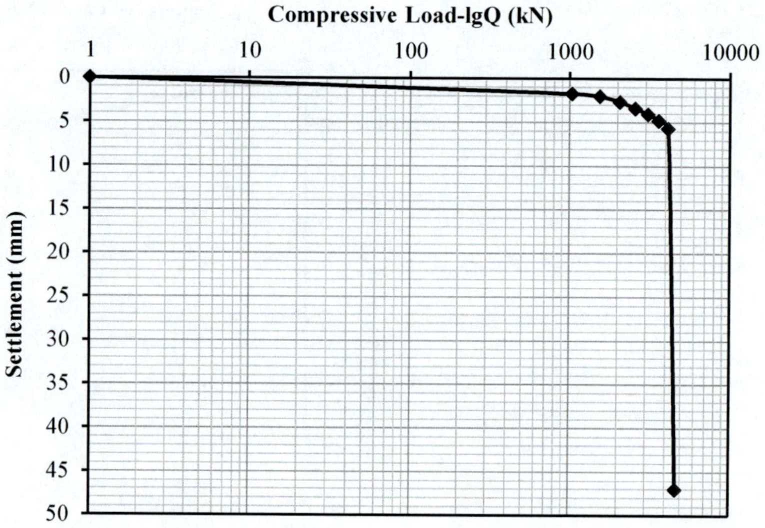

Fig. 7.23 s-lgQ curve of P40

capacity was obtained, which was equal to 1/0.0001 = 10,000 kN. A summary of these pile capacities is provided in Table 7.3.

As shown in Table 7.3. It can be seen that the ultimate bearing capacity obtained from traditional methods (Q-s, s-lgt and s-lgQ curves) and interpretations (double tangent, Davission's and Chin's) are generally close. For the standard pile, comparing with the ultimate bearing capacity of grouted pile as depicted in Table 7.1 (4,577 kN), the tested pile's capacity is overall 4,777 kN, which illustrates the design method provided by JGJ 94-2014 (2014) (Grouting) is ideal. It should be noted that, in this chapter, the diagram based on the Chin's method is determined using the test data before failure occurred (the maximum applied loads with corresponding settlement are not used in the plotted Chin's diagram). The reason to do this is to check if the Chin's method can provide appropriate results under non-failure condition or if this method can be used for capacity determination under Proof Tests (PTs). As shown in Table 7.3, the capacity obtained by Chin's method is close to the results acquired from the other methods, which confirms that the Chin's method can be appropriately used for pile capacity prediction when the PTs being performed. Also, it found that the Chin's method cannot be used when pile suffered from eccentricity.

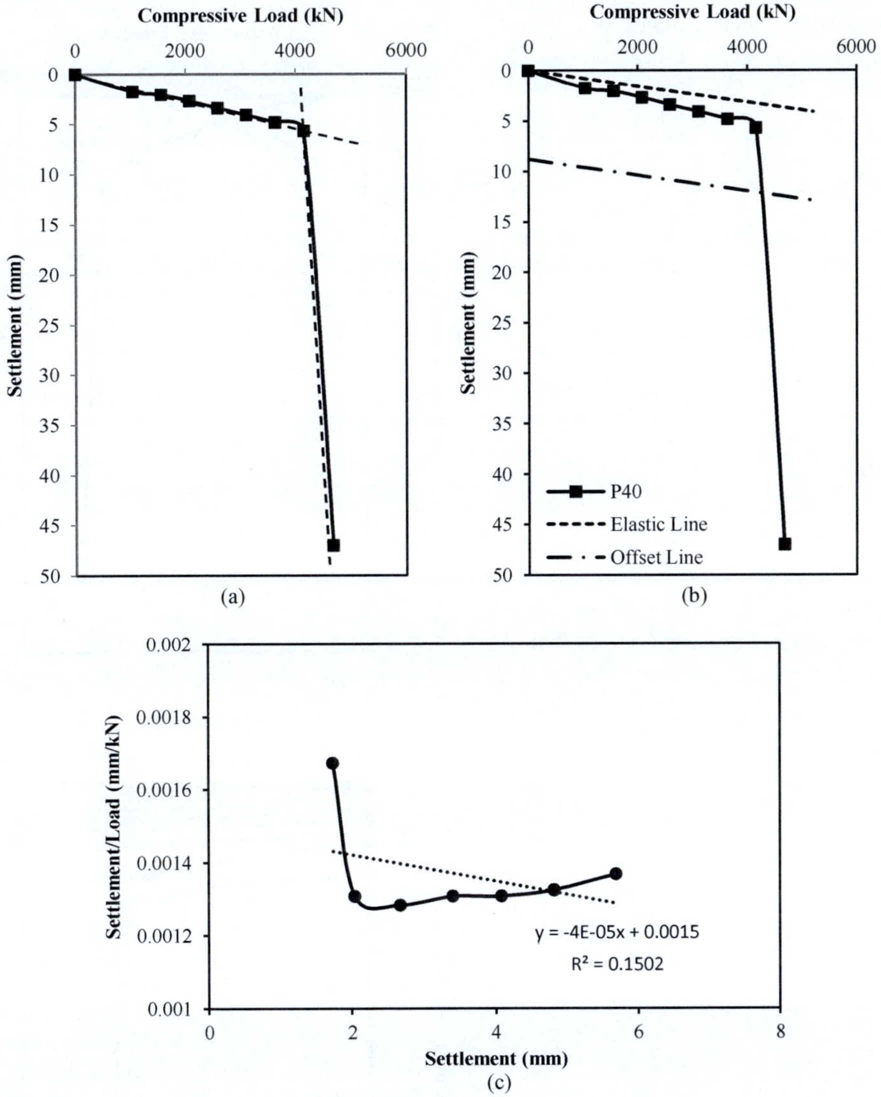

Fig. 7.24 Interpretations **a** Double tangent method **b** Davisson's offset method **c** Chin's method

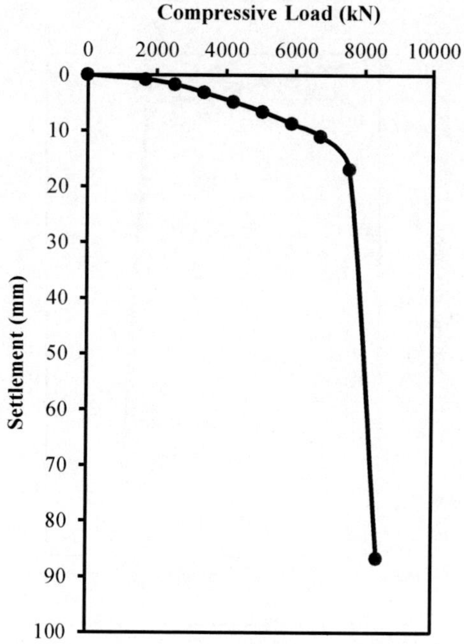

Fig. 7.25 Q-s curve of P12

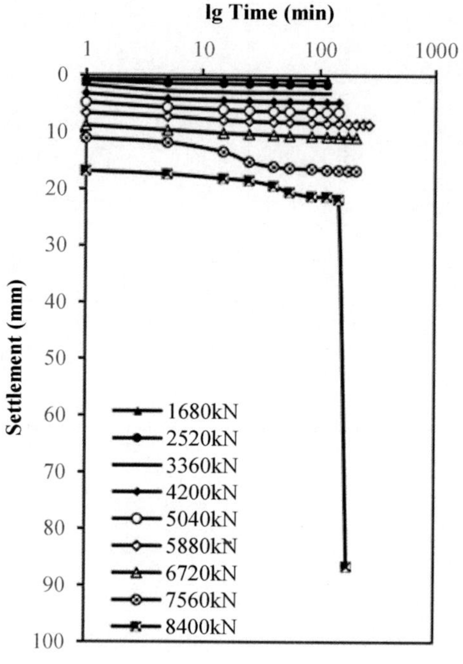

Fig. 7.26 s-lgt curve of P12

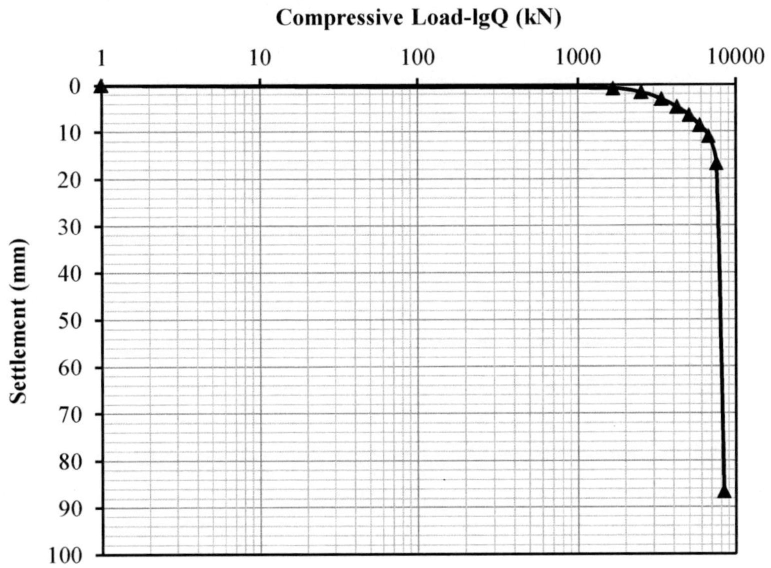

Fig. 7.27 s-lgQ curves of P12

7.8 Concluding Remarks

This chapter has presented a pile investigation for performing compressive SLTs with the ultimate load applied. To determine the pile behaviour under ultimate loading, four piles were cast. Three types of tests that lead to failure were considered. The first test involved loading the pile to failure caused by inadequate concrete strength, the second test involved loading the pile to failure because of eccentricity and the final test involved loading the pile until failure because of the soil being soft (or the pile being overdesigned for its capacity). The behaviours of these three piles were compared with a pile that achieved the design. In this chapter, after the conventional result presentation as well as different data interpretations being provided, the pile capacity of these tested piles was summarised. To check whether the Chin's method can provide appropriate results under non-failure condition or to verify if this method can be used for capacity determination under PTs, the interpreted diagrams by Chin's method was based on the obtained test data before failure occurred (the maximum applied loads with corresponding settlement are not used in the plotted Chin's diagram).

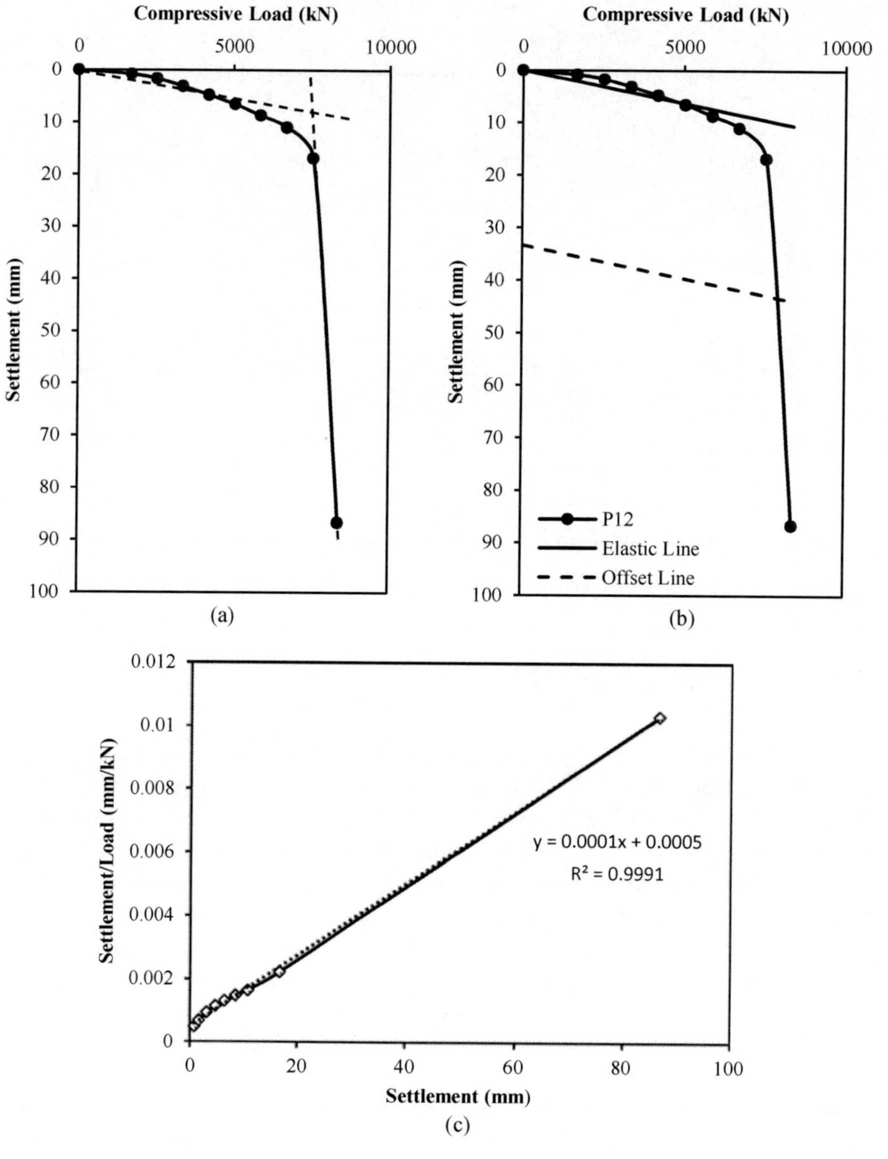

Fig. 7.28 Interpretations **a** Double tangent method **b** Davisson's offset method **c** Chin's method

Table 7.3 Summarisation of the pile capacity

Pile label	Pile condition	Q-s (kN)	s-lgt (kN)	s-lgQ (kN)	Double tangent (kN)	Davisson's (kN)	Chin's (kN)
P80	Standard pile	4,680	5,200	4,680	3,900	>5,200	5,000
P60	Inadequate concrete strength	3,640	3,640	3,640	3,600	>3,600	3,333.3
P40	Eccentricity	4,160	4,160	4,160	4,190	4,250	N/A
P12	Inadequate soil strength	7,560	7,560	7,560	7,560	7,900	10,000

References

American Association of State Highway and Transportation Officials (AASHTO). (2002). *Standard Specifications for Highway Bridges*, Washington, D. C.

Castelli, R. J., & Wilkins, E. (2004). Osterberg load cell test results on base grouted bored piles in Bangladesh. *In GeoSupport 2004: drilled shafts, micropiling, deep mixing, remedial methods, and specialty foundation systems* (pp. 587–602).

Handley, B., Ball, J., Bell, A., & Suckling, T. (2006). *Federation of piling specialists handbook on pile load testing*. Forum Court: FPS.

Ho, C. E. (2003). Base grouted bored pile on weak granite. In *Proceedings of the Third International Conference on Grouting and Ground Treatment* (pp. 716–727).

Li, X., Xie, K., Zeng, G., & Hou, X. (2000). Research of bored pile slurry effect created during construction. *Structural Construction, 30*(5), 21–23.

Ministry of Construction of the People's Republic of China. (1999). *Standard for soil test method (GB/T 50123-1999)*. Beijing, China: Ministry of Construction of People's Republic of China.

Ministry of Construction of the People's Republic of China. (2001). *Code for investigation of geotechnical engineering (GB 50021-2001)*. Beijing, China: Ministry of Construction of People's Republic of China.

Ministry of Construction of the People's Republic of China. (2013). *Standard for test methods of engineering rock mass (GB/T 50266-2013)*. Beijing, China: Ministry of Construction of People's Republic of China.

Ministry of Construction of the People's Republic of China. (2014). *Technical code for testing of building foundation piles (JGJ 106-2014)*. Beijing, China: National Standard of the People's Republic of China.

Nguyen, V. L., Nie, L., & Zhang, M. (2012). Method cement post-grouting to increase the load capacity for bored pile. *Research Journal of Applied Sciences, Engineering and Technology, 5*(19), 4727–4732.

Patel, D., Glover, S., Chew, J., & Austin, J. (2015). The pinnacle-design and construction of large diameter deep base grouted piles in London. *Ground Engineering*, 24–31.

Shi, C. (2005). The application of the pile-end mud-jacking technique in the construction of bored caisson pile. *Sci/Tech Information Development & Economy, 15*(23), 293–296.

Sinnreich, J., & Simpson, R. C. (2013). Base grouting case studies including full scale comparative load testing. In *Seventh International Conference on Case Histories in Geotechnical Engineering*. No. 2.16, pp. 1–8.

Standards Australia. (2009). *Piling—design and installation* (AS 2159) (pp. 1–97).

Zhou, J. L., Zhang, X., Jiang, H. S., Lyu, C., & Oh, E. (2017). Static and dynamic load tests of shaft and base grouted concrete piles. *Advances in Civil Engineering, 2017*, 1–11.

Chapter 8
Capacity and Settlement Analysis

8.1 General Introduction

This chapter presents the results discussion for the post grouted piles, displacement and non-displacement piles, composite piles and piles tested until ultimate load, with consideration of the state-of-art construction technology.

For the grouted piles, the behaviours of piles under various grouting technologies are compared, and the ultimate compressive and uplift bearing capacity is determined though SLTs and dynamic load tests. These traditional results are presented and compared with the extrapolated results. Further, the static and dynamic results are compared and discussed in reference to the ultimate bearing capacity. In addition, the method used for displacement prediction under compressive and uplift loads is provided.

After the investigation of post grouted bored piles, the investigations of precast piles are presented in Sect. 8.3, which include small and large displacement piles and non-displacement piles. For the displacement piles, the pile behaviours driven from different locations are discussed. The behaviours of small and large displacement piles with various lengths are researched with consideration of soil layer analysis. Further, the behaviours of non-displacement piles suffering from uplift and horizontal loading are discussed, and the empirical formula for pile displacement prediction are provided.

This chapter also presents a discussion of the composite piles. For the piles with FRP confinement, the capacity obtained from various interpretation methods is determined and compared, and the improved capacity caused by FRP confinement is examined. For the FRP inside reinforced pile, the horizontal deflection of this pile is compared with traditional pile (steel bar reinforcement). The discussion of composite pile considering the construction process (excavation depth and strut application) is also illustrated.

J. Zhou and E. Oh, *Full-Scale Field Tests of Different Types of Piles*,
Advanced Topics in Science and Technology in China 62,
https://doi.org/10.1007/978-981-33-6183-6_8

Finally, the behaviours of piles applied with ultimate loading are discussed. The result presentations are compared and the interpretation methods are discussed. Moreover, the behaviour of piles with three types of failure conditions are detailed.

After working through the material in this chapter, the reader can analyse the settlement and capacity of the super-long and large diameter pile by themselves using the same procedures. In detail, for the capacity analysis of these piles, the interpretation methods in Sect. 8.2.1.1 should be used, the readers can also base on the methods that are illustrated in Chaps. 2 or 3. For the settlement prediction and empirical formula determination, find the method in Sect. 8.2.2.1.

8.2 Post Grouted Concrete Piles

8.2.1 Capacity Discussion

8.2.1.1 Compressive Static Load Tests

The load settlement results of three post grouted piles (P51, P121 and P126) obtained from compressive SLTs illustrated the settlement behaviours under increment and decrement loading. It was found that the final settlements of P51 (no grouting pile) and P121 (base grouting pile) were 8.11 and 7.36 mm, respectively. This represented a small decrement of total settlement after using base grouting technology, which is highly because the end bearing stratum being the diorite. In other words, the base grouting may effectively decrease the total settlement for a pile that reaches the stratum such as loose sands. P126 showed a small total settlement of 2.8 mm. This demonstrated that the shaft grouting would decrease the total settlement of the pile foundation.

Presenting the load settlement data into logarithmic scale (s-lgQ) as illustrated in Fig. 3.8. Through s-lgQ curvatures, the ultimate compressive bearing capacities of the piles could be determined, albeit with difficulty, to be 6,800, 7,100 and 8,000 kN, respectively. This is because the tests were PTs, which aimed to ensure the pile settlement was acceptable under the maximum loading (equal to twice the design load). The non-failure settlements led to difficulty in determining the capacity from the s-lg Q curves. Thus, determination of the ultimate pile capacity had to be based on interpretation.

Similar to the analysis of s-lgQ curves, as shown in Fig. 3.15 (DeBeer's method interpretation), the ultimate bearing capacities of P51, P121 and P126 obtained from the plotted lgs–lgQ curves were determined, with difficulty, to be 7,600, 7,900 and 8,300 kN, respectively. For Davisson's offset limit method, the intersection between the load–settlement curve and offset line could not be found as shown in Fig. 3.16, which illustrated that the ultimate bearing capacity was over the maximum applied loads. Thus, more comprehensive analysis was still needed. Another common method to plot the failure criteria of a drilled pile is the double tangent method, emphasised by AASHTO (2002) and FHWA (1992) (Samtani and Nowatzki 2006b). As shown in Figs. 3.12, 3.13 and 3.14, through finding two tangent lines from the loading curve, the intersection represents the ultimate bearing capacity and corresponding settlements. The capacities of P51, P121 and P126 were determined to be 7,560, 7,700 and 8,250 kN, respectively. The double tangent method and DeBeer's method both aim to find a point to represent a 'turning trend' through two tangent lines and logarithmic scales; thus, the settlements determined from these two methods were close to each other, as shown in Table 3.4.

As shown in Fig. 3.17, the slopes of these three piles were determined to be 1E-4, 9E-5 and 8E-5, and the inverse of these slopes gave the ultimate capacity. Thus, the capacity of piles P51, P121 and P126 were determined to be 10,000, 11,111 and 12,500 kN. It was found that Chin's method provided the greatest values, compared with the other methods. All results obtained from the different interpretation methods demonstrated that the base grouting and base-and-shaft grouting improved the ultimate bearing capacity of the piles. For instance, based on double tangent method, the capacity of the base-and-shaft grouting pile (P126) was 8,250 kN, and the base grouting pile (P121) was 7,700 kN. These two values were greater than the non-grouted pile (P51), which had a capacity of 7,560 kN. This was because the grouting technique increased the shaft and base area to resist the compressive working loads. Further, the results obtained from Chin's method and the dynamic methods were relatively close to each other.

8.2.1.2 Uplift Static Load Tests

The load displacement behaviours of piles suffering from uplift loading are shown in Fig. 3.18. It can be seen that, during the loading stages from 0 to 3,000 kN, the displacement between two piles were similar, and these two increment load lines were approximately linear. By plotting these data into s-lgQ curves, no evident 'turning point' was found, which aligned with the load settlement results (approximately linear from the loading stages). Through analysing the s-lgt diagrams, as shown in Figs. 3.20 and 3.21, it was found that the upwards displacement was stable overall with time before the next loading was applied. Further, there was no evident increase in upwards displacement between the two loading stages.

As depicted in Fig. 3.22, the interpretation of uplift SLTs through the offset method gave the ultimate bearing capacity of P15 and P16 to be 1,500 and 1,700 kN, respectively. However, as discussed above, all result analyses of the Q-s, s-lgQ and s-lgt curves indicated that the ultimate bearing capacity was greater than the maximum loading of 3,000 kN. Hence, the ultimate load capacity or failure load obtained through the offset method will underestimate the real uplift ability of the pile; thus, the recommended uplift design load (stated by FHWA NHI-16-009), which is equal to half to two-thirds of the failure load, will be conservative. Another interpretation for the uplift loaded tests was using the modified Mazurkiewicz method. As illustrated in Figs. 3.23 and 3.24, two functions were determined, and, when $x = 0$, the ultimate uplift bearing capacities of P15 and P16 were 6,299 and 5,442 kN, respectively. This indicated that the base-and-shaft grouted technology increased the ultimate uplift bearing capacity.

8.2.1.3 Dynamic Load Tests

Through dynamic load tests, the shaft resistances along the pile and the end bearing of the pile could be determined. As provided in Fig. 3.27, the shaft resistance of P121 along the pile length was not uniformly distributed compared with P126 as depicted in Fig. 3.28. This was because the shaft grouting material which has mixed with soil layers resulting in the shaft around the pile demonstrated a relatively uniform property. From the load transfer mechanism diagrams of P121 and P126, it was found that the transferred loads decreased with depth.

As illustrated in Table 3.6, the ultimate bearing capacities of P121 and P126 were determined to be 9,071 and 9,851.5 kN, respectively. Further, the shaft resistances of P121 and P126 were determined to be 4,217.3 and 4,850.0 kN. This represented that the base-and-shaft grouting increased the total pile capacity (9,071 kN < 9,851.5 kN) and shaft resistance (4,217.3 kN < 4,850.0 kN). The phenomenon was caused by the post grouting technique, which increased the total friction area between the concrete surface and soil layers. It can also be seen that the grouting technique could also improve the end bearing capacity of piles (9,851.5–4,850.0 > 9,071–4,217.3). Compared with shaft resistance obtained from dynamic load tests and interpretation by modified Mazurkiewicz's method, this interpretation overestimated the ultimate uplift bearing capacity (4,850 kN < 6,299 kN).

From the ultimate compressive SLT results, the results indicated that: (1) the base-and-shaft grouting pile increased about 9.82% of its capacity without any grouting; (2) the base grouting pile increased approximately 2.89% of its capacity without grouting; (3) the capacity of the shaft-and-base grouted pile increased by 6.1% of the base grouting pile, and this value was close to the results obtained from the dynamic load tests (8.6%). By comparing the ultimate uplift SLT results, it could be determined that there was a 15.7% increment in the ultimate pile capacity after using the base-and-shaft grouting technology. By comparing the shaft resistance capacities of the base grouted piles and base-and-shaft grouted piles obtained from the dynamic load test, a 15.0% increment was demonstrated.

8.2.2 Settlement Discussion

8.2.2.1 Compressive Loaded Piles

The maintained SLTs could provide information about which loading stage was unstable. As shown in Fig. 8.1, during each loading stage, the soil foundation system was unstable within a duration of 30 min, and, later, the soil foundation system became stable with increasing time, during which the load was maintained. It can also

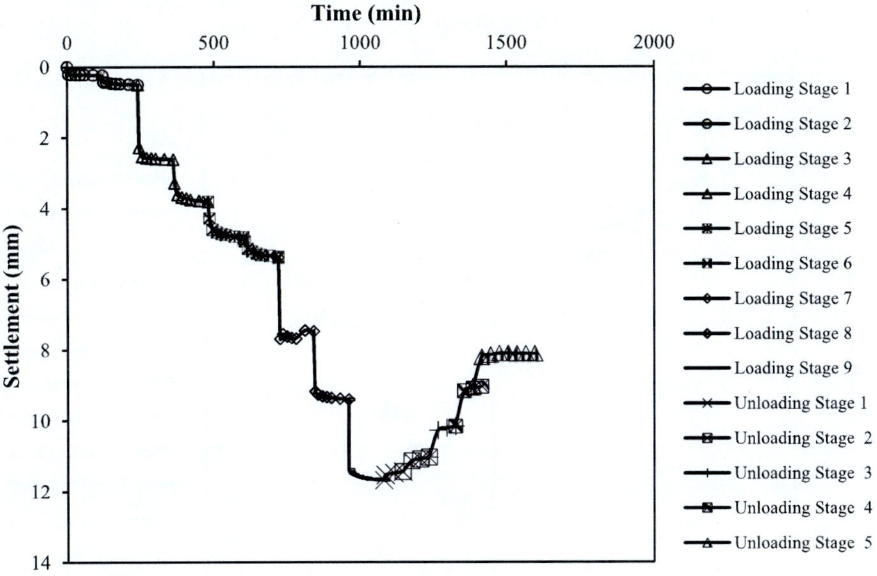

Fig. 8.1 Settlement curves versus maintained time of P51

be seen that, after the loading of Stage 7 was applied, huge settlement was discovered, and the soil foundation system was unstable after a relatively long duration. Thus, the loading of Stage 7 was found to be the critical loading. Similarly, from Figs. 8.2 and 8.3, the critical loading stages of P121 and P126 were determined to be loading Stage 9. From the Figs. 8.1, 8.2 and 8.3, it can also be seen that, during each unloading stage, the soil foundation systems were stable with maintained loading applied. Further, by comparing the final settlements of these three piles, the results indicated that the settlement of the base-and-shaft grouted pile was less than the base grouted pile, and the settlements of these two piles were both less than the traditional pile without grouting application.

Define the settlement ratio $R_s = S_t/s_{(F_i)}$. Under a certain compressive loading, the settlement at a certain time (S_t) divided by the settlement ($s_{(F_i)}$) which represents the final settlement at the end of this loading will be less than 1, but over 0. The value $R_s = 1$ illustrates that the soil is stable. For example, for the pile P126, from a loading stage of 7,200 kN, before a time of 120 min, $S_t/s_{(F_i)}$ is less than 1. After the time of 120 min, $S_t/s_{(F_i)} = 1$, the next loading stage will be applied soon because the settlement is stable at this 7,200 kN load. By plotting the time versus the settlement ratio, R_s, which is obtained from various loading stages, scatter diagrams of P51, P121 and P126 can be obtained, as shown in Figs. 8.4, 8.5 and 8.6. By using the non-linear regression, a settlement equation is proposed, as illustrated in Eq. 8.1:

$$S_t = s_{(F_i)} \times \frac{t^n}{m^n + t^n} \tag{8.1}$$

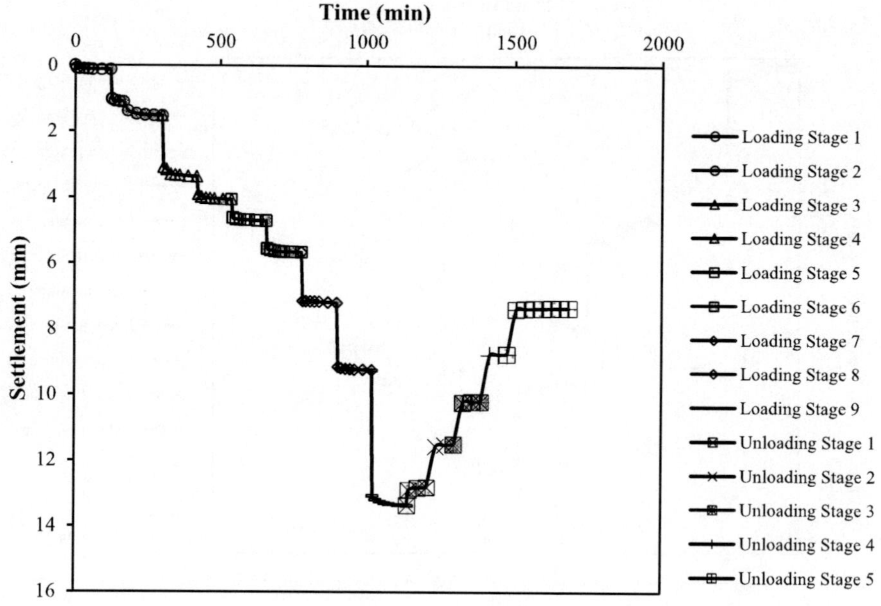

Fig. 8.2 Settlement curves versus maintained time of P121

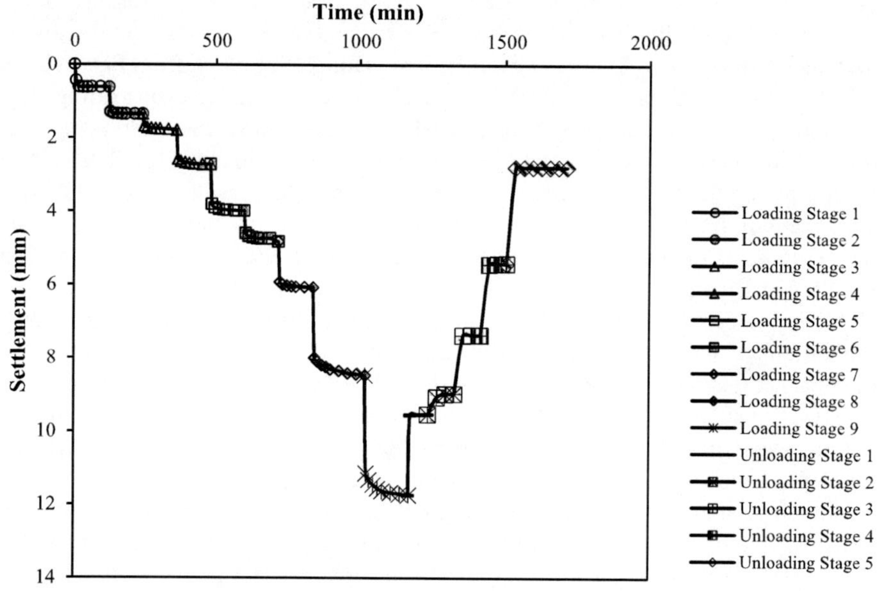

Fig. 8.3 Settlement curves versus maintained time of P126

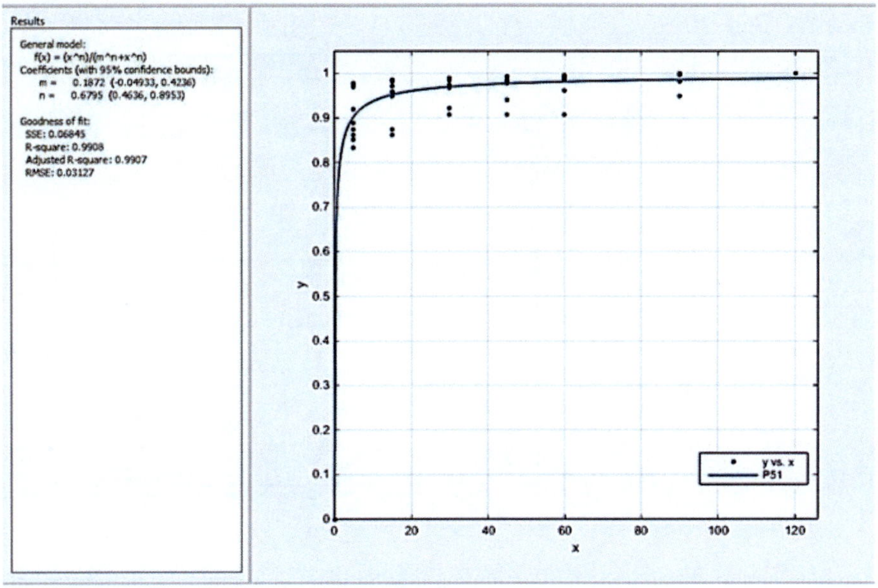

Fig. 8.4 Non-linear regression of P51 (loading stage)

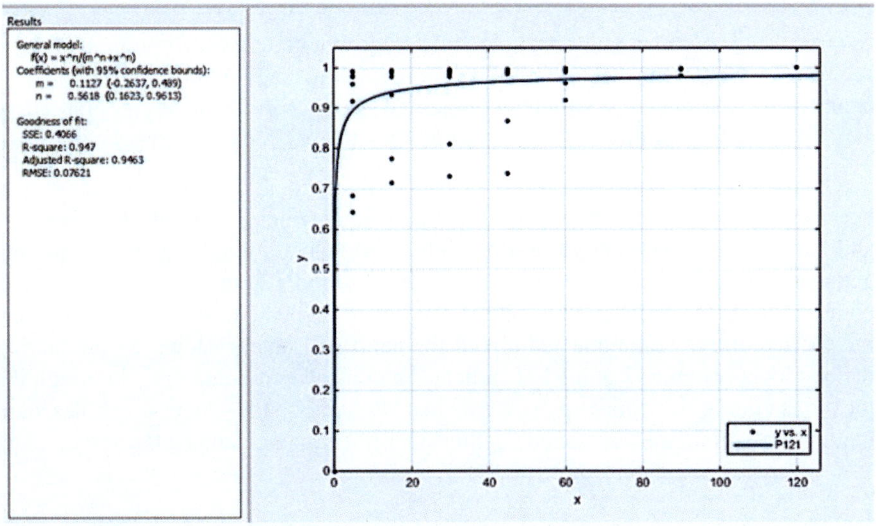

Fig. 8.5 Non-linear regression of P121 (loading stage)

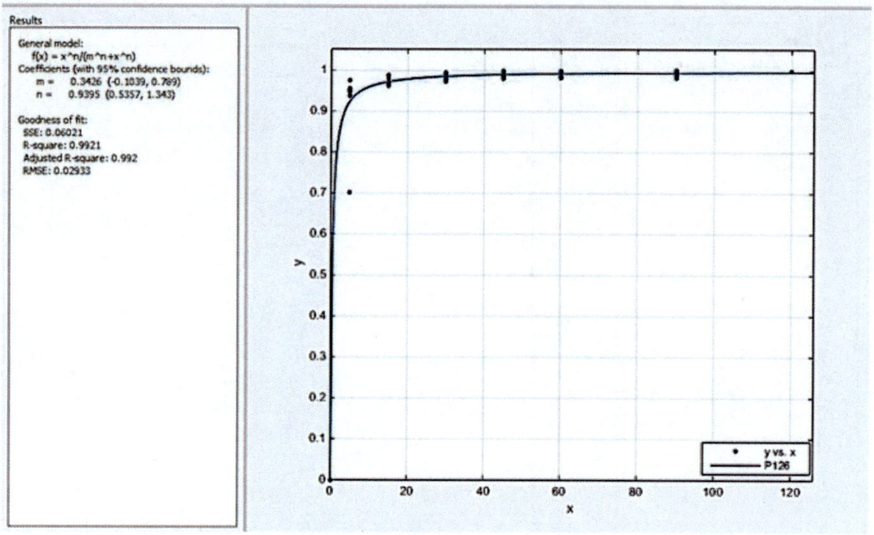

Fig. 8.6 Non-linear regression of P126 (loading stage)

where:

$S_t =$	Settlement at a certain loading stage at a certain time in mm
$s_{(F_i)} =$	Settlement at the end of a certain loading stage in mm
m and $n =$	Settlement coefficients obtained from non-linear regression curve (Table 8.1).

Comparing the scatter diagrams of P51 with P126, P126 shows better results because most points assemble on the regression line, which represents that the shaft-and-base grouting technology reinforced the soil layers around the pile foundation and made the soil layers stable. Comparing P121 with P126 indicates that the points obtained from the base grouted pile (P121) were dispersed, which illustrates that the shaft grouting technique reinforced the shaft soil layers. However, comparing the results between P51 and P121 indicates a controversial outcome. Although the locations among these three piles were close with each other, more boreholes were required to explain the mentioned conflict. This phenomenon could also be caused by

Table 8.1 Settlement coefficients of compressive loaded bored pile

Type of pile	Compressive loading			Compressive unloading		
	m	n	R^2	m	n	R^2
Traditional pile	0.1872	0.6795	0.9908	0.8265	4.814	0.9996
Base grouted pile	0.1127	0.5618	0.947	0.3928	3.268	0.9999
Base-and-shaft grouted pile	0.3426	0.9395	0.9921	0.5395	3.942	0.9999

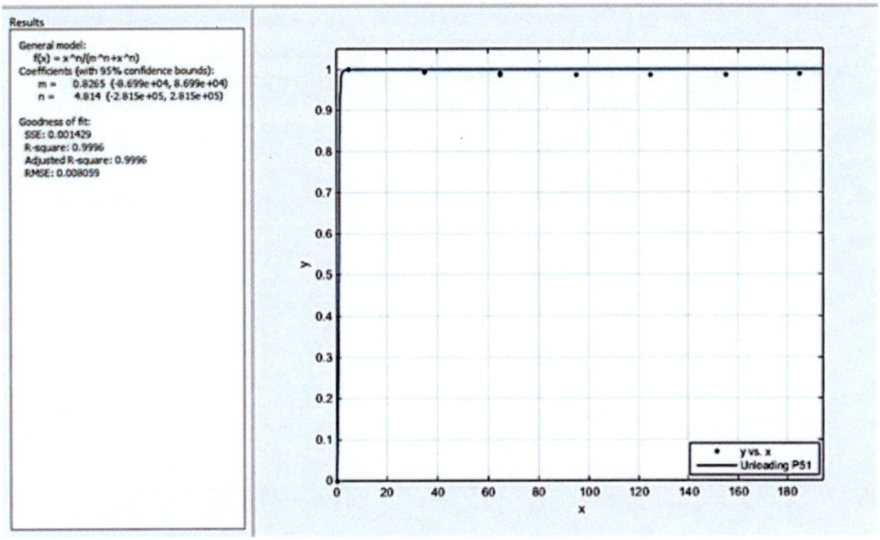

Fig. 8.7 Non-linear regression of P51 (unloading)

the construction process associated with the disturbance of soil layers. For example, the vibrations during grouting may cause the soil to be disturbed.

Equation 8.1 was only used to predict the settlement from loading stages. In order to obtain the empirical equation which was used to predict the settlement during unloading stages, define the settlement ratio $R_s = S_t/s_{(in)}$. Under a certain compressive unloading, the settlement at a certain time (S_t) divided by the settlement ($s_{(in)}$) which represents the initial settlement at the beginning of this unloading will be less than 1, but over 0. The value $R_s < 1$ illustrates that the soil is relatively stable. By plotting the time versus the settlement ratio, R_s, which is obtained from various unloading stages, scatter diagrams of P51, P121 and P126 can be obtained, as shown in Figs. 8.7, 8.8 and 8.9. By using non-linear regression, a settlement equation is proposed, as illustrated in Eq. 8.2:

$$S_t = s_{(In)} \times \frac{t^n}{m^n + t^n} \tag{8.2}$$

where:

$S_t =$	Settlement at a certain loading stage at a certain time in mm
$s_{(In)} =$	Settlement at the beginning of a certain loading stage in mm
m and $n =$	Settlement coefficients obtained from non-linear regression curve (Table 8.1).

These two empirical equations can be used to predict the settlement under a certain loading at a certain time, while the coefficients of m and n can be determined from Table 8.1. Equation 8.1 is similar to the equations proposed by Yang et al. (2012),

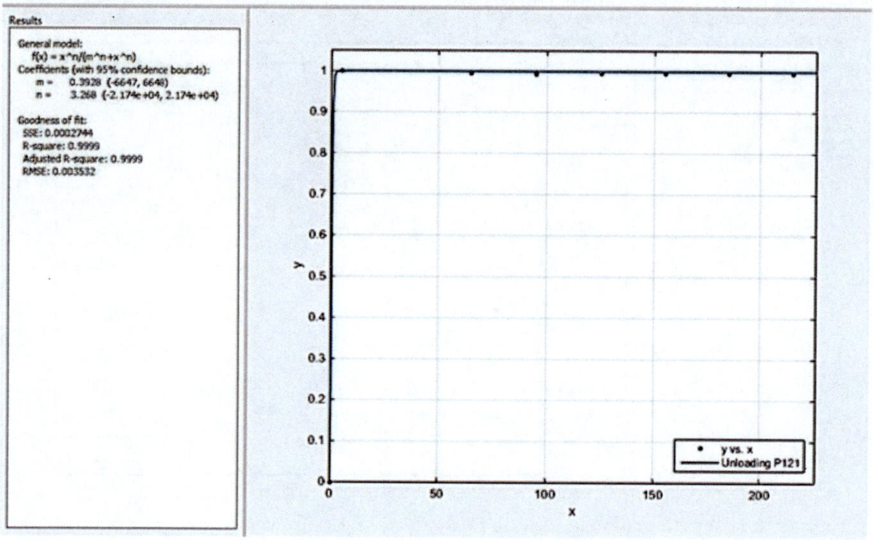

Fig. 8.8 Non-linear regression of P121 (unloading)

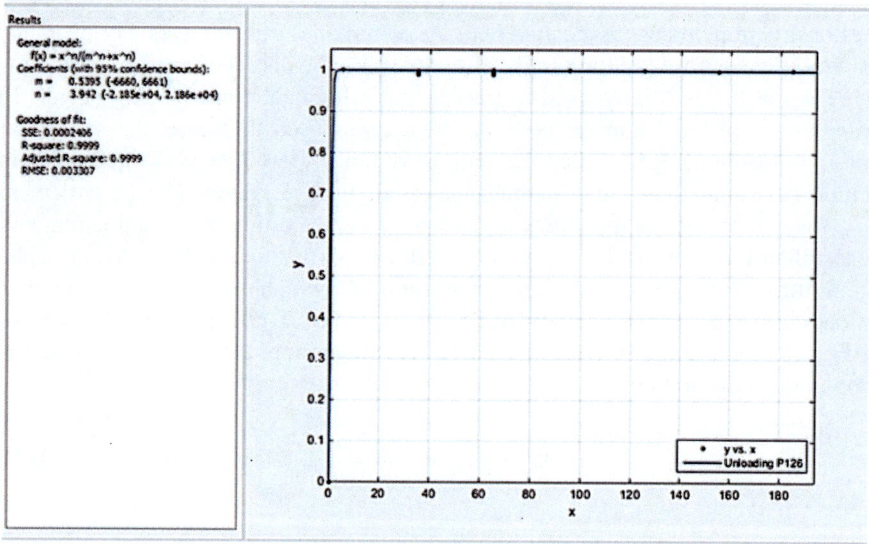

Fig. 8.9 Non-linear regression of P126 (unloading)

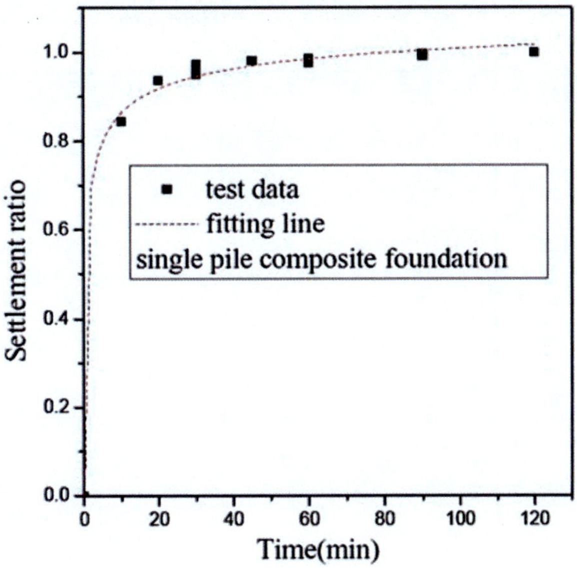

Fig. 8.10 Non-linear regression of single pile (Yang et al. 2012)

with different coefficient values of m and n. As shown in Fig. 8.10, the scatters assembled very close to each other. This was because the tests loaded to a maximum of 200 kPa, where loading was relatively small, and the record frequency (number of times) are lesser under the maintained loads. In contrast, in the case of P51, P121 and P126, where loading was up to 9,000 kN (17,914 kPa), they required much more time to allow the soil settlement to be stable, especially from the beginning of the loading (five to 60 min in each loading stage).

Based on the proposed equations of Eqs. 8.1 and 8.2, the predicted settlement and test settlement of P51, P121 and P126 from the loading stages and unloading stages are provided in Figs. 8.11, 8.12, 8.13, 8.14, 8.15 and 8.16. It can be seen that these two equations provided accurate computed results.

Equation 8.1 was proposed based on the settlement data from time five to 120 min (the settlement difference was less than 0.1 mm when the time was beyond 120 min; thus, these data were sufficient to determine the fitting line). As shown in Table 8.2, for P126, it was found that, beyond a time of 120 min, the predicted settlements at a time of 150 and 180 min were very close to the test settlements. For another example, for P51, from the loading stage of 5,400 kN, when the time was beyond 120 min, the predicted settlement was 5.312 mm, which was very close to the collected test settlement value of 5.395 mm. Similarly, Eq. 8.2 could also predict the settlement value beyond a time of 95 min.

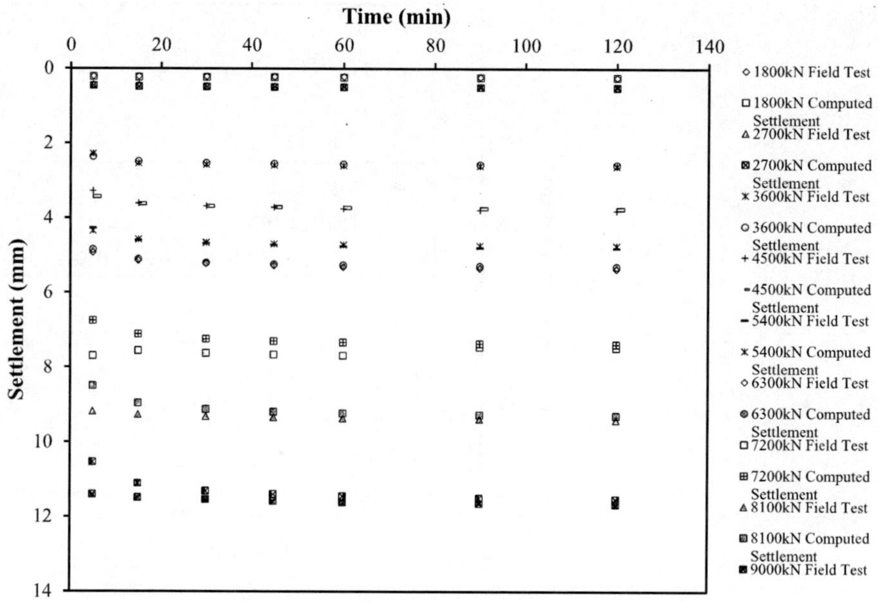

Fig. 8.11 Test and computed settlements from loading stages of P51

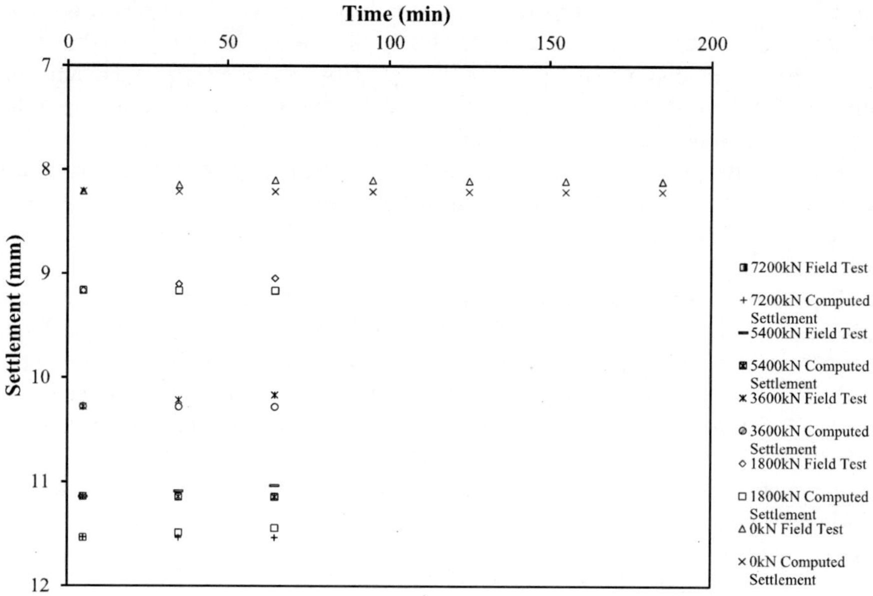

Fig. 8.12 Test and computed settlements from unloading stages of P51

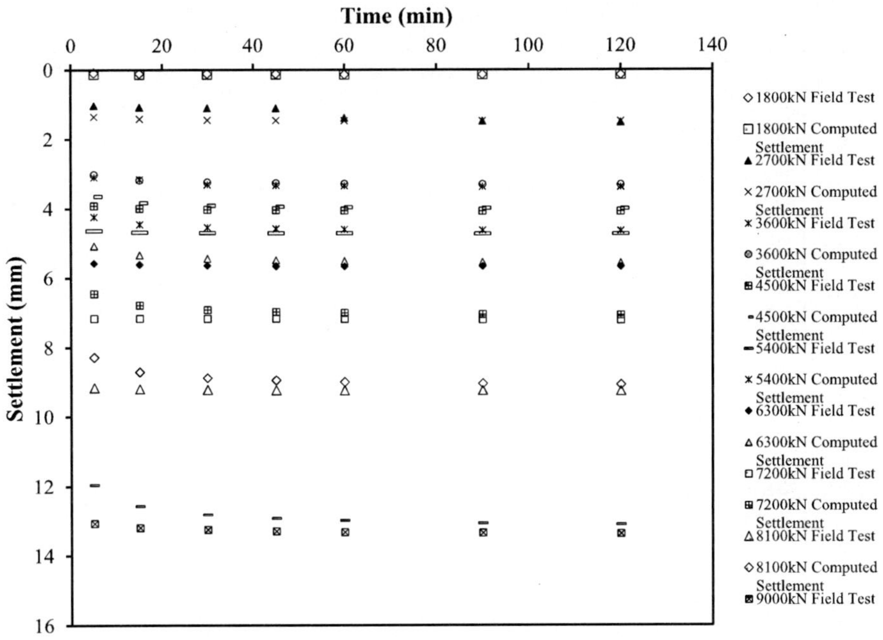

Fig. 8.13 Test and computed settlements from loading stages of P121

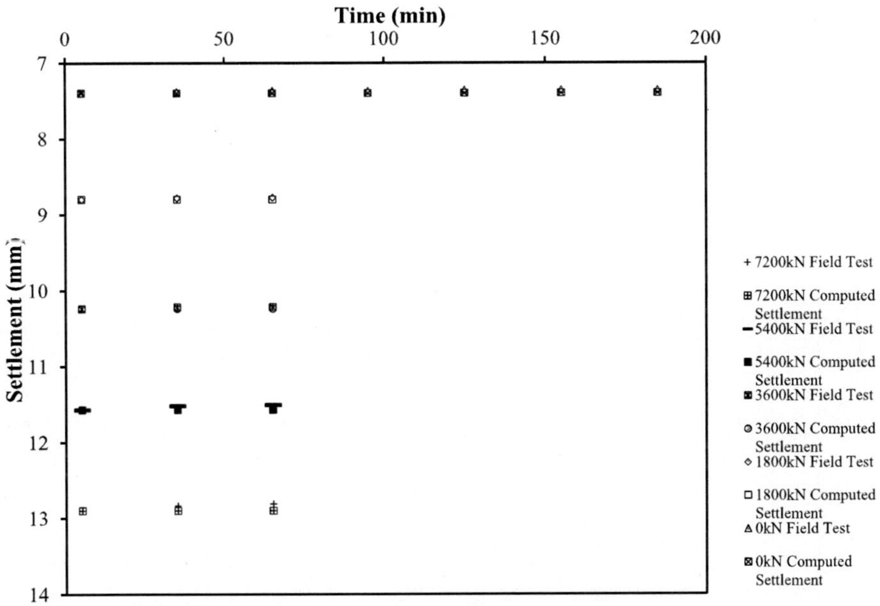

Fig. 8.14 Test and computed settlements from unloading stages of P121

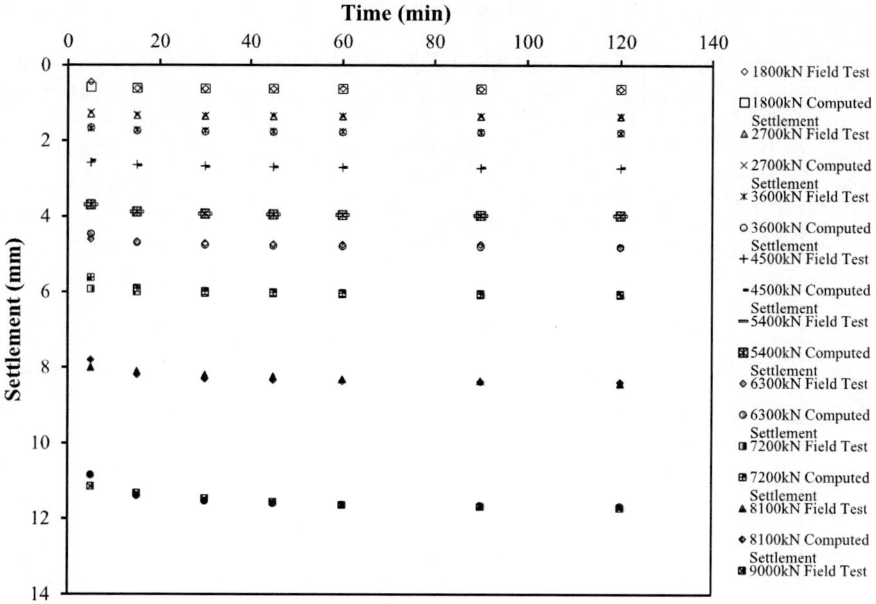

Fig. 8.15 Test and computed settlements from loading stages of P126

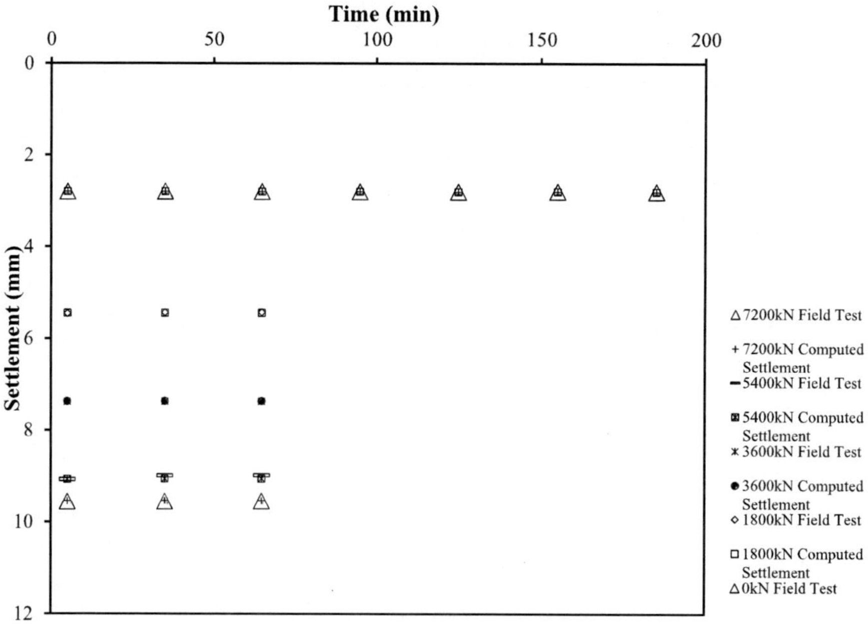

Fig. 8.16 Test and computed settlements from unloading stages of P126

Table 8.2 Test and computed settlements of P126 from loading 8,100 kN

Time (min)	Load (kN)	LVDT1 (mm)	LVDT2 (mm)	LVDT3 (mm)	LVDT4 (mm)	Average (mm)	Computed (mm)	Information
5	8,100	8.12	8.41	6.69	8.77	7.9975	7.998	Data used for non-linear regression
15	8,100	8.17	8.49	6.82	8.91	8.0975	8.098	
30	8,100	8.24	8.6	6.93	9.01	8.195	8.195	
45	8,100	8.29	8.67	6.96	9.06	8.245	8.245	
60	8,100	8.34	8.72	7.06	9.13	8.3125	8.313	
90	8,100	8.38	8.77	7.09	9.16	8.35	8.350	
120	8,100	8.44	8.83	7.17	9.24	8.42	8.420	
150	8,100	8.46	8.84	7.2	9.27	8.4425	8.443	Data used for comparison
180	8,100	8.49	8.87	7.23	9.29	8.47	8.470	

8.2.2.2 Uplift Loaded Piles

The vertical displacements of two piles versus time are provided in Figs. 8.17 and 8.18. As shown in these diagrams, it was found that, during each loading stage, the settlement trends during each maintained loading were almost parallel to the x-axis, and the displacement difference between two loading stages were small and similar, which represented no critical loading information. By comparing these gradients of each loading/unloading stage between two piles, it was discovered that the settlement

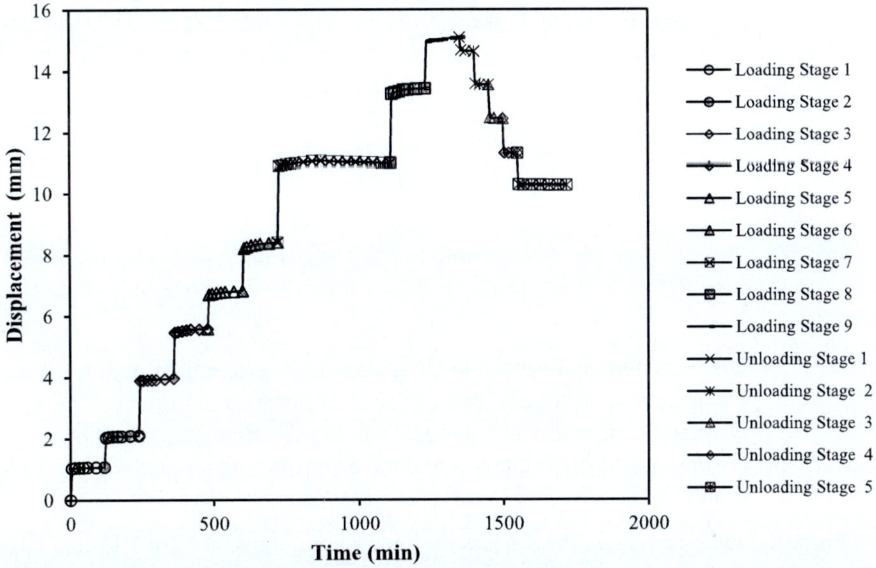

Fig. 8.17 Vertical displacement of P15 under maintained load

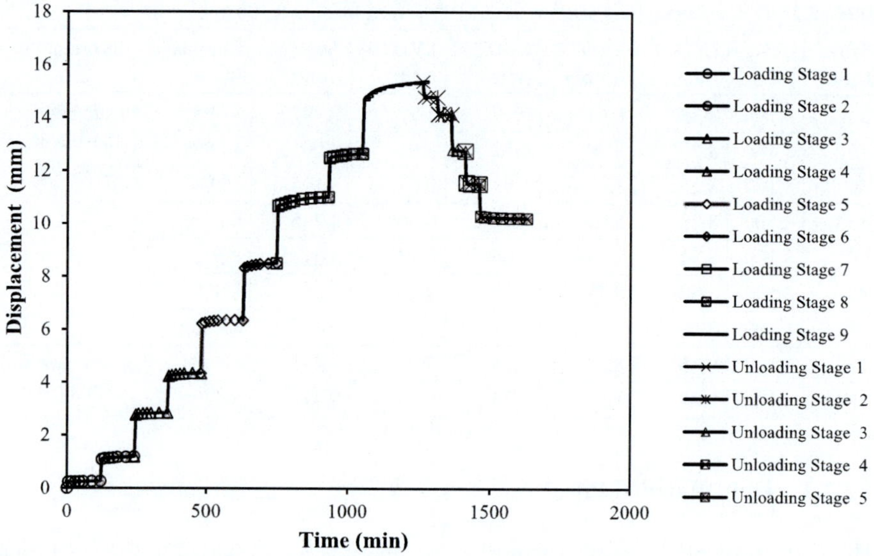

Fig. 8.18 Vertical displacement of P16 under maintained load

of P16 was unstable. In other words, the gradient of P16 was relatively higher than P15, which indicated that the base-and-shaft grouting technology improved the uplift displacement behaviour.

Similar to the method illustrated in the previous section which is used to determine the non-linear regression lines, define the displacement ratio $R_d = D_t/D_{(F_i)}$ and displacement ratio $R_d = D_t/D_{(in)}$ to determine the non-linear regression lines (uplift loaded bored piles) during loading and unloading stages, respectively:

$$D_t = D_{(F_i)} \times \frac{t^n}{m^n + t^n} \tag{8.3}$$

$$D_t = D_{(In)} \times \frac{t^n}{m^n + t^n} \tag{8.4}$$

where:

$D_t =$ Displacement at a certain loading stage at a certain time in mm
$D_{(F_i)} =$ Displacement at the end of a certain loading stage in mm
$D_{(in)} =$ Displacement at the beginning of a certain loading stage in mm
m and $n =$ Displacement coefficients obtained from non-linear regression curve (Table 8.3).

From the loading stages, the regression lines were determined for P15 and P16, as shown in Figs. 8.19 and 8.20, respectively. It was found that the data of P16 scattered, and the points of P15 assembled better and close to the regression line. This illustrated that the grouting technology made the soil layers stable when the

Table 8.3 Settlement coefficients of uplift loaded bored piles

Type of pile	Uplift loading			Uplift unloading		
	m	*n*	R^2	*m*	*n*	R^2
Base-and-shaft grouted pile	0.00702	0.581	0.9997	0.01761	1.359	1.0
Traditional pile	0.01287	0.5059	0.9952	0.4913	3.921	0.999

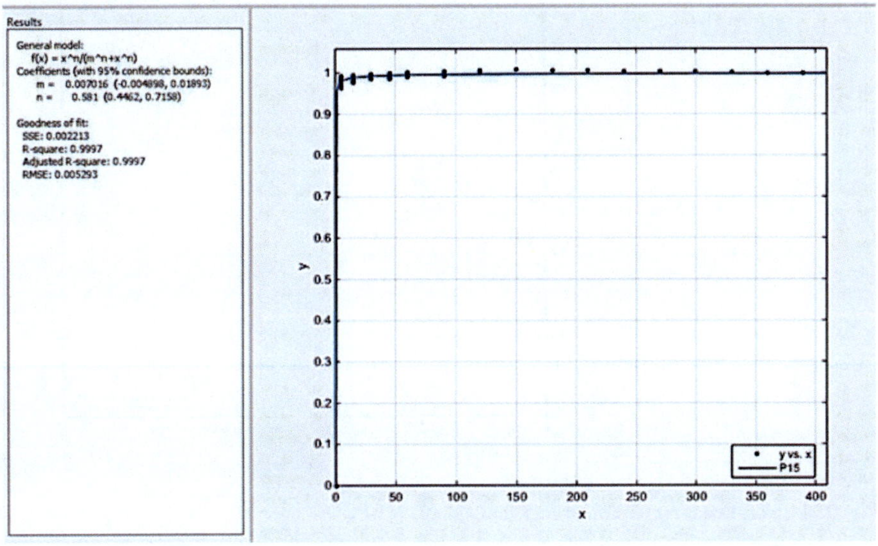

Fig. 8.19 Non-linear regression of P15 (loading stage)

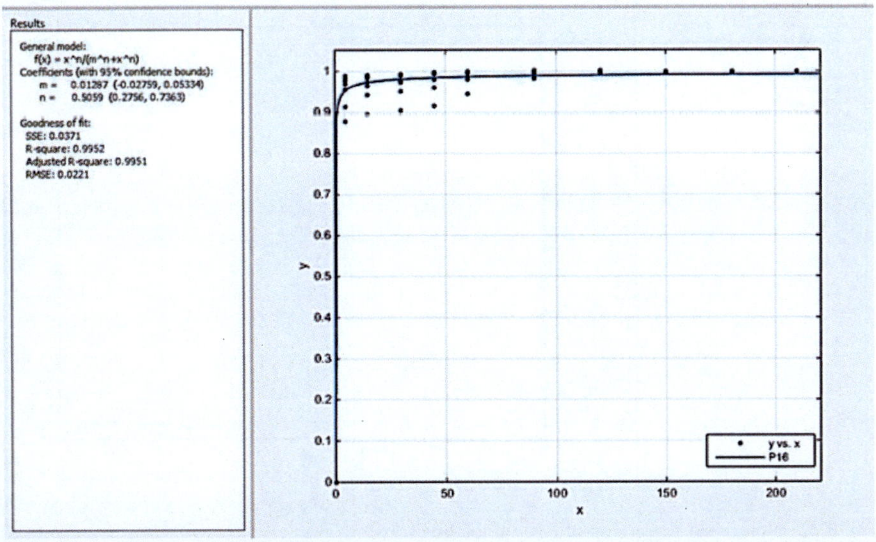

Fig. 8.20 Non-linear regression of P16 (loading stage)

maintained loading was applied. As shown in Figs. 8.21 and 8.22, the data indicated that the displacement of uplifted piles during unloading stages was stable, which was similar to the outcome of the compressive loaded pile during unloading stages.

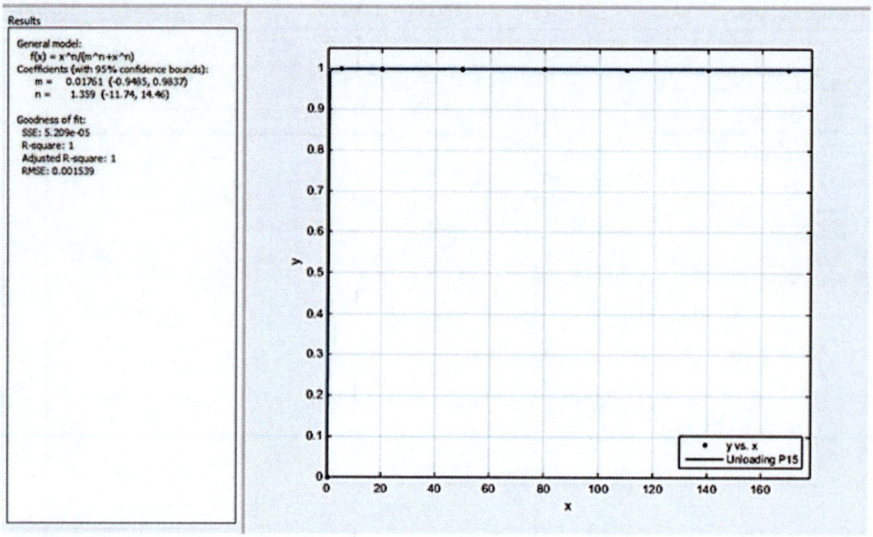

Fig. 8.21 Non-linear regression of P15 (unloading)

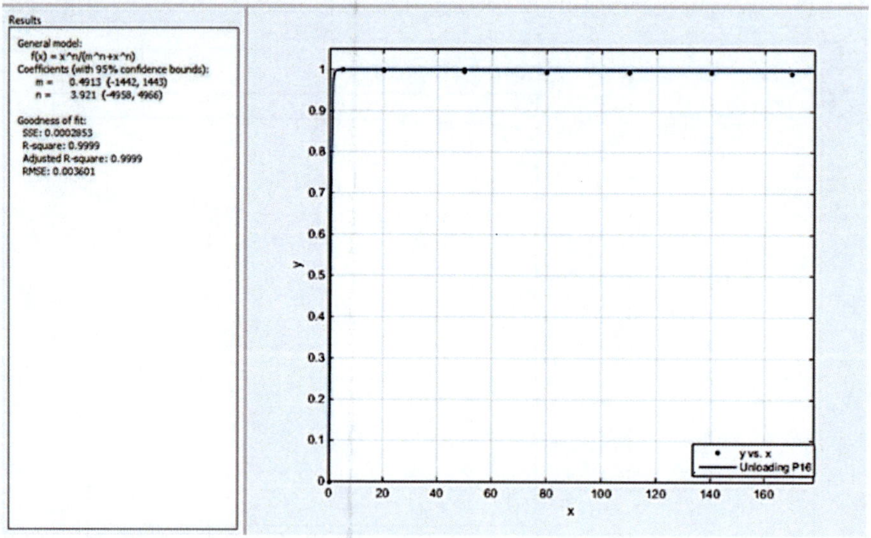

Fig. 8.22 Non-linear regression of P16 (unloading)

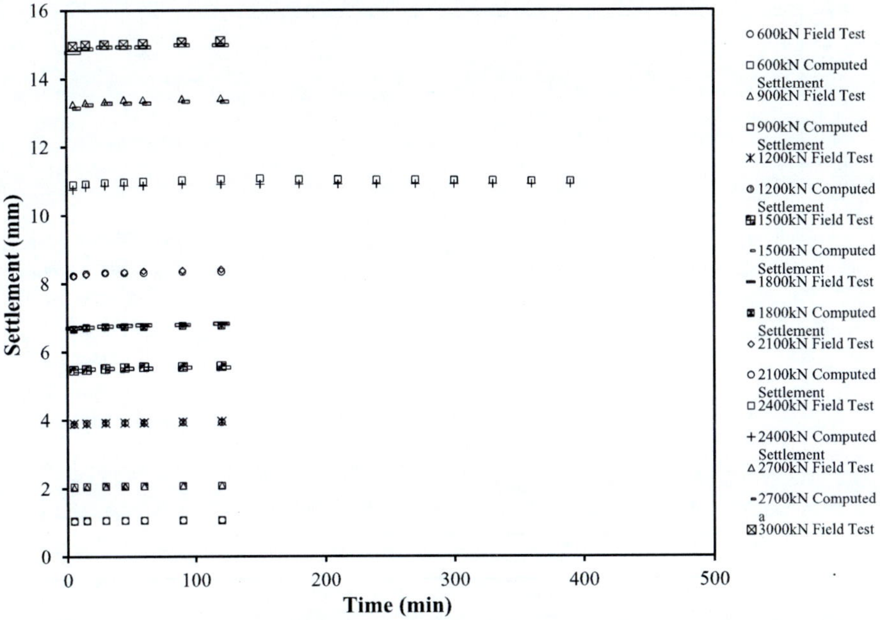

Fig. 8.23 Test and computed settlements from loading stages of P15

Based on these regression lines, the displacement equations were determined, and the *m* and *n* values are provided in Table 8.3.

Based on the proposed Eqs. 8.3 and 8.4, the computed displacements of uplift loaded piles with test data were compared, as shown in Figs. 8.23, 8.24, 8.25 to 8.26. It can be seen that the proposed equations provided accurate results for the piles' displacement.

8.3 Precast Concrete Piles

8.3.1 Small and Large Displacement Piles

8.3.1.1 Capacity Discussion

The capacity design of displacement piles as illustrated in Sect. 4.2, Chap. 4 is detailed in this section. The capacity of a pile foundation can be determined by calculating the shaft and end capacity, and based on the local project standard of Technical Code for Building Pile Foundations, JGJ 94-2008 (2008), the capacity of the precast rectangular piles and pipe piles can be determined as follows:

For solid rectangular piles:

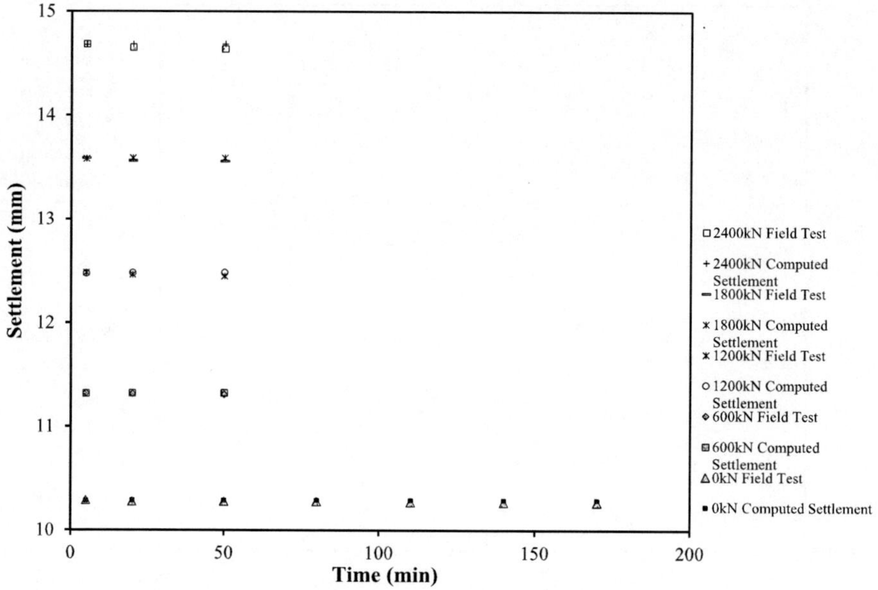

Fig. 8.24 Test and computed settlements from unloading stages of P15

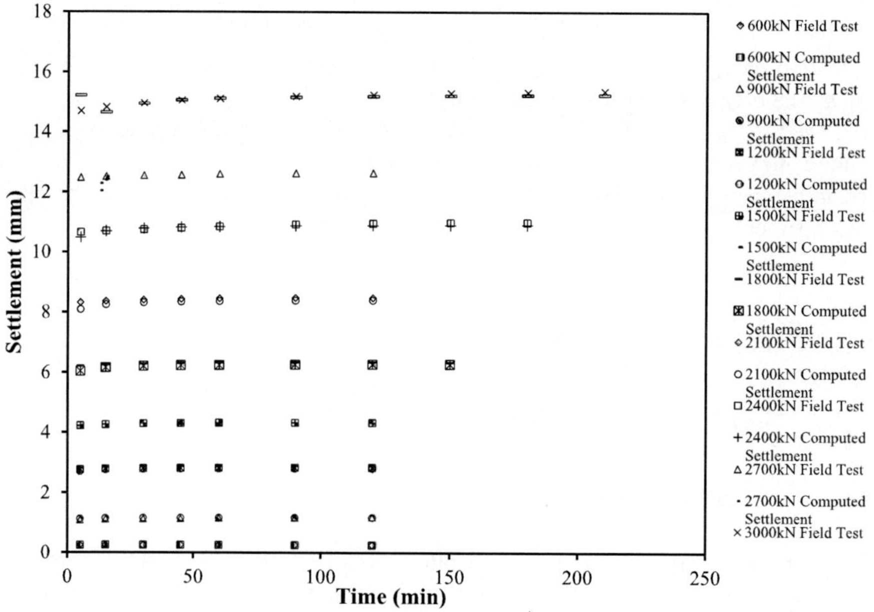

Fig. 8.25 Test and computed settlements from loading stages of P16

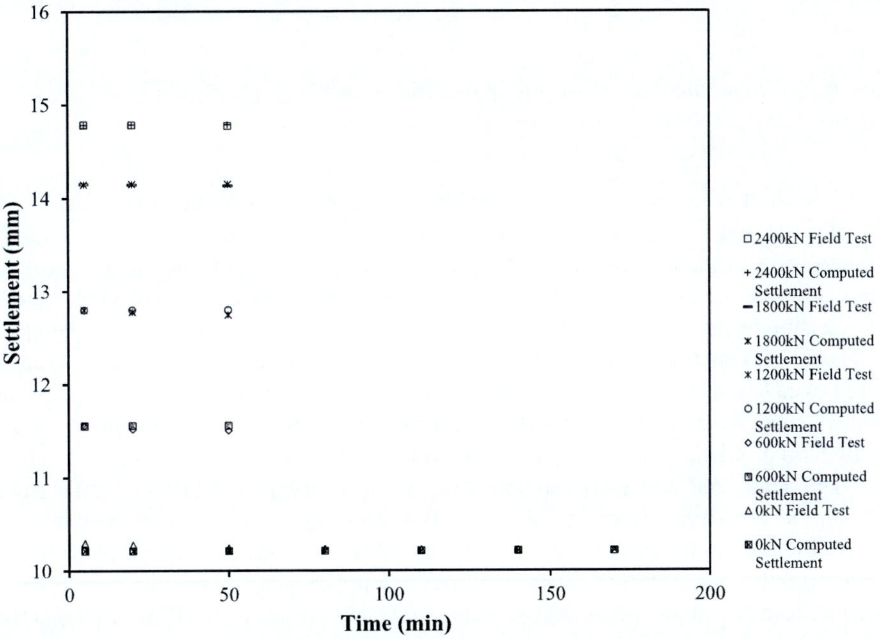

Fig. 8.26 Test and computed settlements from unloading stages of P16

$$Q_{uk} = Q_{sk} + Q_{pk} = u\sum q_{sik}l_i + q_{pk}A_p \tag{8.5}$$

For pipe piles:

$$Q_{uk} = Q_{sk} + Q_{pk} = u\sum q_{sik}l_i + q_{pk}\left(A_j + \lambda_p A_{p1}\right) \tag{8.6}$$

$$\lambda_p = 0.16h_b/d \quad \text{when}\, h_b/d < 5$$

$$\lambda_p = 0.8 \quad \text{when}\, h_b/d \geq 5$$

where:

$Q_{uk} =$ Ultimate bearing capacity of pile
$Q_{sk} =$ Shaft capacity of pile
$Q_{pk} =$ End capacity of pile
$u =$ Cross-section perimeter of pile
$q_{sik} =$ Shaft resistance
$l_i =$ Thickness of ith soil
$q_{pk} =$ End resistance
$A_p =$ Area from pile toe
$A_j =$ Effective area of pile toe and $A_j = \frac{\pi}{4}(d^2 - d_1^2)$

$\lambda_p =$ Plug effect coefficient
$A_{p1} =$ Hollow area of pile toe
$h_b =$ Embedment depth of pile in bearing stratum
$d =$ External diameter of pipe pile
$d_1 =$ Internal diameter of pipe pile.

The shaft resistance, q_{sik}, and end resistance, q_{pk}, can be determined through Tables 8.4 and 8.5, respectively. The pile capacity should be based on the nearest borehole information, instead of the average soil thickness. Further, it is important to determine the elevation of the first soil layer based on the construction elevation. For example, for the concrete rectangular pile near the borehole with a label of BH162, the construction elevation was about +4.4 m; hence, there would be around 1 m excavation, as shown in Fig. 8.27. After the soil thickness near the pile was determined, based on the parameters obtained from laboratory and in situ tests, the calculation of shaft and end capacity could be determined, as shown in Table 8.6.

The shaft, end and total capacities of all the tested pipe and rectangular piles (Chap. 4) are summarised in Table 8.7. Based on the double tangent method, the capacities of the tested concrete rectangular piles were determined. By plotting the interpretation versus the calculated ultimate bearing capacity as shown in Fig. 8.28, it can be seen that all the points were located in the lower part, which illustrated that the double tangent method was conservative in determining the ultimate bearing capacity. The ultimate bearing capacity obtained from the double tangent interpretation can be modified by multiplying a factor. In this case, the factor η_D was determined to be 1.83. By plotting the designed capacity versus the modified double tangent interpretation, as shown in Fig. 8.29, it can be seen that the modified double tangent interpretation could provide a good prediction of the ultimate bearing capacity

Traditionally, the allowable loading, Q_a, is equal to the calculated ultimate bearing capacity divided by the safety factor of 2. By plotting the allowable load versus the double tangent method, as shown in Fig. 8.30, it can be seen that the double tangent method could appropriately predict the allowable loads. Compared with the double tangent method, Chin's method overestimated the ultimate bearing capacity. A reduction factor (η_{Ch}) of 1.35 for a pile with a length less than 25 m and of 1.2 for a pile with a length over 25 m is recommended. As shown in Fig. 8.31, based on the proposed reduction factor, by plotting the modified Chin's interpretation versus the calculated capacity, all points assembled in the angle bisector of the first quartile.

For the precast pipe piles, the calculated ultimate bearing capacity was around 2,500 kN and the allowable load of a single pile was 1,250 kN. Based on the Technical Code for Testing of Building Foundation Piles, JGJ 106-2014 (2014) the maximum load applied during tests should be twice the allowable load—which was 2,500 kN. However, the required capacity of this project was 520 kN, and these SLTs only applied maximum loads of 1,040 kN (This is a very uneconomical design). Further research on precast pipe piles with more loading application is required.

Table 8.4 Shaft resistance, q_{sik}

Soil type	Soil state		Concrete precast pile	Post grouting bored pile	Traditional bored pile
Fill			22–30	20–28	20–28
Sludge			14–20	12–18	12–19
Mucky soil			22–30	20–28	20–28
Cohesive soil	$I_L > 1$		24–40	21–38	21–38
	$0.75 < I_L \leq 1$		40–55	38–53	38–53
	$0.5 < I_L \leq 0.75$		55–70	53–68	53–66
	$0.25 < I_L \leq 0.50$		70–86	68–84	66–82
	$0 < I_L \leq 0.25$		86–98	84–96	82–94
	$I_L \leq 0$		98–105	96–102	94–104
Red clay	$0.7 < a_w \leq 1$		13–32	12–30	12–30
	$0.5 < a_w \leq 0.7$		32–74	30–70	30–70
Silt	Soft	$e > 0.9$	26–46	24–42	24–42
	Medium soft	$0.75 \leq e \leq 0.9$	46–66	42–62	42–62
	Stiff	$e \leq 0.75$	66–88	62–82	62–82
Silty sand	Loose	$10 < N \leq 15$	24–48	22–46	22–46
	Medium loose	$15 < N \leq 30$	48–66	46–64	46–64
	Dense	$N > 30$	66–88	64–86	64–86
Medium sand	Medium dense	$15 < N \leq 30$	54–74	53–72	53–72
	Dense	$N > 30$	74–95	72–94	72–94
Coarse sand	Medium dense	$15 < N \leq 30$	74–95	74–95	76–98
	Dense	$N > 30$	95–116	95–116	98–120
Gravel sand	Loose	$5 < N_{63.5} \leq 15$	70–110	50–90	60–100
	Medium to dense	$N_{63.5} > 30$	116–138	116–130	112–130
Gravel	Medium to dense	$N_{63.5} > 10$	160–200	135–150	135–150
Cobble	Medium to dense	$N_{63.5} > 10$	200–300	140–170	150–170
Weathered soft rock		$30 < N \leq 50$	100–120	80–100	80–100
Weathered rag-stone		$30 < N \leq 50$	140–160	120–140	120–150
Highly weathered soft rock		$N_{63.5} > 10$	160–240	140–200	140–220

(continued)

Table 8.4 (continued)

Soil type	Soil state		Concrete precast pile	Post grouting bored pile	Traditional bored pile
Highly weathered rag-stone		$N_{63.5} > 10$	220–300	160–240	160–260

Source JGJ 94-2008 (Ministry of Construction of the People's Republic of China 2008)

8.3.1.2 Settlement Discussion

As illustrated in Chap. 4, Fig. 4.1 shows the tested piles at Areas A, B and C. The soil profiles of Areas A, B and C are shown in Figs. 8.32, 8.33 and 8.34. It was found that a pile length ranging from 20 to 30 m may be required to reach the bearing stratum (angular gravel) in Area A, a pile length ranging from 24 to 32 mm may be required in Area B, and a pile length ranging from 24 to 39 m may be needed in Area C.

The load settlement results of compressively loaded piles were presented in Chap. 4. As shown in Fig. 4.13, these Q-s curves of all 21 m piles were different. It was found that the maximum settlements of piles ranged from 16 to 24 mm. This was because the bearing stratum was brecciated gravel, and the shaft resistance was different (thickness of soil layers varied). This total head displacement is the sum up of pile deformation and soil deformation. It was also found that, after consecutive unloading back to 0 kN, all piles curves were parallel (unloading curves). This was because all these piles were made of the same concrete with the same moduli. Once the loading was released, the shortened pile would expand back at the same rate (Δsettlement/Δload, unit of mm/kN), and the plastic deformation of soil dominates elastic deformation.

When defining the permanent settlement ratio, R_{P-S}, governed by Eq. 8.7, the average R_{P-S} of small displacement pipe piles was determined to be 52.2% in Area A (strength of concrete of 30 MPa) and 54.9% in Area B (strength of concrete of 50 MPa):

$$R_{P-S} = \frac{s_{max} - s_p}{s_{max}} \times 100 \tag{8.7}$$

where:

$R_{P-S} =$ Permanent settlement ratio;
$s_{max} =$ Settlement under maximum applied load;
$s_p =$ Permanent settlement after releasing load.

As provided in Tables 4.1 and 4.2, the ratios of all these tested piles were close to the average permanent ratio in Areas A and B, respectively. Similar to these pipe piles, as depicted in Table 4.3, the average R_{P-S} of large displacement rectangular piles was determined to be 49.0% (strength of concrete was 40 MPa) and the ratios of rectangular piles were close to the average value of 49.0%.

Table 8.5 End resistance, q_{pk}

Soil type	Soil state		Concrete precast pile				Post grouting bored pile				Traditional bored pile		
	Pile Type		l	l	l	l	l	l	l	l	l	l	l
Cohesive soil	$0.75 < I_L \leq 1$		210–850	650–1,400	1,200–1,800	1,300–1,900	150–250	250–300	300–450	300–450	200–400	400–700	700–950
	$0.5 < I_L \leq 0.75$		850–1,700	1,400–2,200	1,900–2,800	2,300–3,600	350–450	450–600	600–750	750–800	500–700	800–1,100	1,000–1,600
	$0.25 < I_L \leq 0.50$		1,500–2,300	2,300–3,300	2,700–3,600	3,600–4,400	800–900	900–1,000	1,000–1,200	1,200–1,400	850–1,100	1,500–1,700	1,700–1,900
	$0 < I_L \leq 0.25$		2,500–3,800	3,800–5,500	5,500–6,000	6,000–6,800	1,100–1,200	1,200–1,400	1,400–1,600	1,600–1,800	1,600–1,800	2,200–2,400	2,600–2,800
Silt	Medium dense	$0.75 \leq e \leq 0.9$	950–1,700	1,400–2,100	1,900–2,700	2,500–3,400	300–500	500–650	650–750	750–850	800–1,200	1,200–1,400	1,400–1,600
	Dense	$E \leq 0.75$	1,500–2,600	2,100–3,000	2,700–3,600	3,600–4,400	650–900	750–950	900–1,100	1,100–1,200	1,200–1,700	1,400–1,900	1,600–2,100
Silty sand	Medium dense	$10 < N \leq 15$	1,000–1,600	1,500–2,300	1,900–2,700	2,100–3,000	350–500	450–600	600–700	650–750	500–950	1,300–1,600	1,500–1,700
	Dense	$N > 15$	1,400–2,200	2,100–3,000	3,000–4,500	3,800–5,500	600–750	750–900	900–1,100	1,100–1,200	900–1,000	1,700–1,900	1,700–1,900
Fine sand	Medium dense to dense	N > 15	2,500–4,000	3,600–5,000	4,400–6,000	5,300–7,000	650–850	900–1,200	1,200–1,500	1,500–1,800	1,200–1,600	2,000–2,400	2,400–2,700
Medium sand			4,000–6,000	5,500–7,000	6,500–8,000	7,500–9,000	850–1,050	1,100–1,500	1,500–1,900	1,900–2,100	1,800–2,400	2,800–3,800	3,600–4,400
Coarse sand			5,700–7,500	7,500–8,500	8,500–10,000	9,500–11,000	1,500–1,800	2,100–2,400	2,400–2,600	2,600–2,800	2,900–3,600	4,000–4,600	4,600–5,200
Gravel sand	Medium dense to dense	N > 15	6,000–9,500		9,000–10,500		1,400–2,000		2,000–3,200		3,500–5,000		
Gravel sand		$N_{63.5} > 10$	7,000–10,000		9,500–11,500		1,800–2,200		2,200–3,600		4,000–5,500		
Cobble		$N_{63.5} > 10$	8,000–11,000		10,500–13,000		2,000–3,000		3,000–4,000		4,500–6,500		
Weathered soft rock	$30 < N \leq 50$		4,000–6,000				1,000–1,600				1,200–2,000		
Weathered rag-stone	$30 < N \leq 50$		5,000–8,000				1,200–2,000				1,400–2,400		
Highly weathered soft rock	$N_{63.5} > 10$		6,000–9,000				1,400–2,200				1,600–2,600		

(continued)

Table 8.5 (continued)

Soil type	Soil state	Concrete precast pile				Post grouting bored pile				Traditional bored pile		
	Pile Type	l	l	l	l	l	l	l	l	l	l	l
Highly weathered rag-stone	$N_{63.5} > 10$	7,000–11,000				1,800–2,800				2,000–3,000		

Source JGJ 94-2008 (Ministry of Construction of the People's Republic of China 2008)

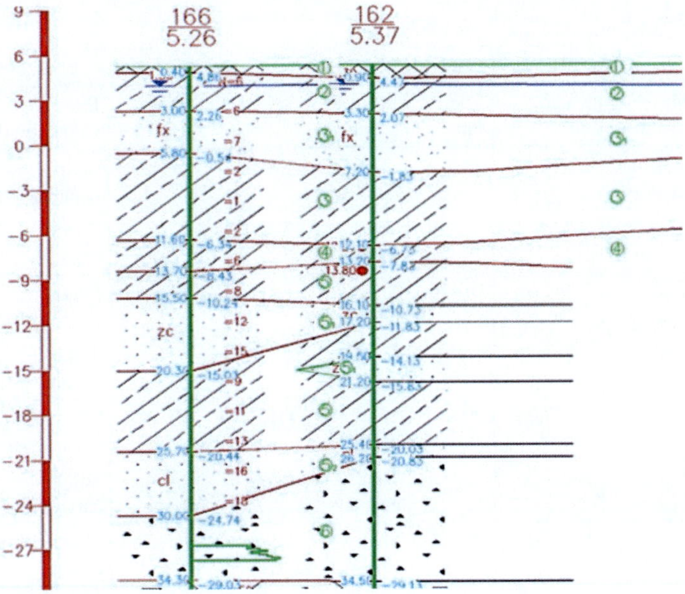

Fig. 8.27 Borehole log 162

Table 8.6 Calculation of pile adjacent to borehole 162

1. Shaft capacity			
Label	Soil thickness (m)	q_{sik} (kPa)	Shaft resistance (kN)
1	1	20	20
1-1	0	20	0
2	3	50	150
3	5	38	190
3-1	3	18	54
4	1.2	65	78
5	3	78	234
5-1	1.2	50	60
5	8	78	624
6	0	150	
Total shaft resistance			2,256
2. End resistance			
End layer	A_p	Toe area	End bearing (kN)
6	9,500	0.16	1,520
3. Ultimate bearing capacity (kN)		3,776	

Table 8.7 Summarization of tested precast piles

Concrete rectangular pile			Concrete pipe pile					
Pile label	Pile capacity	kN	Pile label	Pile capacity	kN	Pile Label	Pile Capacity	kN
P1	Shaft capacity	2,256	P1172	Shaft capacity	1,333.37	P643	Shaft capacity	1,496.65
	Total capacity	3,776		Total capacity	2,329.24		Total capacity	2,492.52
P106	Shaft capacity	2,447.36	P1470	Shaft capacity	1,358.87	P1509	Shaft capacity	1,531.44
	Total capacity	3,967.36		Total capacity	2,354.74		Total capacity	2,527.32
P24	Shaft capacity	1,977.28	P976	Shaft capacity	1,451.94	P1485	Shaft capacity	1,411.87
	Total capacity	3,497.28		Total capacity	2,447.81		Total capacity	2,407.74
P41	Shaft capacity	1,799.04	P991	Shaft capacity	1,373.94	P1475	Shaft capacity	1,483.21
	Total capacity	3,319.04		Total capacity	2,369.81		Total capacity	2,479.08
P91	Shaft capacity	2,495.2	P995	Shaft capacity	1,460.48	P1580	Shaft capacity	1,542.74
	Total capacity	4,015.2		Total capacity	2,456.35		Total capacity	2,538.62
P48	Shaft capacity	2,667.36	P1163	Shaft capacity	1,441.39	P1481	Shaft capacity	1,478.19
	Total capacity	4,187.36		Total capacity	2,437.26		Total capacity	2,474.06
P147	Shaft capacity	2,761.6	P1165	Shaft capacity	1,393.66	P1465	Shaft capacity	1,706.02
	Total capacity	4,281.6		Total capacity	2,389.53		Total capacity	2,701.90
P124	Shaft capacity	2,634.88	P1175	Shaft capacity	1,343.92	S73	Shaft capacity	1,653.40
	Total capacity	4,154.88		Total capacity	2,339.79		Total capacity	2,649.27
P238	Shaft capacity	2,224.64	P639	Shaft capacity	1,490.12	S103	Shaft capacity	1,653.15
	Total capacity	3,744.64		Total capacity	2,485.99		Total capacity	2,649.02
P173	Shaft capacity	2,683.2	P1275	Shaft capacity	1,708.16	S126	Shaft capacity	1,633.30
	Total capacity	4,203.2		Total capacity	2,704.03		Total capacity	2,629.18

(continued)

Table 8.7 (continued)

Concrete rectangular pile			Concrete pipe pile					
Pile label	Pile capacity	kN	Pile label	Pile capacity	kN	Pile Label	Pile Capacity	kN
P137	Shaft capacity	2,611.36	P1287	Shaft capacity	1,714.44			
	Total capacity	4,131.36		Total capacity	2,710.31			
P4	Shaft capacity	2,638.4	P1299	Shaft capacity	1,729.39			
	Total capacity	4,158.4		Total capacity	2,725.26			
P16	Shaft capacity	2,651.36	P1453	Shaft capacity	1,727.13			
	Total capacity	4,171.36		Total capacity	2,723.00			
P85	Shaft capacity	2,748.8	P630	Shaft capacity	1,412.25			
	Total capacity	4,268.8		Total capacity	2,408.12			
P93	Shaft capacity	2,693.12	P1473	Shaft capacity	1,432.34			
	Total capacity	4,213.12		Total capacity	2,428.22			

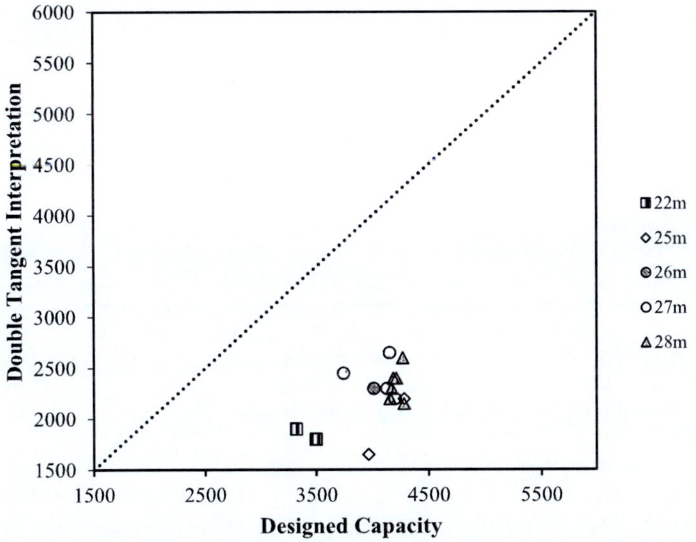

Fig. 8.28 Designed capacity versus double tangent interpretation

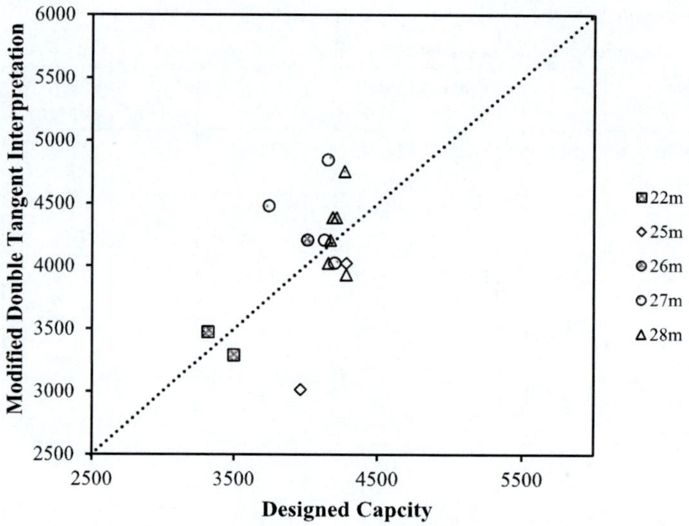

Fig. 8.29 Designed capacity versus modified double tangent interpretation

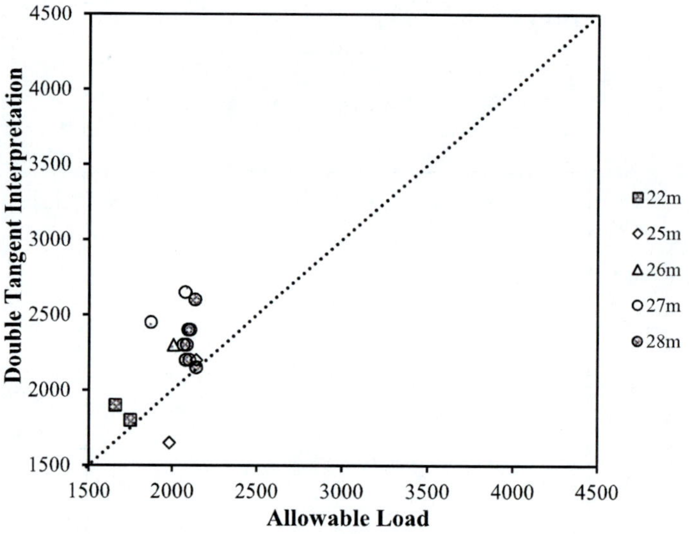

Fig. 8.30 Allowable load versus double tangent interpretation

As illustrated in Tables 4.1, 4.2 and 4.3, the permanent settlement ratio of small displacement piles was greater than the large displacement piles; however, this was not a general outcome, since there was a lot of uncertainty. Thus, further research is required with consideration of the strength of concrete used, pile length, shaft areas and so forth.

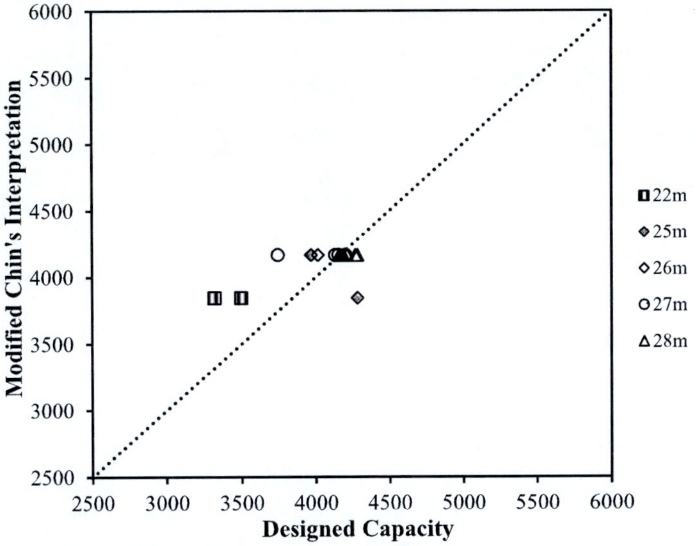

Fig. 8.31 Designed capacity versus modified Chin's interpretation

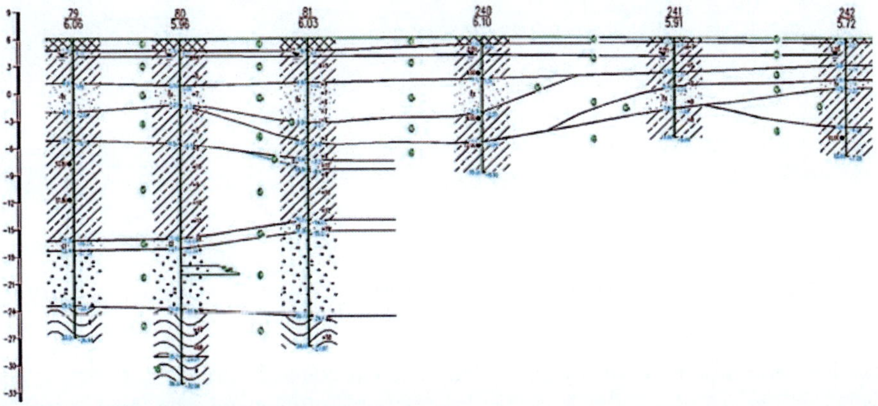

Fig. 8.32 Soil profile of Area A based on borehole logs

Similar to the 21 m piles, the load settlement results of the 22 and 24 m piles were provided in Figs. 4.14 and 4.15. These piles were made of the same concrete material of 30 MPa. It was found that the maximum settlements were different because the shaft resistances of each layer were different. It could also be seen that the curves from the unloading stages were parallel with each other because the same material was being used, and the elongation deformation would develop at the same rate. To further confirm this, the load–settlement curves of piles with the same concrete (50 MPa) were parallel from the unloading stage, as shown in Fig. 4.17. If plotting

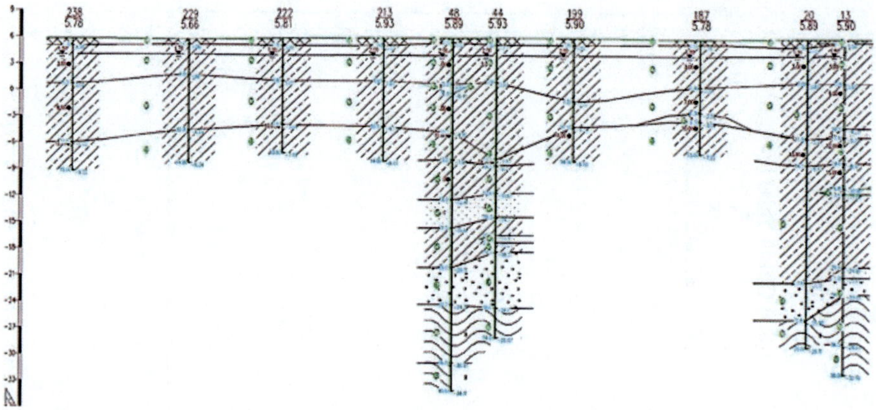

Fig. 8.33 Soil profile of Area B based on borehole logs

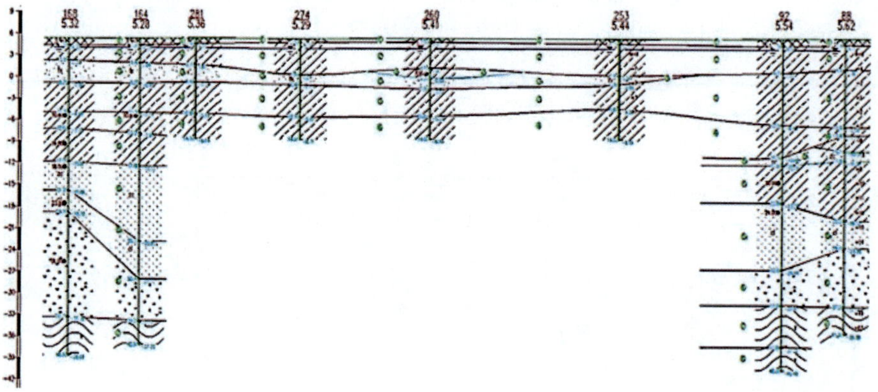

Fig. 8.34 Soil profile of Area C based on borehole logs

the load–settlement curves of piles with different concrete strengths, these curves would not be parallel with each other from the unloading stage. The same outcome was obtained for the 27 and 28 m rectangular piles, as shown in Figs. 4.18 and 4.19.

Under consideration of pipe and rectangular pile lengths, as shown in Figs. 4.16 and 4.21, respectively, the maximum settlement of pipe and rectangular piles were different, yet the R_{P-S} values of the pipe and rectangular piles were close to the same type of pile. In detail, the R_{P-S} of P1172 (20 m), P1165 (21 m), P1485 (22 m) and P1299 (24 m) was 55.14, 53.72, 51.86 and 54.7%, respectively, for the pipe piles, and the R_{P-S} of P14 (22 m), P106 (25 m), P91 (26 m), P124 (27 m) and P147 (28 m) was 49.02, 49.09, 54.37, 48.14 and 49.03%, respectively, for the rectangular piles. This indicates that, for the piles that reached the hard soil stratum, during loading and unloading, the structure deformation was a dominant factor. Thus, the permanent settlement ratios of one type of pile will be close to each other.

To further prove this outcome with consideration of loading value, the results of P1 and P106 were compared. These piles were made of the same concrete and were 25 m. As shown in Fig. 4.20, different loading increments were applied, and this demonstrated that the two curves were parallel with each other during unloading stages. The R_{P-S} of P1 and P106 was found to be 51.05% and 49.09%, respectively, which was close to each other.

8.3.2 *Non-displacement Precast Piles*

8.3.2.1 Capacity Discussion

As discussed in literature review of Chap. 2, there has been limited previous research on non-displacement precast piles—the construction process in which soil is first removed and then the precast pile is lowered. Thus, this study investigated the behaviours of six piles under uplift and lateral loading, and the results were provided in Chap. 4.

For the horizontal load movement results obtained from the lateral SLTs, the critical load and ultimate load could be preliminarily determined from Figs. 4.22, 4.23 and 4.24 for P15 (24.7 m), P163 (26.6 m) and P152 (28 m), respectively. However, this is an empirical method that is highly dependent on practical experience. For example, the ultimate horizontal capacity of P163 was difficult to determine. Further, it was found that, before the ultimate loading being applied, the piles' behaviours were similar. As shown in Figs. 4.22 and 4.23, the maximum horizontal movements of P15 and P163 were 5.56 and 5.125 mm, respectively, when 400 kN was applied. In addition, the residual movements of P15 and P163 were 0.59 and 0.57 mm, respectively. To further prove this outcome, the maximum horizontal movements of P15 and P163 were 11.28 and 11.91 mm, and the residual movements were 1.51 and 2.66 mm, respectively, when 500 kN was applied.

These three typical presentations preliminarily illustrated that the critical loads and ultimate loads increased with pile length. However, the ultimate load of P163 was difficult to determine. The method of horizontal SLT interpretation involved plotting horizontal loading versus nominated gradient settlement. As shown in Figs. 4.25, 4.26 and 4.27, theoretically, two 'turning points' could be determined that represented critical and ultimate loading conditions. Through determining these points, it was found that critical load and ultimate load increased with pile length, as illustrated in Table 4.6.

The load–settlement curves of the three uplift loaded piles were provided in Fig. 4.28. It can be seen that the curves from the loading stages were relatively linear, and the maximum uplift displacements were small. Through interpretation of Mazurkiewicz's method, the ultimate uplift capacities of the three piles were determined in Fig. 4.29. It can be seen that the capacity increased with pile length. The load transfer mechanism of the non-displacement precast pile was presented in Figs. 4.30 and 4.31. This behaviour was similar to the compressive loaded piles, where loading was transferred into piles resisted by shaft force developed by soil and pile shaft surface, and the there is no load transferred from the pile end.

8.3.2.2 Displacement Discussion

The proposed empirical Eqs. 8.1 to 8.2 and 8.3 and 8.4 can be used to determine the compressive and uplift displacement prediction respectively, with various coefficient m and n values which can be obtained from non-linear regression lines. Similar to these equations, for the lateral loaded piles, during the loading and unloading stages, the horizontal displacement ratio was defined as $R_h = H_t/H_{(F_i)}$. Under a certain lateral loading, the horizontal displacement at a certain time (H_t) divided by the displacement ($H_{(F_i)}$) which represents the final displacement at the end of this loading or unloading will be less than 1, but over 0. By plotting the time versus the horizontal displacement ratio, R_h, from the loading and unloading stages, as shown in Figs. 8.35, 8.36, 8.37 and 8.38, the non-linear regression lines were determined. Finally, the horizontal displacement equation was proposed, as illustrated in Eq. 8.8:

$$H_t = H_{(F_i)} \times \frac{t^n}{m^n + t^n} \tag{8.8}$$

where:

$H_t =$ Settlement at a certain loading stage at a certain time in mm;
$H_{(F_i)} =$ Settlement at the end of a certain loading/unloading stage in mm;
m and $n =$ Settlement coefficients obtained from non-linear regression curve.

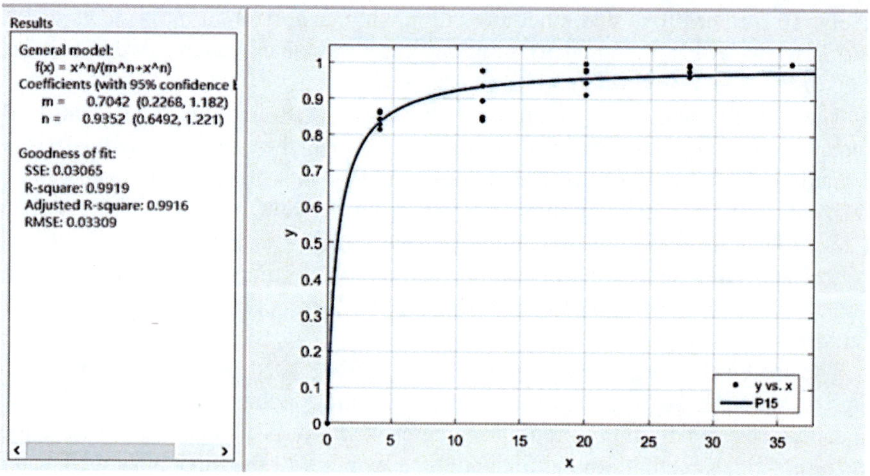

Fig. 8.35 Non-linear regression of P15 (loading stage)

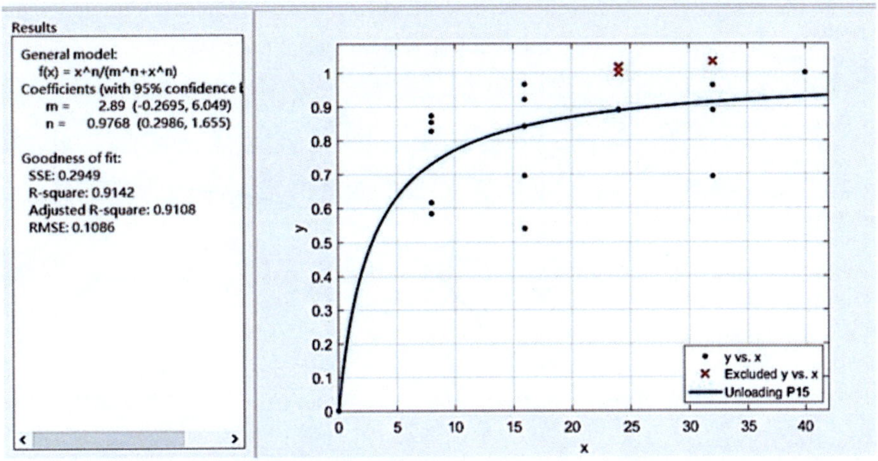

Fig. 8.36 Non-linear regression of P15 (unloading stage)

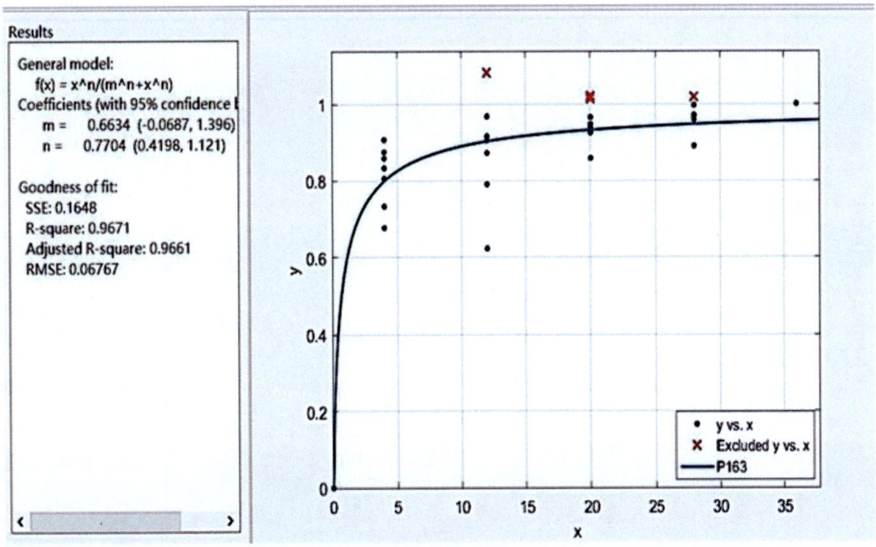

Fig. 8.37 Non-linear regression of P163 (loading stage)

As shown in Fig. 8.36, 8.37 and 8.38, for the piles during loading and unloading stages, some points should be excluded because the horizontal displacement ratio value was defined as 0 to 1. This was mostly because of the error caused by dial gauges. Also, for the pile during unloading stages, when the pile foundation deflects back to its initial location, the soil layers may collapse and consequently lead to the inaccurate reading of horizontal displacement. As shown in Fig. 8.36, this regression line was relatively trustworthy and could provide good displacement predictions,

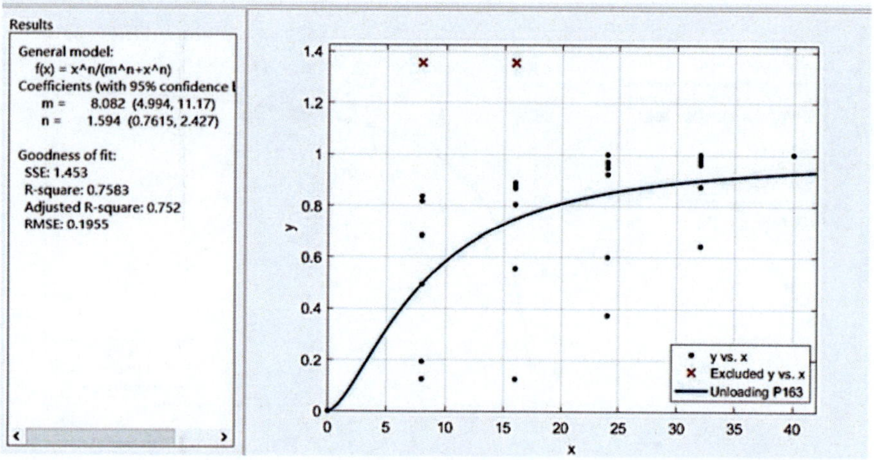

Fig. 8.38 Non-linear regression of P163 (unloading stage)

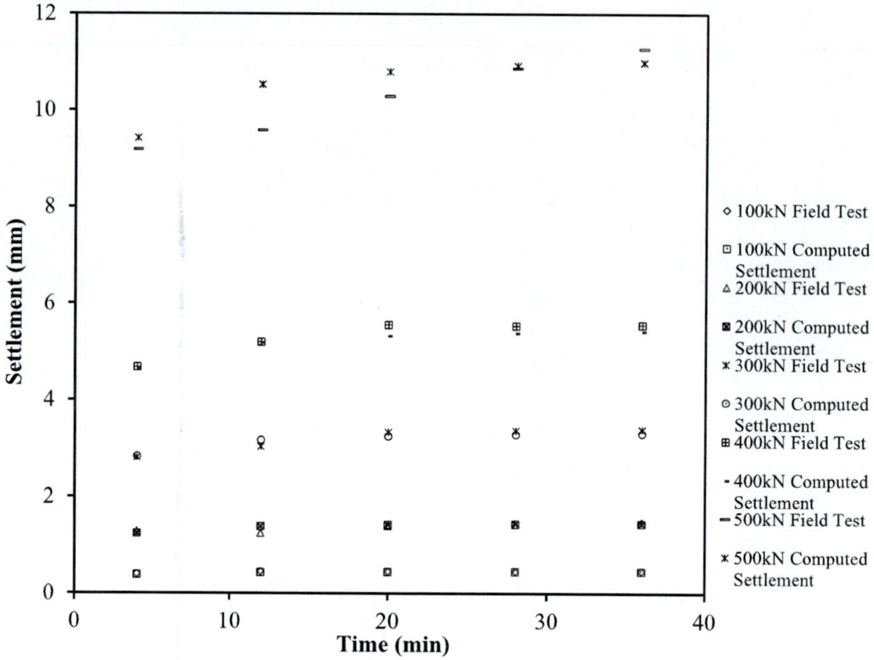

Fig. 8.39 Test and computed settlements from loading stages of P15

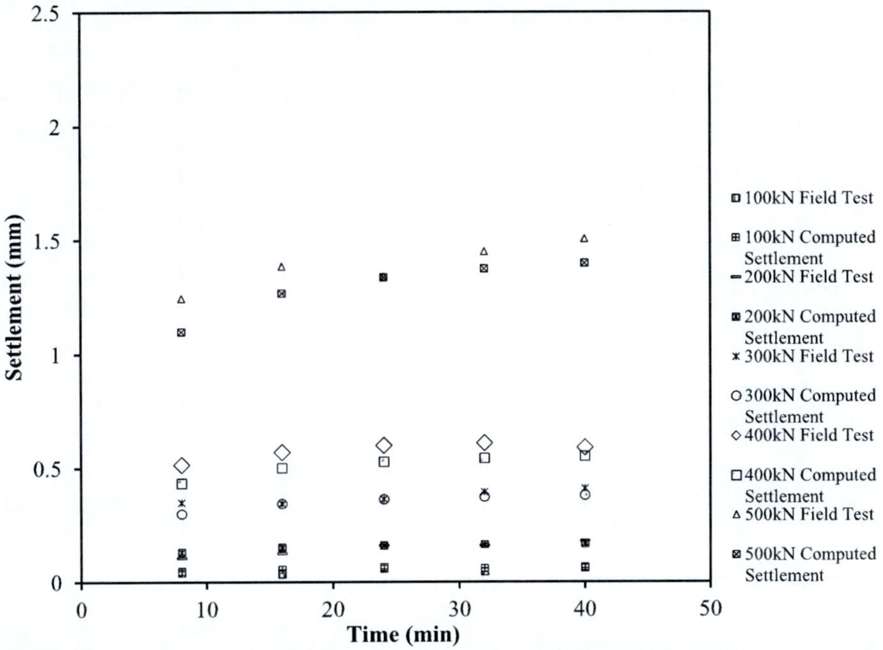

Fig. 8.40 Test and computed settlements from unloading stages of P15

as illustrated in Fig. 8.40. The regression line with an R-squared value of 0.7583, as depicted in Fig. 8.38, is not recommended for use. From the loading stages, as shown in Figs. 8.35 and 8.37, these regression lines could provide accurate values of coefficients *m* and *n*; hence, the predicted horizontal displacement was trustworthy, as illustrated in Figs. 8.39 and 8.41. The limitation of this research was that there is no comparison between this non-displacement pile and displacement driven precast pile. Hence, the effect caused by removing soil first is unknown. Only the behaviours of these non-displacement piles have been investigated; thus, further research is required.

8.4 Concrete Composite Piles

8.4.1 FRP Laminar–Confined Composite Piles

By comparing the Q-s curves of CFRP laminar–confined composite piles with traditional concrete piles, as depicted in Figs. 5.8 and 5.9, it was found that the settlements of the CFRP-confined pile were less than the traditional piles during the application of loading. After the loads were unloaded, the permanent settlements of these two

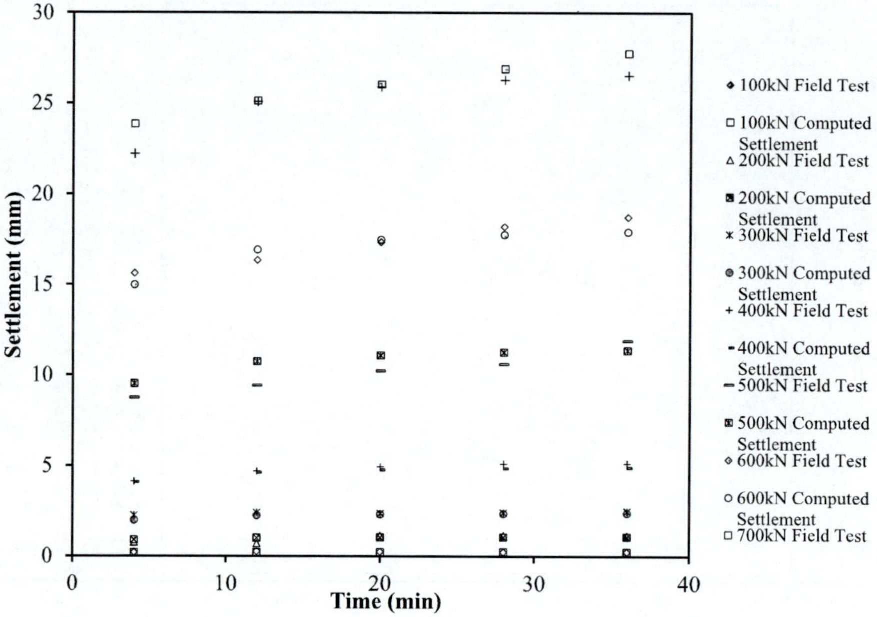

Fig. 8.41 Test and computed settlements from loading stages of P163

composite piles were less than the traditional concrete piles. Since the geotechnical condition was the same (the soil layers were the same around the tested piles) and all piles reached the bearing stratum, the settlement was primarily due to the concrete shortening. Under the same loading value, the shortening of the CFRP composite pile was less than the normal pile because of the circumferential stress caused by CFRP confinement.

Through the s-lgQ curves, as shown in Fig. 5.10, it was difficult to find a turning point that could represent the failure criteria; hence, the determination of the ultimate bearing capacity of these tested piles needed to be based on interpretation. The s-lgt curves provided the same information as the Q-s curves, and indicated that the settlements of the CFRP-confined piles were less than the traditional piles. Similar to the s-lgQ curves, DeBeer's method also provided the same outcome because there was no turning point, as shown in Figs. 5.17 and 5.18.

As displayed in Figs. 5.11, 5.12, 5.13 and 5.14, there was no point showing a dramatic increasing of vertical settlement in each loading line, and no extreme 'drop' illustrating dramatic settlement increase under one specific loading stage. The ultimate bearing capacity could not be determined via these four diagrams, yet provided valuable information that these four piles' capacity should be more than the maximum applied loads. Through determining the elastic lines and offsetting these lines based on Eq. 2.38, Davisson's method also provided the same outcome. Given that the offset line does not intersect with the Q-s curves obtained from the loading curves, the ultimate bearing capacity of these piles should be more than the maximum loads applied from the SLTs, as shown in Figs. 5.19 and 5.20.

Based on the double tangent method, the pile capacities of P116, P115, P2108 and P2054 were determined to be 2,600, 2,400, 850 and 760 kN, respectively, as shown in Figs. 5.15 and 5.16. However, as discussed above, the capacity of these four piles should be greater than 4,800, 4,800, 1,300 and 1,300 kN, respectively. The reason is that the double tangent method is a conservative method; thus, the determined ultimate load would be less than the maximum applied loads, especially when used for obtaining the pile capacity under non-plunging SLTs and proof tests. Following the method provided by Yang and Xiao (2011), which is based on Chin's method, the functions were used to evaluate the capacities of piles, as depicted in Figs. 5.21, 5.22, 5.23 and 5.24.

The values of ultimate bearing capacity, shaft resistance and toe resistance were summarised in Table 5.1. It was found from Chin's method that the CFRP-confined pipe pile had a greater capacity than the traditional pipe pile, and that this confinement increased the shaft and toe resistance simultaneously. From the double tangent method, it was also found that the CFRP confinement increased the pipe pile capacity. Further, the double tangent method illustrated that the CFRP confinement could increase rectangular pile capacity.

This investigation focused on the behaviour of piles under CFRP confinement. However, the limitation was that the loading applied was not sufficiently large, which led to difficulty in determining the capacity. Further research is required with consideration of the types of CFRP material used, quantity of laminar used, FRP orientation arrangement and so forth.

8.4.2 FRP Bar–Reinforced Composite Piles

The deflection behaviour of GFRP-reinforced concrete piles was researched in comparison with traditional steel-reinforced concrete piles in Chap. 5. These two piles suffered from lateral loading caused by changed earth pressure during excavation construction, and the deflection behaviours were monitored with consideration of support strut application, as shown in Figs. 5.39 and 5.40.

Analysis of the lateral displacement diagrams of these piles showed that the deflection increased from pile head along the pile to a certain depth, and this maximum horizontal movement would increase along the pile. Also, it could be seen that the horizontal movements along the pile shaft are small, thus, the project was relatively overdesigned. The GFRP composite pile demonstrated a maximum lateral deflection of −6.02 mm at the depth of −12.5 m, while the maximum lateral deflection of the steel-reinforced concrete bored pile was −10.56 mm at the depth of 12.5 m. Further, it was found that, after each support strut was applied, the maximum horizontal deflection of the GFRP composite pile was less than the normal reinforced concrete pile, and the maximum horizontal movement of piles was approximately 12 to 13 m underneath, as depicted in Table 5.5.

It cannot be concluded that the GFRP bar–reinforced concrete pile demonstrated better than the traditionally reinforced concrete pile, since the ultimate tensile strength of GFRP and steel are different (GFRP > 500 MPa; steel = 400 MPa). However, it can be concluded that the GFRP bar–reinforced concrete bored pile could be a suitable alternative to replace the traditionally reinforced concrete bored pile in deep excavation. Further research is required to determine the stresses along the pile, and more advanced analysis is needed, such as a bending moment study along the pile length.

8.5 Piles Under Ultimate Load

Chapter 7 provided a pile behaviour investigation with consideration of the inadequate concrete strength of piles, punching failure of piles and piles applied with eccentric loads. SLTs were conducted and the results of the Q-s, s-lgt and s-lgQ curves were analysed for capacity determination. Chapter 7 also presented the double tangent method and Davisson's offset method interpretations of these failed piles. By using the data before failure condition occurred and selecting Chin's method, prediction of these piles' capacities was also presented.

Through observation of these piles under ultimate loading, it was discovered that, for the pile with low concrete strength caused by inappropriate construction, the concrete will crack at the pile head, which may extend up to 30 mm, and the concrete will crush at a certain depth. The reinforcement in the crushed concrete zone will be forced outwards. In the case of the pile which suffered from eccentricity, concrete will crush in the compressive zone, and, in this zone, the reinforcement will be forced outwards. The steel in the tensile zone, however, resists loads after the concrete has ruptured, and the pile consequently fails. In the case of the pile suffering from punching failure, there were no concrete cracks discovered, but the soil crack could extend up to 50 mm.

By comparing the Q-s and s-lgQ curves, the ultimate bearing capacities were the same as those summarised in Table 7.3, this is because these methodologies are to find a turning point, which illustrated a huge settlement increase. It can also be seen that, for these piles under ultimate loading, the double tangent method can provide good results, although they may be slightly conservative (3,900 kN < 4,680 kN of P80). Further, it can be seen that Davisson's method can provide a good interpretation, although it may sometimes overestimate the ultimate capacity (5,200 kN > 4,680 kN). In addition, it was found that Chin's method can provide acceptable prediction of ultimate bearing capacity, although it may sometimes overestimate the capacity (10,000 kN > 7,560 kN). Further, it can be seen that Chin's method was not fit to predict the pile capacity under eccentric loading.

For the results of the s-lgt curves, as shown in Fig. 7.18, the behaviour of the pile with inadequate concrete strength could be determined. It can be seen that, during the applied loading of 3,640 kN, there was a decreasing curve. This increase in settlement was caused by the concrete being crushed. Given that there was only compressive loading, this failure in the concrete structure did not lead to failure of the pile foundation. It was not until the loading of 4,160 kN was applied that the soil-pile system failed. The behaviour of the pile with eccentric loading can be seen in Fig. 7.22. It was found that the concrete structure failed when a loading of 4,680 kN was applied. Two downwards trends were determined. The first downwards trend was caused by the total failure of the concrete structure, as proven by analysing Fig. 7.11 (After the field test was performed, the steel in the tensile zone yielded). The second downwards trend was caused by the failure of the pile soil system. As shown in Fig. 7.26, for plunging failure pile, the pile foundation was unstable under the loading of 7,560 kN, and resisted the loading of 8,400 kN, but finally failed.

8.6 Concluding Remarks

This Chapter has detailed analysis and discussion referring to the capacity and the settlement of those different types of pile foundations. In terms of the capacity, traditional result presentations were discussed and the extrapolations of different types of piles were compared. Most importantly, this Chapter provides a method for the reader to derive the empirical formula, which can be used for settlement prediction. This method considered three loads orientations, namely the compressive, uplift and horizontal directions. Also, the empirical formula considered the loading and unloading stages.

References

American Association of State Highway and Transportation Officials (AASHTO), *Standard Specifications for Highway Bridges*, Washington, D. C., 2002.

Ministry of Construction of the People's Republic of China. (2008). *Technical code for building pile foundations (JGJ 94-2008)*. Beijing, China: National Standard of the People's Republic of China.

Ministry of Construction of the People's Republic of China. (2014). *Technical code for testing of building foundation piles (JGJ 106-2014)*. Beijing, China: National Standard of the People's Republic of China.

Samtani, N. C., & Nowatzki, E. A. (2006b). *Soils and foundations—Volume II* (No. FHWA-NHI-16-009). US Department of Transportation.

Yang, P., Hu, H.-S., & Xu, J.-F. (2012). Settlement characteristics of pile composite foundation under staged loading. *Procedia Environmental Sciences, 12*, 1055–1062.

Yang, H., & Xiao, D. (2011). Back analysis of static pile load test for SPT-based pile design: A Singapore experience. In *Advances in Pile Foundations, Geosynthetics, Geoinvestigations, and Foundation Failure Analysis and Repairs* (pp. 144–152).

Chapter 9
Conclusions and Recommendations

9.1 General Introduction

This book provides the theoretical background, technical concepts and extrapolation of pile foundations related to state-of-the-art field tests. Based on the literature, Chap. 2 reviewed the up-to-date research and determined its limitations, the general principles and practices are detailed. Following this, based on the observed limitations, further research was provided in Chaps. 3 to 7 though testing a total number of 64 piles. After discussion of these results which was provided in Chap. 8, the current chapter presents the study conclusions and recommendations with reference to the tested pile foundations.

9.2 Post Grouted Piles

To determine the capacity of piles with different grouting technology applications, a total number of 5 piles were selected to test. Compressive and uplift loaded SLTs were conducted and two dynamic load tests were performed on two of these piles. Though these tests, the behaviours of these piles under different load directions were identified, and the ultimate bearing capacity was determined based on interpretation. The results led to the following conclusions:

For these bored piles under compressive loads:

1. From the SLT results, the base-and-shaft grouted pile increased about 9.82% of its capacity (compared with the pile without any grouting).
2. From the SLT results, the base grouting pile increased approximately 2.89% of its capacity (compared with the pile without any grouting).
3. From the SLT results, the base-and-shaft grouted pile increased around 6.1% of its capacity (compared with the base grouting pile).

J. Zhou and E. Oh, *Full-Scale Field Tests of Different Types of Piles*,
Advanced Topics in Science and Technology in China 62,
https://doi.org/10.1007/978-981-33-6183-6_9

4. From the dynamic load test results, the base-and shaft grouted pile increased around 8.6% of its capacity (compared with the base grouting pile).
5. An empirical formula for compressive settlement prediction was proposed (including loading and unloading stages, Chap. 8, Sect. 8.2.2.1).

For piles under uplift loads:

1. The base-and-shaft grouted pile increased about 15.7% of its capacity (compared with the pile without any grouting).
2. The shaft resistance of the base-and-shaft grouted pile increased about 15.0% of its capacity (compared with the pile with base grouting only).
3. An empirical formula and uplift displacement prediction were proposed (including loading and unloading stages, Chap. 8, Sect. 8.2.2.2).

The interpretation methods indicated that the double tangent method was conservative and Chin's method slightly overestimated the pile capacity for compressively loaded piles. Mazurkiewicz's method overestimated the capacity for uplift loaded piles. Further, the offset method for uplift loaded piles was too conservative. After the uplift SLTs were performed, the recommended uplift design load should be equal to two-thirds to four-fifths of this failure load.

9.3 Precast Concrete Piles

For the precast displacement piles, a total number of 40 piles were selected. Twenty-two piles with various lengths were tested in Area A, three piles with various lengths were tested in Area B, and fifteen piles with various lengths were tested in Area C. By defining the permanent settlement ratio, $R_{P\text{-}S}$, it was found that:

1. The $R_{P\text{-}S}$ from all piles with concrete strength of 30 MPa was close to 52.2%, and the Q-s curves from the unloading stages were parallel with each other in Area A.
2. The $R_{P\text{-}S}$ from all piles with concrete strength of 50 MPa was close to 54.9%, and the Q-s curves from the unloading stages were parallel with each other in Area B.
3. The $R_{P\text{-}S}$ from all piles with concrete strength of 40 MPa was close to 49.0%, and the Q-s curves from the unloading stages were parallel with each other in Area C.

The other conclusions were as follows:

1. For the precast piles, if the tested piles are the same type, the Q-s curves from the loading stages will be different (shaft resistance varied) and the Q-s curves from the unloading stages will be parallel with each other.
2. For the proof test result, if using double tangent method, the ultimate bearing capacity is the interpreted value multiplying by factor η_D. If using Chin's method,

the ultimate bearing capacity is the interpreted value divided by reduction factor η_{Ch}. In the case of the project in China, $\eta_D = 1.83$, $\eta_{Ch} = 1.35$ ($L < 25$) and $\eta_{Ch} = 1.2$ ($L > 25$) are recommended.

If in one area, the Q-s curve of a pile from the unloading stages are not parallel to the curves of other piles which are made of same materials (gradient of the unloading trend line is evidently different), it represents that the soil layers from this location may be very different to the others, or the pile are damaged. Thus, more borehole observations are recommended, and a low stain integrity test is suggested. It seems that the $R_{P\text{-}S}$ value of large displacement piles is smaller than that of small displacement piles; however, there is much variables such as concrete strength and geometric configuration being different among these tested piles, further research is required.

For the behavior determination of non-displacement precast piles, 6 piles were selected with different design lengths. Three uplift and three horizontal SLTs were performed, with conclusions as follows:

1. The ultimate uplift capacity of the pile increased with pile length.
2. The ultimate horizontal capacity of the pile increased with pile length.
3. The transferred loads from the pile decreased along the pile length, and shaft resistance increased along the pile.
4. Shaft resistance will develop under increased loads, and particular soil layers provide dominant shaft resistance.
5. An empirical formula for the horizontal displacement prediction were proposed (including loading and unloading stages). The proposed equation from the loading stages is recommended; however, the equation from the unloading stages may provide inaccurate results. Hence, it is only used for optional analysis (Chap. 8, Sect. 8.3.2.2).

Similar to the interpretation of grouted piles illustrated in the previous section, the offset method of uplift loaded piles is not recommended. For the horizontal SLTs, the nominated gradient settlement method is recommended. Given that there was no driven precast pile being tested, the capacity comparison between the non-displacement and displacement pile is unknown; hence, the effect caused by removing the soil first is uncertain. Thus, further research is required.

9.4 Concrete Composite Piles

To determine the effect of CFRP confinement, two types of precast piles were selected. For this research, a total number of 4 piles were tested, including one pipe pile and one pipe pile with CFRP application, one rectangular pile and one rectangular pile with CFRP application. Through results analysis and interpretation, it was concluded that:

1. The CFRP confinement increased the total capacity of rectangular pile by about 8.3%.

2. The CFRP confinement increased the total capacity of pipe pile by about 11.8% and 25% based on the double tangent method and Chin's method, respectively.
3. For the pipe pile, the CFRP confinement increased the shaft resistance by 27.2%.
4. For the pipe pile, the CFRP confinement increased the end resistance by 23.7%.

The limitation of this research was that the applied loads were insufficient, which led to the small effect of CFRP confinement. Further research is also required with consideration of the types of FRP material selected, quantity of laminar applied and FRP orientation arrangement. This new type of pile with FRP application is recommended for use in coastal conditions if the pile is fully covered and confined by FRP (compressive loaded pile). From the structural point of view, this FRP confined pile should perform very well when the pile toe reaching strong bearing strum such as rocks.

For the study of FRP inside reinforcement application in geotechnical conditions, 2 piles were monitored by inclinometer. The deflection behaviors of these two piles were determined and the results were compared. It could not be concluded that the GFRP bar–reinforced concrete pile demonstrated better than the traditionally reinforced concrete pile, since the ultimate tensile strength of GFRP and steel are different (GFRP > 500 MPa; steel = 400 MPa). However, it could be concluded that the GFRP bar–reinforced concrete bored pile could be a suitable alternative to replace the traditionally reinforced concrete bored pile in deep excavation. Further research is required with determination of stresses along the pile, and more advanced analysis is needed, such as a bending moment study along the pile length.

9.5 Super-Long and Large Diameter Piles

To research the load transfer, shaft resistance development and distribution of the super-long and large-diameter piles, 3 drilled shaft concrete piles were tested. From the conventional diagram presentation, it was concluded that:

1. By increasing the pile length, especially when reaching a harder bearing stratum, it can effectively increase the ultimate bearing capacity of the single pile. Also, increment of the pile diameter and pile length can improves the total capacity of a single pile. By comparison of these piles, the 83 m pile with 1800 mm diameter was conservatively designed.
2. The exist of compressible silty sand from the pile tip will led to large pile toe displacement, and thus presented a small concrete compression. These pile toe and pile head curves (Q-s curves) also illustrate that the 83 m length pile with 1800 mm diameter was conservatively designed. Moreover, the trend of Q-s curves from the loading stages of this 83 m pile was discovered relative linear (close to the elastic deformation line), which illustrated the deformation of concrete element is the main factor referring to the ultimate bearing capacity, thus by applying post-grouting technology would not be recommended.

3. From the load transfer curves (load versus depths), the principle finding was that the transferred load decreased along the pile length during each applied load, but the gradients were different, this is due to different pile shaft and the exist of different soil layers. Also, when condition of the soil being soft and very soft (clay), and loose to very loose (sand), the transfer rate of axial force will be smaller.
4. By comparing the shaft and end bearing percentage, it was found that the increase of the pile length can make contribution to the increase of the shaft resistance and hence reduce the transferred loads to the bearing stratum. Also, by comparison of the proportion of shaft and toe resistance under working load, the 83 m pile with 1800 mm diameter was a shaft dominated pile (the total capacity of pile is mainly contributed by shaft resistance).
5. From the load-shaft resistance curves, it was found that the shaft resistance of the most of soil layers developed from zero to a maximum value and then maintained when increasing the applied loads. Also, the shaft resistance from the upper layers developed firstly. The shaft resistance softening and hardening were discovered from some soil layers. Most importantly, the 'Mutual Compensation' phenomenon was discovered, that is when shaft softening occurred from one soil layer, the shaft hardening of the other soil will occur simultaneously.
6. From the Shaft Resistance Distribution Curves, It Was Found that the Shaft Resistance of Each Soil Layer Does not Develop Simultaneously. The Shaft Resistance Distribution Illustrated an 'R' Shape.

From a practical engineering point of view, the soil layers around these three piles were mostly sand, which possesses a relatively high voids; thus, post-grouting technology is highly recommended to improve the ultimate bearing capacity. However, it is recommended that shaft-and-base grouting be applied to P66 and P12, while only shaft grouting be applied to P105. This is because P105 is a shaft-dominated pile or a pure-friction pile; thus, using base grouting to reinforce the pile toe cannot effectively increase the ultimate bearing capacity. Further, for P105, because the transferred load from the pile toe was observed to be relatively small, increasing the pile length would be inappropriate to increase the ultimate bearing capacity; thus, increasing the pile diameter is recommended.

9.6 Piles Under Ultimate Load

To determine the pile behavior under ultimate loading, 4 piles were cast. Three types of tests that lead to failure were considered. The first test involved loading the pile to failure caused by inadequate concrete strength, the second test involved loading the pile until failure because of the soil being soft (or the pile being overdesigned for its capacity) and the final test involved loading the pile to failure because of eccentricity. The behaviors of these three piles were compared with a pile that achieved the design, and it was concluded that:

1. For the pile with inadequate rigidity, the pile will suffer from structural failure. Under particular loads, the concrete with inadequate strength will be crushed, and the steel reinforcement will try to resist the vertical loading, but will be 'squeezed out' immediately. This structure can still resist extra loading, and soil-pile system failure will occur soon.
2. For the pile applied with eccentricity load, the pile suffered compression and tension in different parts. Under a certain loading, the concrete crushes and there will be a dramatic settlement increase. Later, the tension part will elongate so the increase of downward settlement may change to small settlement decreasing; Lastly, the soil-pile system will fail which will lead to increase of settlement again.
3. For the pile suffering from plunging failure, it is believed that the shaft resistance will develop under increased loads, and, after the total shaft resistance is fully developed, the loads will transfer to the pile end, thereby leading to the failure of the bearing stratum.
4. For the piles suffering from ultimate loading, s-lgt curve is recommended for analysis because this method can provide information to check where loading has caused the pile foundation to become unstable. Similar to the previous conclusion, the double tangent method was conservative and Chin's method overestimated the ultimate capacity. It is also suggested that Chin's method should not be used for pile capacity suffering from eccentric load.

9.7 Concluding Remarks

This book provides research of different types of pile foundations which includes cast-in-situ piles (grouted pile and non-grouted piles), precast piles (displacement and non-displacement piles) composite piles (CFRP laminar confined piles and GFRP bar reinforced piles) and super long and large diameter piles. The research considers recent popular topics including grouting technique, non-displacement precast pile, fibre material application, shaft resistance of super-long and large-diameter pile and failure behaviour of pile foundation. Moreover, various project case studies are introduced including high-rise building project and tunnelling project. This book also covers all field tests of pile foundation in terms of static load tests with load orientation consideration (compressive, uplift and horizontal) and dynamic load test as well as inclinometer tests.

Overall, this book has provided practical knowledge for geotechnical and foundation engineering. It is expected that this book will provide academic guidance for both foundation and geotechnical engineers. Further, the presented methods and test data can be used to confirm the suitability of pile soil systems to satisfy the designed load with an appropriate safety factor. This book may also provide information for the implementation of new test analysis methods or procedures. Additionally, this book has provided recommendations for consistent use in geotechnical practice.